Springer Series in Optical Sciences   Volume 61

Editor: Herbert Walther

Springer Series in Optical Sciences

Volumes 1–41 are listed on the back inside cover

Nigel G. Douglas

Millimetre and Submillimetre Wavelength Lasers

A Handbook of cw Measurements

With 15 Figures

Springer-Verlag Berlin Heidelberg GmbH

Dr. Nigel G. Douglas

Kapteyn Sterrenwacht, Rijksuniversiteit te Groningen,
NL-9301 KA Roden, Netherlands

ISBN 978-3-662-14492-3

Library of Congress Cataloging-in-Publication Data. Douglas, Nigel G., 1955-. Millimetre and submillimetre wavelength lasers : handbook of cw measurements / Nigel G. Douglas. p. cm.-(Springer series in optical sciences : v. 61) Includes index.
ISBN 978-3-662-14492-3   ISBN 978-3-540-46095-4 (eBook)
DOI 10.1007/978-3-540-46095-4

1. Far infrared lasers-Measurement-Hand-
books, manuals, etc. 2. Plasma waves-Measurement-Handbooks, manuals, etc. I. Title. II. Title: Millimeter and submillimeter wavelength lasers. III. Series.
TA1696.D68 1989   621.36'6-dc20   89-11399

2154/3150-543210 – Printed on acid-free paper

to Yonas

Preface

The optically pumped laser has made an enormous contribution to research in the part of the electromagnetic spectrum known as the far infrared, or submillimetre region. I hope that this book will be useful to both practising and prospective workers in the field, since it contains an up-to-date catalogue of measurements of the main properties of submillimetre lasers as well as an introductory review of the measurement techniques themselves.

Wavelength and frequency measurements have been exhaustively compiled (in Part II of this book) along with molecule and pump identification. Part I contains a short review of the relevant measurement techniques in each of these areas and, in addition, a review of power measurements.

Working in this field, as in any other, one's satisfaction is determined largely by the colleagues one has and the friends one makes along the way. I am very grateful to Dr G. Dodel, Dr L.C. Robinson and Dr G.F. Brand for introducing me to the field. Dr I.S. Falconer and Dr P.A. Krug have been good colleagues and friends. For this book in particular I am grateful to Dr Dodel, Dr K.M. Evenson, Dr H. Figger, Prof. M. Fourrier, P. Kempf, Dr K.J. Siemsen and Dr M.S. Tobin for their comments, and to Dr D.J.E. Knight for a great deal of help, including data from unpublished or obscure sources which he had gathered for his own compilation. My brother, Neil Douglas of Melbourne University, assisted with the chemical nomenclature. I am indebted to the Observatory of Leiden University for their hospitality. Theo Jurriens of the University of Groningen assisted with LaTeX, especially with the preparation of large data tables. I take this opportunity to express my immense gratitude to Y.M. ten Wolde for her careful and patient work.

Roden, The Netherlands   Nigel Douglas
February 1989

Bohr erzählte, daß er einmal zusammen mit Kramers am Strand eine aus der Kriegszeit stammende angetriebene Mine gefunden hätte, und sie hätten dann um die Wette versucht, die Zündkapsel zu treffen. Nach vergeblichen Versuchen sei ihnen aber klar geworden, daß sie ja dabei die Freude, getroffen zu haben, doch nicht mehr ins Bewußtsein hätten aufnehmen können, da die explodierende Mine ihrem Leben vorher ein Ende gesetzt hätte; und daraufhin hätten sie sich einem anderen Ziel zugewandt

Werner Heisenberg, *Schritte über Grenzen*

Contents

Part I

Measurement of the Properties of Millimetre and Submillimetre Wavelength Lasers

1. Introduction

Far-infrared physics is becoming a mature discipline. By this I mean that the applicable technology is so far developed, and the techniques sufficiently routine, that an increasing number of applications of far-infrared physics are being found. Research has broadened from an investigation of sources and detectors to their use in other fields. Development work is of course continuing, and especially in the area of detectors much improvement can still be made before sensitivities are comparable to those of optical detectors. Yet, as well as the long-standing use of far-infrared techniques in plasma diagnostics, we are now seeing a vast increase in the pursuit of astronomy in this part of the spectrum and the use of the far-infrared in metrology. There has also been a corresponding, gradual commercialization of the necessary components.

Without question, the *optically pumped far-infrared laser* has provided the main technological impetus for this phase of maturity. Notwithstanding the importance of backward-wave oscillators and the gradual development of high-frequency solid-state sources, the lasers have made possible a plethora of investigations otherwise too difficult or too costly to perform.

The operating principle of this type of laser is rather simple and is illustrated in Fig. 1.1. The incoming radiation is produced by an infrared laser, usually a CO_2 laser, and is absorbed on vibrational-rotational transitions in a molecular vapour. The vapour is thereby excited to states normally almost totally empty. The resulting population inversions provide several possibilities for laser emission, and as shown in Fig. 1.1 these can be either in the upper or lower vibrational level. It may be reasoned that, since stimulated absorption and emission have equal probabilities, the maximum quantum efficiency for the optically pumped laser is 0.5; in practice the value is seldom as high as 0.1. In terms of energy, the efficiency is lower, of course, by the ratio of the far-infrared to the infrared wavelengths, not to mention the efficiency of the infrared pump laser. Nevertheless the "cleanness" and simplicity of the system make it very attractive.

Optical pumping is not the only way to excite a molecular vapour. Historically the technique was preceded by that of electrical excitation, which is responsible for several very important far-infrared lines, notably in H_2O, HCN and DCN vapour [1.1]. In view of the present dominance of optical pumping one should not underrate these early achievements with electrically excited lasers. The im-

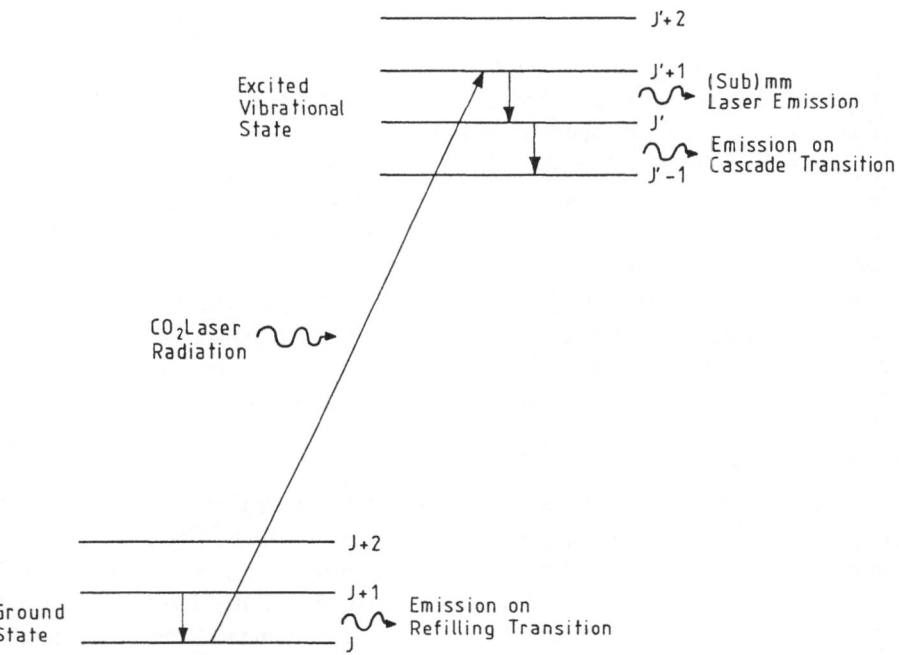

Fig. 1.1. Diagram of the molecular energy levels involved in the absorption and emission processes in the optically pumped laser. The refilling transition is in the lower vibrational state, the others are in the higher vibrational state. Adapted from [1.9]. Reproduced with the permission of the Harry Diamond Laboratories, Adephi

portance of "cascade" transitions (see Fig. 1.1) was shown before the molecular levels responsible for laser action had been identified [1.2] and by 1967 the first heterodyne frequency measurements were being made (Chap. 3). The HCN laser is still one of the most powerful of submillimetre sources (Table 4.1) and even atomic species have been made to "lase" by electrical excitation [1.3].

The first optically pumped far-infrared laser was set up by *Chang* and *Bridges* in 1970 [1.4, 5]. They obtained pulsed emission from methyl fluoride vapour. The first true cw results with optical pumping followed in the same year [1.6]. It quickly became apparent that the technique would be far more prodigious in producing new emissions than electrical excitation had been. The present number of known optically pumped lines is 4218, and of electrically excited lines it is only 337. These numbers represent the results of an extensive body of work involving scores of researchers and over three hundred published articles. Over the last four years, an average of over 400 new lines were reported each year. Beyond this there have been hundreds of papers on the properties of lasers, their applications and so on.

Both pulsed and cw (continuous-wave) emission are possible, although *this book only deals with lasers which operate cw* (or nearly so, as explained in Part II).

3

What is this book? It is *not* a review of the field of far-infrared lasers: there have been several of these over the years, including a general review of optical pumping [1.7], a review of electrically excited lasers [1.8], and a very complete treatment of optically pumped lasers [1.9].

In the first part of this book, I have attempted to fill a gap in the literature by reviewing the *experimental and theoretical techniques* used to acquire the fundamental data on millimetre and submillimetre lasers. That is, I have devoted sections to: molecular spectroscopy and line identification (Chap. 2), wavelength and frequency measurement (Chap. 3), and power (i.e. intensity) measurement (Chap. 4), including a listing of those lines for which power has indeed been reliably measured. I conclude Part I with some notes on pump lasers (Chap. 5). I hope that even experienced readers will enjoy these sections, but their primary purpose is to provide an easy introduction to the literature for those less familiar with the subject.

The main purpose of this book however is to bring the literature on laser lines up-to-date with *a new compilation of all the known cw lines*. The early compilations of this nature [1.10–12] were quickly outstripped by new discoveries, and the last comprehensive update of cw lines appeared in 1981 [1.14] (it contained 1350 lines). Although a review of frequency measurements appeared recently [1.13], only 914 of the lines in the present compilation of 4555 have had their frequencies measured (50 of them since publication of [1.13]). The catalogue which appears in Part II of this book was commenced several years ago because of the need for an up-to-date source of information on laser lines suitable for use in astronomical heterodyne receivers (see for example [1.15]). In common with a number of applications, these receivers require the use of cw lasers at specific frequencies. Spectral coverage is now good enough that the publication of a new, exhaustive catalogue seems opportune.

I have chosen to include only those lines whose production requires what may be termed "standard techniques". *Electrically excited* and *optically pumped* lasers are included, except those involving multi-photon systems, tunable pump lasers, or stark-shifted [1.16] far-infrared lasing. These are all useful techniques but beyond the purposes of this book. Gas-dynamic lasers [1.17], chemical lasers and molecular beam masers [1.18] are also excluded.

Part II contains both *wavelength sorted* and *molecule/pump sorted* listings. The latter will be of interest for researchers already investigating a particular vapour, or for those who may want to locate genuinely monochromatic emissions with no known "partners" on the same pump transition. Along with the various listings and the database of articles from which they were derived, Part II contains information on exactly which lines were accepted into the lists, and on how erroneous entries were detected and eliminated. I hope thereby to be providing a useful up-to-date and durable summary of far-infrared laser lines.

A word about terminology: over the years many phrases have been used to describe the parts of the spectrum we are dealing with here. Pump laser wavelengths are often referred to as "infrared" and the pumped emission as "far-infrared". One problem with that is that the vast majority of "real" infrared workers regard *both* parts of the spectrum as "far-infrared", although "mid-infrared" is sometimes used for wavelengths from ten to a few tens of micrometres (μm). So the term "submillimetre" was coined; but the fact that lasers (in particular) operate out to several millimetres wavelength without change in the essential technology has evidently caused such discomfort that several acronyms such as "MSM", "NMMW", and "SMNSM" have been invented, although none of them has really caught on.

My contribution to the nomenclature is this: pump laser lines are of "infrared" wavelength. Pumped emission is "submillimetre" or "millimetre" (abbreviated to *submm* and *mm*), whichever is applicable, if the wavelength is known. If the wavelength is not known, or is unimportant, or if a whole group of submm and mm lines is being referred to, or in other circumstances where a general term is needed, I use the expression "(sub)millimetre" (abbreviated to *(sub)mm*). This is fairly compatible with older usage, it should keep the purists happy, and I have found it to be not too intrusive.

References

1.1 W.M. Müller, G.T. Flescher: Appl. Phys. Lett. **8**, 217–218 (1966)

1.2 M.A. Pollack, T.J. Bridges, W.J. Tomlinson: Appl. Phys. Lett. **10**, 253–256 (1967)

1.3 J.S. Levine, A. Javan: Appl. Phys. Lett. **14**, 348–350 (1969)

1.4 T.Y. Chang, T.J. Bridges: Opt. Commun. **1**, 423–426 (1970)

1.5 T.Y. Chang, T.J. Bridges: "Submillimeter Wave Laser Action in Optically Pumped CH_3F", in *Proc. Symp. Submill. Waves*, Microwave Research Institute Symposia Series Vol. XX, ed. by J. Fox (Polytechnic Press, New York 1970)

1.6 T.Y. Chang, T.J. Bridges, E.G. Burkhardt: Appl. Phys. Lett. **17**, 249–251 (1970)

1.7 T.Y. Chang: "Optical pumping in Gases", in *Nonlinear Infrared Generation,* Topics in Applied Physics Vol. 16, ed. by Y.-R. Shen (Springer, Berlin, Heidelberg 1977)

1.8 F.K. Kneubühl, Ch. Sturzenegger: "Electrically Excited Submillimeter-Wave Lasers", in *Infrared and Millimeter Waves* **3**, (Academic, New York 1980)

1.9 M.S. Tobin: Proc. IEEE **73**, 61–85 (1985)

1.10 M. Rosenbluh, R.J. Temkin, K.J. Button: Appl. Opt. **15**, 2635–2644 (1976)

1.11 J.J. Gallagher, M.D. Blue, B. Bean, S. Perkowitz: Infrared Phys. **17**, 43–55 (1977)

1.12 H.P. Röser, G.V. Schultz: Infrared Phys. **17**, 531–536 (1977)

1.13 M. Inguscio, G. Moruzzi, K.M. Evenson, D.A. Jennings: J. Appl. Phys. **60**, R161–R192 (1986)

1.14 D.J.E. Knight: NPL Report Qu 45 (1979), revised Feb 1981. An update (references only) is to appear in "Handbook of Lasers and Technology, Supplement 1: Gas Lasers", ed. by M.J. Weber (C.R.C. Press, Florida)

1.15 G. Chin: Int. J. Infrared Mmwaves **8**, 1219–1234 (1987)

1.16 H.R. Fetterman, H.R. Schlossberg, C.D. Parker: Appl. Phys. Lett. **23**, 684–686 (1973)

1.17 B.L. Wexler, T.J. Manuccia, R.W. Waynant: Appl. Phys. Lett. **31**, 730–732 (1977)

1.18 F.C. de Lucia, W. Gordy: "Millimeter and Submillimeter Wave Molecular Beam Masers", in *Proc. Symp. Submill. Waves,* Microwave Research Institute Symposia Series Vol. XX, ed. by J. Fox (Polytechnic Press, New York 1970)

2. Molecular Spectroscopy and Line Identification

(Sub)millimetre lasers are useful in a number of applications, but in addition are studied for the insight they give into the physics of the lasing molecule. The frequency of both the pump and laser transition can often be measured with great accuracy. Comparison with calculated frequencies allows the molecular constants to be corrected, assuming that these are first known well enough to identify the transitions concerned. Information on the relative polarizations of the pump and laser line is useful for reducing the number of candidate lines in the identification.

In this section I review briefly the nature of the vibrational and rotational levels of interest and the corresponding selection rules. My purpose is to make the literature, especially that on which the catalogue of laser lines is based, more accessible. Therefore, although physical principles are discussed, the emphasis is on simple formulae and nomenclature. I restrict the discussion to material commonly encountered in the context of molecular lasers — very good broader accounts of the subject of molecular spectroscopy are widely available [2.1–4].

Speaking generally, it may be stated that optical pumping of the molecules we will consider takes place via vibrational transitions (with associated small rotational energy shifts) while the more closely spaced rotational transitions themselves are responsible for the (sub)mm laser action. The exception to this is provided by a class of "mid-infrared" lasers whose emission, as well as absorption, takes place on vibrational-rotational transitions [2.5]. In only one case, to my knowledge, has a vibrational band been proposed as the mechanism for (sub)mm laser action [2.6].

Diatomic molecules are relatively unimportant in the present context since there are very few whose vibrational bands are fortuitously located with respect to pump laser frequencies. This does not exclude the possibility of pumping via other mechanisms, such as electrical or chemical excitation, and a number of (generally mid-infrared) emissions have been observed from such molecules as HCl, HF and OH radical [2.7]. None has been observed so far in true cw mode. However the case of the diatomic molecule will be treated as it introduces most of the main concepts. To a first approximation the diatomic molecule is a rigid rotor with a negligible moment of inertia about one axis and a value of

$$I_B = \mu r^2 \tag{1}$$

7

about the others. Here μ is the reduced mass of the molecule, r the bond length and, by convention, the moments of inertia I_A, I_B and I_C are designated to be in increasing order of size. In the case of the diatomic molecule I_A is nearly zero and $I_B = I_C$. Now the classical energy of a rotating body is $E = \frac{1}{2}I\omega_c^2$ and such a rotating body, if it possesses a permanent dipole moment, will radiate at the frequency of rotation, ω_c. A "typical" diatomic molecule has a bond length of 0.1 to 0.3 nm. For $^{12}C^{16}O$ for example it is 0.113 nm. Given the atomic masses, one can calculate I_B and hence show that, if the rotational energy of the molecule equals its kinetic energy at room temperature, $\omega_c \sim 7.55 \times 10^{12}$ rad/sec. This corresponds to emission at 250 μm wavelength. The intention here is not to dwell on the usefulness of classical arguments. I wish only to suggest that it is intuitively reasonable that typical light molecules should have rotational transitions in the (sub)millimetre part of the spectrum. As we can see from the equations which follow, the value of the rotational quantum number (J) which corresponds most nearly to this value of the rotational energy is 10, and for the appropriate transition (10 → 9) the calculated wavelength is 259 μm, so the classical calculation is not too bad.

Analogy with the classical case also suggests that symmetric molecules (e.g. $^{14}N_2$), which have no permanent electric dipole moment, should exhibit no pure rotational electric dipole transitions.

Defining the rotational constant

$$B_e = \frac{h}{8\pi^2 I_B} \tag{2}$$

we can express the (quantum mechanically derived) **energy levels** of the rigid rotor as [2.1]

$$E_J = hB_e J(J+1) \tag{3}$$

where J is the rotational quantum number ($J = 0, 1, 2 \ldots$). The subscript e associated with B stands for "equilibrium", since as we will see the effective value for B differs from this in some cases. Not all authors use this subscript. The **selection rule** applicable for the case of pure rotational transitions considered here is

$$\Delta J = \pm 1 \tag{4}$$

– this applies for all but a few special cases (see text following (7) below).

It should be mentioned that *Raman-effect* transitions are occasionally observed in lasers [2.8], being those in which the polarizability of the molecule changes during the transition. Spectroscopists distinguish between Raman and "infrared" transitions and note that different selection rules apply in the two cases [2.2]. This selection rule in the Raman case applies to the overall change of state of the molecule: there may not even be an "upper level" through which the molecule passes. In the context of cw lasers however we are always dealing with the resonantly enhanced Raman effect in which there is a near-lying upper level.

Equation (4) applies, even in the Raman case, to the (sub)millimetre transition involving this upper level.

The CO molecule in low vibrational states (dealt with below) and fairly low rotational levels (J below about 50) is a reasonable approximation to the rigid rotor. According to (3) and (4) above, the transition frequencies are those of the "rotational ladder"

$$\nu(J + 1 \to J) = \frac{E_{J+1} - E_J}{h} = 2B_e(J + 1) \tag{5}$$

where for $^{12}C^{16}O$ $B_e = 57.898$ GHz. Thus one has (in an obvious shorthand often seen in astronomy) CO(4–3) = 460 GHz, CO(1–0) = 110 GHz and so on.

Before we leave the subject of the rotational spectra of diatomic molecules, mention should be made of the interesting case of the oxygen molecule. The most common form of oxygen, $^{16}O_2$, is homonuclear, so as mentioned earlier it has no pure rotational electric dipole spectrum. It is, however, one of a small number of molecules not in the ground state of *electronic* spin angular momentum. We have neglected electrons up to now because any change in electronic state implies a very great change in energy, normally corresponding to visible or ultra-violet radiation, and because the characteristics of vibrational-rotational transitions within the ground electronic state are generally unaffected by the detailed characteristics of that state. In O_2 however the effect of the (unpaired) electrons is to give the molecule a net electron spin $S = 1$ and thus a permanent *magnetic* dipole moment. The resulting allowed transitions are very important, not directly in the context of the production of laser radiation but because of the many atmospheric absorption lines which they represent. For the angular momentum coupling case appropriate here [2.9] the rotational energy of the O_2 molecule is given by an expression like (3) above but with J replaced by N, where the vector relation $J = S + N$ holds. This means that J can take only the values $N - 1, N$, and $N + 1$. Owing to the high symmetry of $^{16}O_2$ the rotational levels with even values of N are missing [2.10]; the selection rule for N is $\Delta N = 0, \pm 2$. Ignoring for a moment the fine-structure interaction between J and N we therefore obtain a rotational ladder of lines like that of CO but with "rungs" at $B_e(4N + 6)$, where $N = 1, 3, \ldots$. The first of these transitions, $\Delta J = 0$, $N = 1 \to 3$, is at 706 μm (424.76 GHz [2.9]) and subsequent lines appear at shorter and shorter wavelengths across the submillimetre range appearing for example at around 105 μm for $N = 15 \to 17$.

Consideration of the allowed values for N and J, and the selection rule $\Delta J = 0, \pm 1$ (O_2 is one of the exceptions to (4) mentioned earlier), leads one to conclude that each transition in the rotational ladder should be accompanied by *two* allowed transitions for which $\Delta J = +1$. Their offset from the $\Delta J = 0$ transition is around 60 GHz for most values of N, and they appear experimentally as somewhat weaker lines above and below the $\Delta J = 0$ line.

The final set of allowed rotational transitions for the oxygen molecule are the fine-structure lines, $\Delta J = +1$, $\Delta N = 0$. As can be deduced from the foregoing, they occur around 60 GHz for most values of N [Ref. 2.1, Fig. 7–6] and accumulate to form an atmospheric absorption band of considerable strength. The one such line outside this band is the $N = 1$, $J = 0 \rightarrow 1$ transition at 118.75 GHz [2.11]. Together the oxygen fine-structure lines form the boundaries of one of the atmospheric windows of millimetre wavelength radio astronomy.

To account for oxygen lines in the red and near-infrared requires a discussion of the vibrational-rotation spectrum, but here we will leave our special case and return to the general discussion.

The **vibrational energy** of a diatomic molecule is given by

$$E_\nu = h\omega_e \left(\nu + \frac{1}{2} \right) \tag{6}$$

where ν is the vibrational quantum number ($\nu = 0, 1, 2 \ldots$) and ω_e is a frequency related to the force constant k of the molecular bond (to be precise $\omega_e = 1/2\pi c\sqrt{k/\mu}$) [2.3]. For CO for example ω_e, in the units usually quoted, is 2170 cm^{-1}. The fact that the rotational constants and vibrational frequencies are quoted in different units is historical in nature and has to do with the fact that their respective sizes are such that while rotational transitions often lie in the millimetre or microwave region, vibrational transitions tend to appear as infrared or Raman bands. It is worth pointing out that ω_e in the above equation is to be regarded as a true frequency, not an angular frequency.

For a harmonic potential, the **vibrational selection rule** is

$$\Delta\nu = 0, \pm 1 \tag{7}$$

although anharmonicities can cause other, weaker, transitions to appear. In vibrational transitions the selection rule for J, rule (4) above, is maintained in all but a few special cases [Ref. 2.2, Sect. IV.1]. These include molecules, like O_2 considered above, which are not in the ground electronic state. The case $\Delta\nu = 0$ is included to allow for pure rotational spectra, possible only if the molecule has a permanent dipole moment.

One has to depart from the rigid rotor model to handle vibrational-rotational interactions properly, indeed strictly speaking to account for vibrations at all. As J increases the average bond length changes due to centrifugal distortion. Conversely, the vibrational state of the molecule affects the rotational constants due to the fact that the interatomic forces are anharmonic, so that the equilibrium configuration of the molecule is affected. Exactly how one handles these effects depends upon one's model of the potential function — the morse function being a common first step [2.4].

Linear polyatomic molecules, like diatomic molecules, have only one important moment of inertia and the **rotational energies** have the same form as (3). The **selection rule** for pure rotation transitions likewise has the same form as rule (4) above.

The **vibrational energy** component can be evaluated by summing expressions like (6) over all vibrational modes together with cross-terms to allow for anharmonicities [2.3]. A more explicit treatment of vibrational energies for linear and other polyatomic molecules is beyond the scope of this book, but the effect of those vibrational modes on the rotational spectra requires some attention. Nor can we neglect mentioning centrifugal corrections. To a certain approximation these two effects can be taken into account by adjusting the rotational constant thus:

$$B = B_e - \sum \alpha_i \left(\nu_i + \frac{1}{2} \right) - J(J+1)D \qquad (8)$$

where α_i and D are (small) constants and i runs over the vibrational modes of the molecule. A molecule of N atoms has in principle $3N$ degrees of freedom, of which 3 are accounted for by translation of the molecule as a whole. A further 3 are accounted for by free rotation, except in the case of linear molecules presently considered. They have effectively 2 rotational degrees of freedom, leaving $3N-5$ vibrational modes. For a linear triatomic molecule such as OCS or CO_2 there are therefore four modes, although two are degenerate (i.e. they have the same frequency). The three distinct modes are shown in Fig. 2.1. Conventionally, the bending mode is labelled ν_2 and of the stretching modes ν_1 is the one of lower frequency. ν_2 is doubly degenerate since it can occur in the plane shown or perpendicular to it. Any particular vibrational state is designated by the three quantum numbers plus a superscript l. The latter is introduced to allow for the fact that when the molecule is rotating, an axis is defined which breaks the degeneracy of the mode. l can take the values $\nu_2, \nu_2 - 2, \nu_2 - 4, \ldots, -\nu_2$ but it turns out that the frequency only depends on l^2. Consequently one designates the mode with only the magnitude of l.

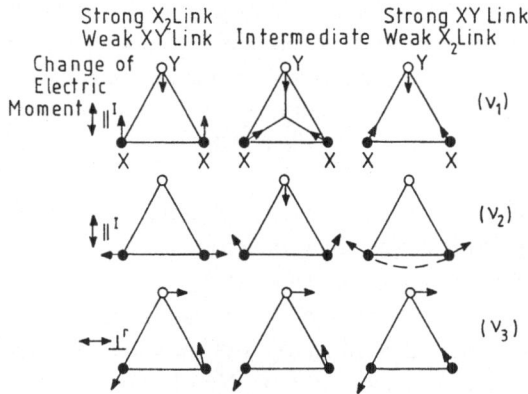

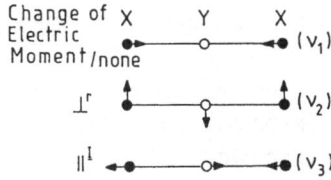

Fig. 2.1. Examples of the sets of normal modes of vibrations of molecules. (a) The three modes of vibration of a triatomic molecule YX$_2$; (b) The special case of a linear triatomic molecule such as CO_2 : ν_1 symmetric stretching mode, ν_2 bending mode, ν_3 asymmetric stretching mode. After [2.4]

The **vibrational selection rule,** rule (7) above, which is applicable again here, implies that only one vibrational mode can change, and by at most one unit, in any allowed transition. Transitions in which this is obeyed are called *fundamentals* and they indeed tend to represent the strongest transitions. As a result of harmonicities however both *overtone* ($|\Delta\nu| > 1$ for a single mode) and *combination* bands ($\Delta\nu \neq 0$ for several modes) may appear, and are quite important. A discussion of which particular overtone and combination bands will be active in infrared and Raman transitions, a property which stems from more fundamental "selection rules" governed by the symmetry of the molecule, is beyond our purposes here and can be found in, for example, [Ref. 2.2, Sect. III.3].

In certain types of vibrational transitions $\Delta J = 0$ may occur in addition to $\Delta J = \pm 1$, which is always allowed. Again this is beyond the scope of our discussion but can be found in, for example [Ref. 2.2, Sect. IV.1].

To give a concrete example of the foregoing, consider the molecule CO_2 for which $\omega_1 = 1388.3$ cm^{-1}; $\omega_2 = 667.3$ cm^{-1}; $\omega_3 = 2349.3$ cm^{-1} [Ref. 2.2, Table 56]. The well-known laser bands have as the upper level 00^01 and as the lower levels 10^00 and 02^00 (see Fig. 5.1). Note the superscript l, mentioned earlier. Spectroscopists would designate these sets of transitions as $\nu_3 - \nu_1$ and $\nu_3 - 2\nu_2$ combination bands respectively. The measured frequencies of the band centres are displaced from the values expected from the above data, being at 961 cm^{-1} and 1064 cm^{-1}. This is a consequence of a phenomenon known as *Fermi resonance* which is discussed briefly in Chap. 5.

The fact that CO_2 is linear and symmetric leads, just as with homonuclear diatomic molecules, to certain simplifications in its spectra. CO_2 has no permanent dipole moment and therefore no pure rotational spectrum. Furthermore, transitions involving only a change in ν_1 are infrared inactive, since this mode of vibration maintains the symmetry (see Fig. 2.1). In the case of the laser bands mentioned above, transitions between vibrational levels are accompanied by a change in rotational quantum number $J = \pm 1$, but those terminating on odd values of J are not allowed, also for reasons of symmetry. None of these remarks would apply to N_2O, for example, because this molecule has the structure N-N-O and is therefore not symmetric.

Non-linear molecules can be handled as more-or-less symmetric tops. The molecules NH_3, PCl_3 and AsF_3 are all examples of symmetric tops with a permanent dipole moment because of their pyramidal shape. BCl_3 and BF_3 on the other hand are planar, owing to the valence of the Boron atom, and lack a pure rotational spectrum. (We assume here that all isotopes of the same element have equal mass. NH_2D of course is an asymmetric top). Non-linear triatomic molecules of the form X_2Y, such as H_2O and NO_2, are asymmetric tops, as are, perhaps surprisingly, such triangular molecules as O_3. Let us examine these categories a little more closely.

The **symmetric top** molecule has two identical moments of inertia (which we will call I_B) and one different (which we will call I_C to be consistent with [2.1]; it may well be called I_A instead [2.3], especially if it is known to be *smaller* than

I_B). Simple examples are ammonia, NH_3, (I_B= 298 GHz), phosphine, PH_3, and methyl chloride, CH_3Cl, but considerably heavier examples exist—for example $C_8H_{13}Cl$ is a symmetric top (I_B= 1.1 GHz). **Rotational energy levels** for the symmetric top can be characterized by two quantum numbers J, K such that

$$E_{JK} = hB_e J(J+1) + h(C_e - B_e)K^2 \qquad (9)$$

where B_e is defined in (2), and C_e in an analogous fashion, and

$$K = 0, \pm 1, \ldots \pm J \quad . \qquad (10)$$

We have written the rotational constants with the subscript e not only to be consistent with (2) but also to emphasize that corrections for centrifugal and vibrational effects have not been included. As to the former, and mostly for the sake of familiarity with the nomenclature, we give the terms of next greatest importance on the right-hand-side of (9) :

$$-hD_J J^2(J+1)^2 \quad \text{and} \quad -hD_{JK} J(J+1)K^2$$

where D_J and D_{JK} are further rotational constants. As to vibrational corrections, these can be attempted as indicated in (8) where now the summation runs over $3N - 6$ vibrational modes. In general, however, the α's are difficult either to calculate or to extract from experimental data so that one is generally dealing with "effective" values of the rotational constants B, C derived directly from spectral measurements of the relevant vibrational band.

As can be seen from (9), the energy depends only on the magnitude of K, so there are $J + 1$ different values. In considering **selection rules** we have to distinguish between *parallel* and *perpendicular* bands, which may be defined according to whether the quantum number K changes:

parallel band—
$$\Delta K = 0, \ \Delta J = 0, \pm 1$$
$$(\Delta J = 0 \text{ forbidden for } K = 0)$$

perpendicular band—
$$\Delta K = \pm 1, \ \Delta J = 0, \pm 1 \quad . \qquad (11)$$

For true (as opposed to "accidental") symmetric tops perpendicular bands can only arise as transitions between vibrational states at least one of which is degenerate [2.2]. Furthermore, transitions within a particular vibrational band (i.e. pure rotational transitions), are always of the parallel type. Both of these rules are relaxed for less symmetric molecules but for the present case the restriction to $\Delta K = 0$ means that the rotational "ladder" has the same simple form as given before for the diatomic molecule, (5), again ignoring centrifugal distortions.

The transitions of a symmetric top are designated using the codes P, Q, R to represent $\Delta J = -1, 0, +1$ respectively and a superscript for the corresponding ΔK term. Thus $^P Q$ means $\Delta J = 0, \Delta K = -1$. A transition designated as

$^{R}R_5(15)$ would have as the lower level $K = 5$ and $J = 15$ (and therefore in this case $K = 6$ and $J = 16$ as the upper level). I chose this particular example because *Graner* [2.12], investigating millimetre emission in the CH_3I laser, proposed this transition in the $\nu_6 = 0 \rightarrow 1$ band to explain the optical pumping by the CO_2 laser. If this is correct, the subsequent millimetre wavelength emission ($J = 16 \rightarrow 15$, $\Delta K = 0$) would be, from (5), at $12B_e$ or around 240.0 GHz [Ref. 2.1, Table 3–3]. The correct value turned out later to be 239.1 GHz. In fact heterodyne measurements were able to resolve the millimetre line into four components separated only by a few MHz (see Table A in Part II). This is the effect of hyperfine coupling between the nuclear spin of the Iodine nucleus ($I = \frac{5}{2}$) and the rotational states (here $J = 16$ or $J = 15$). The interested reader is referrred to [Ref. 2.1, Chap. 6].

Asymmetric tops have all three rotational constants unequal ($A > B > C$). Limiting cases are the prolate ($B \approx C$) and oblate ($B \approx A$) tops. **Rotational energy levels** are specified by three quantum numbers J, K_{-1} and K_{+1} (for example $5_{32}, 7_{17}$) which are sometimes written in a shorter, but equivalent form J_τ where $\tau = K_{-1} - K_{+1}$ (in this case $5_1, 7_{-6}$). The shorter nomenclature contains the same information since the sum of K_{-1} and K_{+1} may only adopt the values J or $J + 1$.

The energy levels for the prolate top are:

$$E_{JK} = \frac{B + C}{2} J(J + 1) + \left(A - \frac{B + C}{2} \right) w \tag{12}$$

where w is rather close to K_{-1}. It is obvious how to transpose this equation for the oblate case. For intermediate cases [Ref. 2.1, Appendix III] gives tables enabling w to be evaluated.

In contrast to the case of more symmetric molecules, the **selection rules** for asymmetric tops are more complicated and depend upon which axis is parallel to the dipole moment. If this is the A axis, then the parity of K_{-1} is conserved, and that of K_{+1} changed, in all allowed transitions. (For example $\Delta K_{-1} = 2$ and $\Delta K_{+1} = -1$ would in such a case be a combination satisfying both of the above rules). If it is the C axis which is parallel to the dipole moment, the reverse is true. If the dipole moment is parallel to the B axis, the parity of both K_{-1} and K_{+1} changes in all allowed transitions [2.1]. The degree of change in K_{-1} and K_{+1} is otherwise not restricted, although the line strengths of the transitions fall as one tends to large integer changes. The selection rule for J is

$$\Delta J = -1, 0, +1 \tag{13}$$

and as before the designation P-, Q- and R-branch applies to these three cases, respectively.

Beyond this short summary it has to be said that each molecule has its own peculiarities and has to be analyzed separately. The "good" quantum numbers will depend on the characteristics of the molecule. For example, the molecule CH_3OH is an asymmetric top in which the CH_3 group can rotate freely, but in which the angle of the CH₃-O-H bond gives rise to a "hindered" rotation of the OH group, best expressed as a torsional mode of oscillation. Quantum numbers (n, τ) are assigned to this torsional mode, which is a kind of intermediate case between vibration and rotation and accounts in part for the rich rotational spectrum of CH_3OH [2.13, 14]. In this particular case the C-O stretch vibrational mode accounts for the infrared spectrum. An instructive analysis of this molecule, largely based on (sub)millimetre data, has recently been given [2.15].

A useful means of reducing the number of possibilities for a line identification is the polarization rule [2.13, 16, 19]. This links the rotational quantum numbers with the observed polarizations, i.e. whether the polarization of the (sub)millimetre emission is parallel or perpendicular to the polarization of the pump field. The polarization rule states that:

if $\Delta J_{\text{pump}} + \Delta J_{\text{emission}}$ is even the polarizations are parallel and

if $\Delta J_{\text{pump}} + \Delta J_{\text{emission}}$ is odd the polarizations are perpendicular. (14)

The rule, which has been called "Chang's Rule", was accounted for by *Henningsen* [2.13] by expressing the oscillator strengths for various types of transitions in terms of the integers J and K, which we have seen, and M, which represents the projection of the angular momentum vector onto a vector in a fixed direction, in this case the direction of the pump field. Evidently [2.17] the rule applies also to cascade transitions in which the pump and emission transitions do not share a common level.

A word of caution concerning polarization measurements is in order. As discussed and shown experimentally in [2.18], rule (14) is not absolute but arises from a gain difference between the two possible polarizations which may not be very large. If the preferred mode is deliberately suppressed, laser action will switch to the other mode if the gain is not too low compared with prevailing losses. For the same reason, unintentional polarization-sensitive elements in lasers, such as mirrors at non-normal incidence, or waveguide imperfections, have given rise to many conflicting measurements of the polarizations of laser lines. Some examples can be found in the notes and references given in Part II of this book.

References

2.1 C.H. Townes, A.L. Schawlow: *Microwave Spectroscopy* (Dover, New York 1975)

2.2 G. Herzberg: *Molecular Spectra and Molecular Structure—II. Infrared and Raman Spectra of Polyatomic Molecules* (Van Nostrand, New York 1945)

2.3 L. Herzberg, G. Herzberg: "Molecular Spectra", in *Fundamental Formulas of Physics,* Vol. 2, ed. by D.H. Menzel (Dover, New York 1960)

2.4 R.C. Johnson: *An Introduction to Molecular Spectra* (Methuen and Co., London 1949)

2.5 C. Rolland, J. Reid, B.K. Garside: Appl. Phys. Lett. **44**, 380–382 (1984)

2.6 N.M. Lawandy: Infrared Physics **21**, 235–239 (1981)

2.7 R. Beck, W. Englisch, K.Gürs: *Table of Laser Lines in Gases and Vapors*, Springer Series in Optical Sciences Vol. 2, (Springer, Berlin, Heidelberg, Third Edition 1980)

2.8 T.Y. Chang: "Optical pumping in Gases", in *Nonlinear Infrared Generation*, Topics in Applied Physics Vol. 16, ed. by Y.-R. Shen (Springer, Berlin, Heidelberg 1977)

2.9 L.C. Robinson: *Physical Principles of Far-Infrared Radiation* (Academic, New York 1973)

2.10 V.E. Zuev: *Propagation of Visible and Infrared Radiation in the Atmosphere* (Wiley, New York 1974; Russian original 1970)

2.11 C.H. Townes: I.A.U. Symp. **4**, 92–103 (1957)

2.12 G. Graner: Opt.Commun. **14**, 67–69 (1975)

2.13 J.O. Henningsen: IEEE J. **QE-13**, 435–441 (1977)

2.14 M. Inguscio, F. Strumia, J.O. Henningsen: "Far-Infrared Laser Lines from Optically Pumped CH_3OH", in *Reviews of Infrared and Millimeter Waves*, Vol. 2 ed. by K.J. Button, M. Inguscio, F. Strumia (Plenum, New York 1984)

2.15 J.O. Henningsen: Int. J. Infrared Mmwaves **4**, 707–732 (1983)

2.16 O.I. Baskakov, S.F. Dyubko, M.V. Moskienko, L.D. Fesenko: Sov. J. Quantum Electron. **7**, 445–449 (1977)

2.17 D. Dangoisse, P. Glorieux: "The Optically Pumped Formic Acid Laser", in *Reviews of Infrared and Millimeter Waves*, Vol. 2 ed. by K.J. Button, M. Inguscio, F. Strumia (Plenum, New York 1984); compare their Table I and Fig. 5.2.

2.18 L.B. Whitbourn, J.C. Macfarlane, P.A. Stimson, B.W. James, I.S. Falconer: Infrared Physics **28**, 7–20 (1988)

2.19 T.Y. Chang, T.J. Bridges: Opt. Commun. **1**, 423–426 (1970)

3. Wavelength and Frequency Measurements

To obtain laser action it is necessary not only to have an appropriate population inversion but also to ensure that the rate of stimulated emission, which depends on the intensity of the local radiation field, is high compared with loss mechanisms. This condition can often be met in pulsed systems even in the absence of a resonator, in so-called "mirrorless" lasers. In cw lasers, however, a resonator is invariably necessary to increase the intensity of radiation which is "circulating" in the laser medium; the intensity of the circulating radiation can be calculated by multiplying the output intensity of the laser by the inverse of the transmission coefficient of the output coupling optics.

To obtain maximum output power from the laser it is necessary to adjust the resonator to give the appropriate phase shift after one round trip. All cw (sub)mm lasers provide some means of adjusting the resonator length. *Steffen* and *Kneubühl* [3.1] pointed out that, as the length is adjusted, relatively distinct and well-defined output power maxima would be seen, and that these could be used to determine the wavelength with useful precision. They coined the term "Laser Resonator Interferogram" for such a record of the output power as a functon of cavity length, although the term *cavity scan* is perhaps now more usual. The technique is by no means self-evident—it is only possible because the profile of the laser transition, whose width is expressed by the "gain bandwidth", is (normally) much narrower than the frequency separation of axial cavity modes. Optical wavelength lasers for example do not usually satisfy this condition and laser action will occur for all cavity lengths. Pulsed (sub)mm lasers may have a gain bandwidth of a gigahertz or so which means that, unless the laser cavity is made very short, there will again always be laser action on at least one mode. In the present case, that of cw (sub)mm lasers, the cavity scan can invariably be used to deduce the wavelength, and indeed the majority of wavelength measurements are made this way. However the precise interpretation of the scans depends upon the type of resonator in use and there are, as discussed below, certain pitfalls to be avoided.

Figure 3.1 shows a typical open resonator laser set-up. Here the resonator is formed by two mirrors which I shall refer to as M_1 and M_2. The term "open resonator" is often used for such a cavity because the sides of the cylindrical resonator, even if they are physically present, play an extremely minor role in

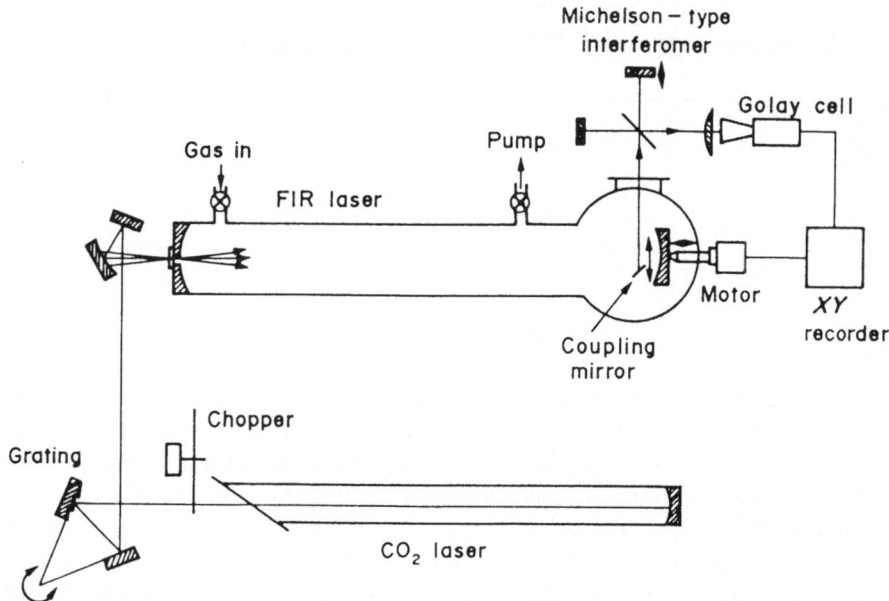

Fig. 3.1. Experimental arrangement described by *Kachi* and *Kon* [3.33]. The CO_2 pump laser cavity is formed by a mirror and a grating at the Littrow angle (autocollimation). Output coupling occurs via the zeroth order reflection from the grating. Note the hinge mount which ensures that the output beam is in a fixed direction for all wavelengths. The submillimeter laser resonator used here is of the Fabry-Perot type. A small amount of the radiation is coupled out to the detection system with a Michelson interferometer used to select the desired line when two or more lines oscillate. From [3.33]

determining the resonant frequencies. One sometimes speaks, rather loosely, of a "Fabry-Perot type" resonator, whereas the true Fabry-Perot is merely one member of the set of open resonators.

There are some circumstances in which one wants the losses of the resonator to be high, in which case one chooses an "unstable" design. We are dealing here however with stable resonators, in which case nearly all the radiation is reflected between the mirrors and the field intensity drops to a very low value at the edges of the mirrors, which can therefore be regarded as infinite in extent. Computations of the actual field patterns for certain cases were given in [3.2] and in subsequent papers by the same authors, and for the case (such as that shown in Fig. 3.1) where a perturbing hole is present, in [3.3] and [3.4]. With the approximation that the mirrors are infinite, the resonant frequencies are given by the expression [3.5]:

$$\frac{\nu}{\nu_0} = (q+1) + \frac{1}{\pi}(2p+l+1)\cos^{-1}\left[\left(1 - \frac{d}{R_1}\right)\left(1 - \frac{d}{R_2}\right)\right]^{1/2} \qquad (1)$$

where $R_1(R_2)$ is the radius of curvature of $M_1(M_2)$; d is the mirror separation; p, l, q are the number of field variations in the azimuthal, radial and axial directions, respectively; $\nu_0 = c/2d$ and c is the speed of light.

The sign of the term in brackets is to be retained after the square root is taken [3.6], although it is positive for all stable resonators. Equation (1) presupposes negligible diffraction loss, so it does not apply, for example, to plane-parallel resonators of Fresnel number less than about 20 (The Fresnel number is the mirror radius squared divided by d times the wavelength).

As the mirror spacing d is changed various modes, which we designate as TEM_{plq}, are brought into resonance with the lasing 0transition. An example of the resulting variation in output power is shown in Fig. 3.2. To compare these results with (1) above, we should in the latter keep the frequency ν fixed, this being determined after all by the lasing medium, and investigate the relationship between d on the one hand and the mode numbers p, l and q on the other. p and l may be said to characterize the *type* of mode while q, usually much larger, is the axial mode number. Henceforth all modes of the same type will simply be referred to as *a mode*, since this is the usual jargon. For example, four modes are labelled in Fig. 3.2, including TEM_{00} and TEM_{01}. The TEM_{11} mode, if it is present, cannot be seen because it falls almost exactly on the next TEM_{00} mode. In this example q can be estimated from the data given in [3.33] to be about 4000.

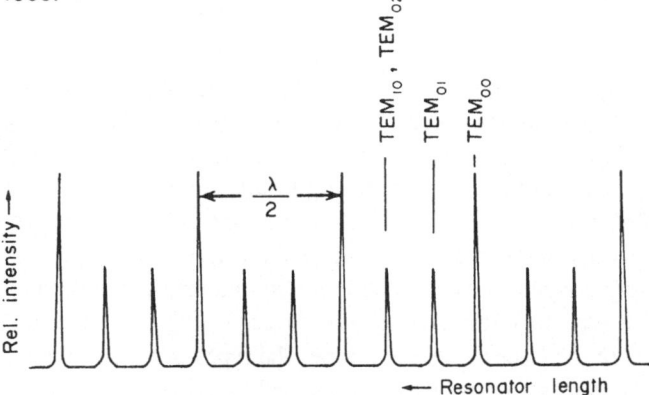

Fig. 3.2. Tuning curve of a submillimeter line observed in the laser of Fig. 3.1, formed as one of the resonator mirrors is shifted in position. The labels are explained in the text. From [3.33]

For a given mode p, l, the change in resonator length Δd between successive axial modes q, $q+1$ is *very nearly* equal to $c/2\nu$ or $\lambda/2$. This is the basis for the labelling in Fig. 3.2 of one half-wavelength between similar modes. The approximation can be checked using (1) and under most circumstances it is good to about 0.05% This is true even if p and l are not equal to zero, provided of course that the choice of p and l is fixed. A more serious problem is that a relatively large number of axial modes needs to be scanned to obtain this accuracy in practice, and this causes modes of different mode numbers p, l to slowly change their relative positions within the group. If such modes are not completely resolved the effect can be a change in the perceived positions of maxima. This becomes very serious when certain degeneracy conditions are met [3.7]. For example, the

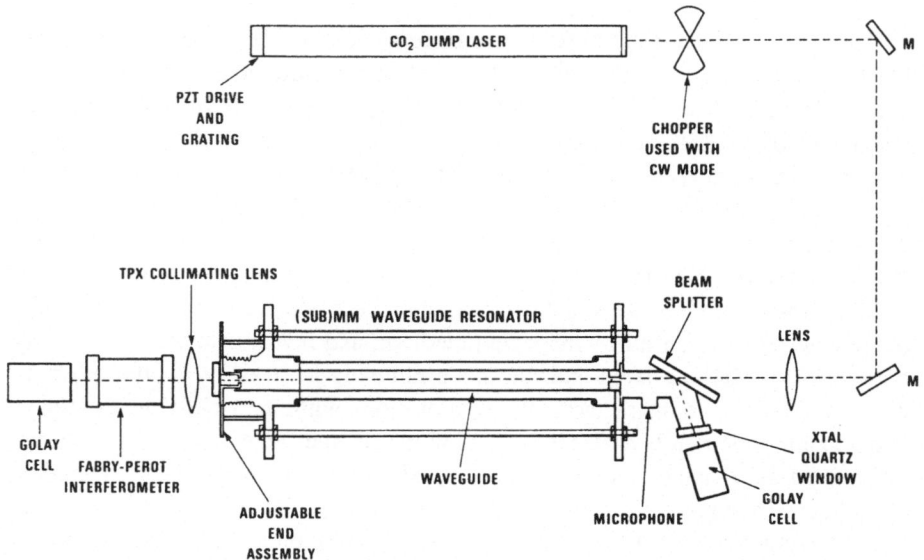

Fig. 3.3. Representative waveguide-type (sub)millimeter laser. In this case the resonator is of length 1 m. Adapted from [3.34]. Reproduced with the permission of the Harry Diamond Laboratories, Adephi

symmetric *confocal resonator* ($d = R_1 = R_2$) has only two distinct sets of modes in its cavity scan, depending on whether l is odd or even. Therefore both sets are highly degenerate. As d is varied, so as to break the confocal condition, the sets of modes slowly split up and could lead to a confusing result. Moreover, the l-odd modes appear *exactly half way* between the l-even modes and would, if present with equal strength, suggest a wavelength one-half of the true wavelength.

In (sub)millimetre work nowadays it is less common to find the open resonator laser than the *waveguide* type of laser such as that shown in Fig. 3.3. Here the transverse shape and size of the laser vessel determines the types of modes, which are of the same general nature as those of the corresponding waveguide. Cavity scans are generated with the aid of a flexible section, such as a bellows, which houses the end mirror. Periodicities in the output on a given mode correspond to one half of the guide wavelength λ_g in that mode. This is related to the cutoff wavelength λ_c, the longest wavelength at which radiation can propagate in that mode, by [3.8]

$$\lambda_g = \frac{\lambda}{\sqrt{1 - (\lambda/\lambda_c)}} \quad . \tag{2}$$

The guide wavelength can differ markedly from the free-space wavelength λ and so wavelengths measured on this basis must be carefully corrected. This requires an absolute identification of the mode in question, which is not so in the Fabry-Perot resonator case, where modes with fixed values of p and l reappear with the same periodicity (very nearly) as do modes with any other fixed values of p and l. For example in Fig. 3.4, which is a cavity scan made with a circular metallic waveguide laser of 19mm diameter, many different periodicities

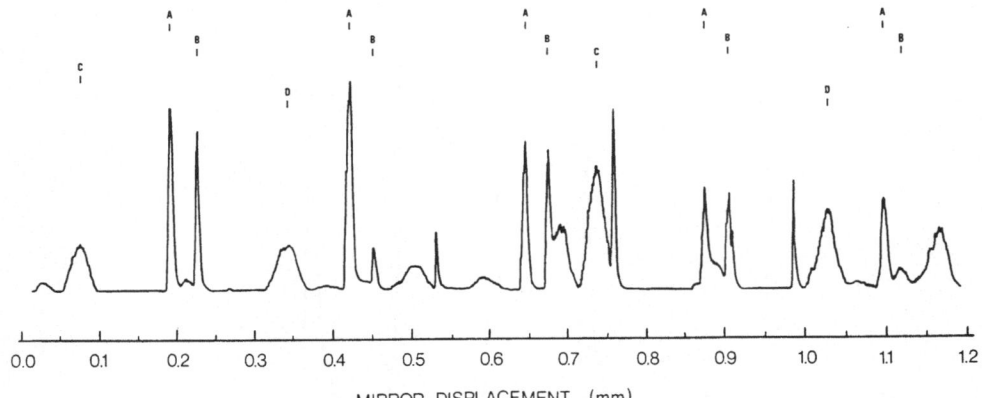

Fig. 3.4. Segment of a cavity scan, or "Laser Resonator Interferogram", in this case for a waveguide laser. Despite the many peaks only two distinct laser transitions are involved — at 447 μm and 1306 μm free space wavelength. The marked features have the following guide wavelengths: A and B — 447 μm, C — 1310 μm, D — 1346μm. The gas was ethyl chloride and the pump line 10R28. Adapted from [3.9] by the author.

are seen, including "wavelengths" of 1310 and 1346 μm. In fact these are due to the same line, with a free-space wavelength near the smaller value. Guide wavelengths are always longer than the free-space wavelength. Inspection of Fig. 3.4 also shows that features appear and reappear as a result of competition, partly between different laser lines (in this case a 447 μm line is present) and partly between modes of the same wavelength. This can lead to fairly gross errors in measurement as pointed out in [3.9].

The use of an external Fabry-Perot interferometer is preferable to cavity scans, for the reasons given above. A wavelength accuracy of better than 10^{-3} is typical; at shorter wavelengths even higher accuracy is possible (if the finesse is the same) because more scans can be observed with a given maximum plate displacement. Moreover, at longer wavelengths the deviation from free-space wavelength has to be taken into account if the desired accuracy is to be much above this figure — in a Fabry-Perot resonator whose diameter is twenty (or less) times the wavelength, the resonator wavelength is greater than the free-space wavelength by 7×10^{-4} (or more) [3.10]. This fact is often overlooked. At shorter wavelengths the problem is usually that of attaining sufficient finesse — if reflectors of metal-mesh [3.11] are used, as is common [3.12], then a finer mesh has to be chosen at shorter wavelengths to attain the same finesse. A segment of an external Fabry-Perot interferogram is shown in Fig. 3.5.

As well as the Fabry-Perot, other types of interferometers have of course been used. Noteworthy was the measurement, as early as 1969, of the wavelength of laser lines to an accuracy of 10^{-6} using a long path-length Michelson interferometer [3.13]. Frequencies had already been measured for some of the lines involved so that a calculation of the speed of light was possible (see below). Noteworthy also is the use of the CO_2 laser radiation, whose wavelength is accurately

21

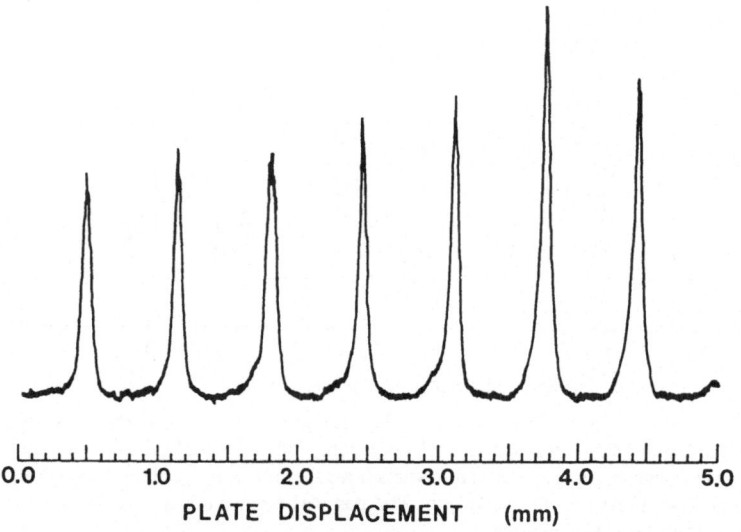

Fig. 3.5. Segment of a Fabry-Perot interferogram for the 1306 μm ethyl chloride line. The laser was the same as that which gave rise to the tuning curve in Fig. 3.4 but was now set to one of the output maxima of type d. The changing amplitude of the peaks is due to a slow drift in the laser cavity, causing a change in the output power of the laser. From [3.9]

known, to calibrate the scan of either an external Fabry-Perot interferometer or of the (sub)mm laser cavity [3.14]. If carefully calibrated, a grating spectrometer can also achieve 0.05% accuracy, as indicated for example in the measurements reported in [3.15].

The property of a Josephson-junction whereby the frequency of the illuminating radiation is converted into a voltage, with a conversion factor based on fundamental constants, means that it too can be used as a spectrum analyzer well suited to (sub)millimetre lasers [3.16].

For a number of purposes, the accuracy obtainable by the interferometric measurement of wavelength is not adequate. The most obvious of these purposes is molecular spectroscopy of the lasing molecule itself, which was discussed in Sect. 2. When used as a local oscillator in an astronomical receiver one would also like to know the laser frequency to within a few megahertz so as to know the radial velocity of the observed objects to within a few km/s. In metrology too, where the laser might be used in a chain to link microwave measurements with those made in the optical, high precision is necessary. For such purposes heterodyne measurements, which yield the frequency directly, are to be preferred, and these are now discussed.

Frequency measurements on (sub)mm lasers had been made even before *Chang* and *Bridges'* paper [3.17]. In 1967 *Frenkel* et al. achieved an accuracy of about 10^{-8} in their measurement of the frequency of the 118.6 μm water-vapour line by phase locking the microwave oscillator used in their experiment to a stable crystal frequency source [3.18]. At about the same time the frequency of the 337 μm line of the HCN discharge laser was measured [3.19]. By mixing with the fourth harmonic of the HCN line *Hocker* et al. [3.20] reported in 1969 that they had measured the frequency of the 84 μm D_2O line. Their result (3 557 143 MHz) could be combined with the best interferometric measurements of the wavelength at the time to calculate a value for the speed of light of 299 792.7 $\pm$ 2.0 km/s. In 1970 *Evenson* et al. [3.21] reported that they had extended the technique to the 28 μm water-vapour line, whose frequency is around 10.7 THz.

Chang and *Bridges* themselves, reporting the first (pulsed) optically pumped submm laser [3.17], CH_3F pumped by the $9P20$ line of CO_2, measured the frequency of six lines. Two of these have since been remeasured in cw lasers. To date about a fifth of known lines have yielded to frequency measurements, the vast majority of these being the work of just a few groups. It can be concluded from this that frequency measurements remain relatively difficult in practice, although the main principles involved can be summarized briefly. All direct frequency measurements require that the radiation be referenced, directly or indirectly, to one or more oscillators whose own frequencies can be counted electronically. Such counting is presently possible up to a few tens of gigahertz. The three ways of referencing a (sub)mm laser to this low microwave regime may be crudely described as "from below", "from above", and "sideways". They all rely on the existence of fast, non-linear detectors which will generate sum and difference signals of the radiation incident on them as well as, where necessary, significant signals at the harmonics thereof. When this requirement is met and two signals ν_1 and ν_2 are present, then detectable signals at $f = n_1\nu_1 \pm n_2\nu_2$ may also be present. ν_1 or ν_2 has to be varied, and the harmonic numbers n_1 and n_2 chosen so that the resulting frequency ν_3 is countable. A review of suitable detectors was given in [3.22].

Harmonic mixing, or mixing "from below", involves the *generation of large multiples of a low frequency microwave signal*. For convenience one often chooses to use the highest frequency that can be conveniently coupled coaxially to the detector — this is usually considered to be about 18 GHz. The klystron usually used to produce this signal is itself referenced, in fact usually locked, to some stable reference frequency. Strong laser lines up to a few hundred gigahertz have been measured this way. Higher frequencies can be reached by using oscillators in the range 75–90 GHz, which is a band where strong sources are available commercially. At these frequencies the signal has to be coupled optically. Harmonics up to the 33rd have been generated in Schottky diodes [3.23] and up to the 401st in Josephson-junctions [3.24]. This type of arrangement is shown in Fig. 3.6, which depicts the apparatus used by the National Physical Laboratory in England [3.25, 26].

Mixing "from above" is a technique pioneered by National Institute of Standards and Technology (N.I.S.T), formerly the National Bureau of Standards. It involves the use of oscillators whose frequencies, higher than those of the lines to

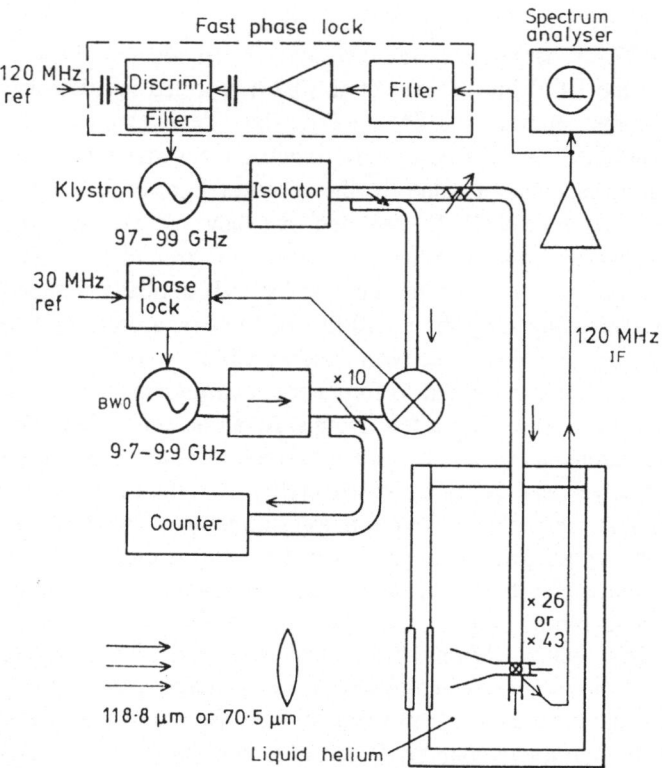

Fig. 3.6. Schematic diagram of a frequency-measurement system using a Josephson junction in a waveguide mount. The apparatus has been set up to measure the frequency of 118.8 or 70.5 μm radiation entering via the lens near the bottom of the drawing. From [3.26]. National Physical Laboratory, UK Crown Copyright

be measured, are known. Possibilities would then include measuring multiples of the (sub)mm frequency by beating with the higher frequency oscillator (another laser) or adding a high frequency signal to the (sub)mm one to produce a tone in the vicinity of a yet higher, but known, frequency. Such higher frequencies may have been measured by techniques not directly applicable to the (sub)mm lasers. In the application chosen by the N.I.S.T., the *difference frequencies between two* CO_2 *lasers* are used [3.27]. The CO_2 laser frequencies themselves are known from previous work (see Chap. 5). The two lasers, together with the (sub)mm radiation to be measured, illuminate a point-contact diode, such as a W-Ni contact, in which the difference frequencies are generated. Figure 3.7 shows the experimental arrangement. For example, the $9P(14)$ and $10R(8)$ lines differ by 2532.9 GHz, which is just 10.1 GHz away from the frequency of the 118.834 μm CH_3OH laser line. To measure the CH_3OH line, the beat note between the three lasers was shifted down to a measurable 54 MHz with the aid of an additional oscillator [3.27]. One complication is to establish whether the difference between the (sub)mm and the synthesized frequencies is positive or negative, which can

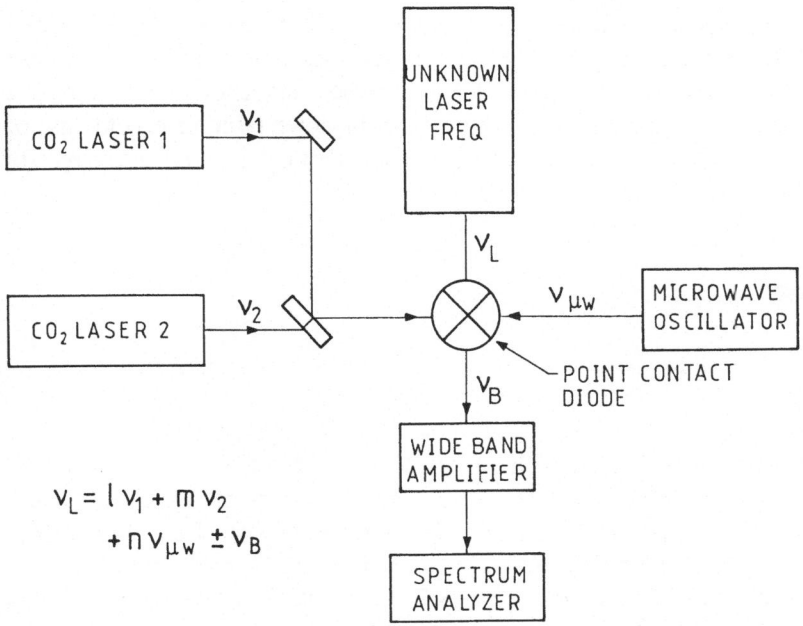

Fig. 3.7. Schematic diagram of the two-CO_2 laser frequency measurement system. In many instances, such as that described in the text, the integers l, m and n take on the values ± 1. Adapted from [3.27]

be done by detuning either CO_2 laser slightly. Two normal CO_2 lasers yield over 7000 difference frequencies, although in practice an additional microwave signal of a few tens of GHz is necessary to bring the combination close enough to the (sub)mm line of interest.

One should add in passing that the usefulness of CO_2 laser difference frequency generation as a source of (sub)mm radiation had been recognised for some time before this (see the review [3.28]) but the N.I.S.T. technique is particularly elegant, since it does not require any complicated phase matching. The accuracy of these techniques is limited by the resetability of the (sub)mm laser used—in the case of the N.I.S.T., an uncertainty of about 2×10^{-7} is claimed [3.29].

A good and recent review of optical frequency measurement, which expands somewhat on the above discussion of (sub)mm heterodyne techniques, can be found in [3.30].

Once a number of (sub)mm lines have been measured it becomes possible to extend the set "sideways" by forming the *difference frequency between the unknown laser line and one close to it in frequency* which has been measured by one of the other techniques. The group at the Dipartimento di Fisica in Pisa, Italy, has been particularly active in applying this technique [3.29] which they attribute to *Epton* et. al. [3.31], although it had been used some time earlier by *Hocker* and *Javan* [3.32]. It is considerably simpler than "direct" measurements and the slight decrease in accuracy is acceptably small, being of order $\sqrt{3}$ [3.29]. Since the difference in frequency between the two lasers is small, it is of course necessary to determine which laser has the higher frequency. This requires a fairly accurate wavelength measurement of the unknown laser, or else the use of a "trick", such as a slight detuning of either laser. This can be done mechanically unless the lasers are highly stabilized [3.35].

References

3.1 H. Steffen, F.K. Kneubühl: IEEE J. **QE-4,** 992–1008 (1968)

3.2 A.G. Fox, T. Li: Bell Sys. Tech. J. **40,** 453–488 (1961)

3.3 D.E. McCumber: Bell Sys. Tech. J. **44,** 333–363 (1965)

3.4 G.T. McNice, V.E. Derr: IEEE J. **QE-5,** 569–575 (1969)

3.5 H. Kogelnik: "Modes In Optical Resonators", in *Lasers — A Series Of Advances,* Vol. 1, ed. by A.K. Levine (Arnold and Dekker, New York 1966)

3.6 G.D. Boyd, H. Kogelnik: Bell Sys. Tech. J. **41,** 1347–1369 (1962)

3.7 D.J. Harris, T.M. Teo, R.J. Batt, S.C. Luk: Int. J. Infrared Mmwaves **1,** 339–349 (1980)

3.8 E.A.J. Marcatili, R.A. Schmeltzer: Bell Sys. Tech. J. **43,** 1783–1809 (1964)

3.9 N.G. Douglas, P.A. Krug: IEEE J. **QE-18,** 1409–1410 (1982)

3.10 F. Westermann, W. Maier: Z. Phys. **179,** 507–524 (1964); a companion article appeared in Z. Phys. **179,** 244–255 (1964)

3.11 Buckbee-Mears Co., 245 E.6th St., St.Paul, Minnesota 55101, U.S.A. As far as I am aware, the company is almost oblivious to the great service it provides to the (sub)mm community by producing inductive mesh!

3.12 E.A.M. Baker, B. Walker: J. Phys. E: Sci. Instrum. **15,** 25–32 (1982)

3.13 V. Daneu, L.O. Hocker, A. Javan, D. Ramachandra Rao, A. Szöke, F. Zernike: Phys. Lett. **29A,** 319–320 (1969)

3.14 J-M. Lourtioz, J. Pontnau, M. Morillon-Chapey, J-C. Deroche: Int. J. Infrared Mmwaves **2,** 49–63 (1981)

3.15 L.E.S. Mathias, A. Crocker: Phys. Lett. **13,** 35–36 (1964)

3.16 T.G. Blaney: J. Phys. E: Sci. Instrum. **4,** 945–948 (1971)

3.17 T.Y. Chang, T.J. Bridges: Opt. Commun. **1,** 423–426 (1970)

3.18 L. Frenkel, T. Sullivan, M.A. Pollack, T.J. Bridges: Appl. Phys. Lett. **11,** 344–345 (1967)

3.19 L.O. Hocker, A. Javan, D. Ramachandra Rao, L. Frenkel, T. Sullivan: Appl. Phys. Lett. **10,** 147–149 (1967)

3.20 L.O. Hocker, J.G. Small, A. Javan: Phys. Lett. **29A,** 321–322 (1969)

3.21 K.M. Evenson, J.S. Wells, L.M. Matarrese, L.B. Elwell: Appl. Phys. Lett. **16,** 159–162 (1970)

3.22 D.J.E. Knight, P.T. Woods: J. Phys. E: Sci. Instrum. **9,** 898–916 (1976)

3.23 H.R. Fetterman, B.J. Clifton, P.E. Tannenwald, C.D. Parker: Appl. Phys. Lett. **24,** 70–72 (1974)

3.24 D.G. McDonald, A.S. Risley, J.D. Cupp, K.M. Evenson, J.R. Ashley: Appl. Phys. Lett. **20**, 296–299

3.25 T.G. Blaney, D.J.E. Knight, E.K. Murray Lloyd: Opt. Commun. **25**, 176–178 (1978)

3.26 T.G. Blaney, N.R. Cross, D.J.E. Knight, G.J. Edwards, P.R. Pearce: J. Phys. D: Appl. Phys. **13**, 1365–1370 (1980)

3.27 F.R. Petersen, K.M. Evenson, D.A. Jennings, J.S. Wells, K. Goto, J.J. Jiménez: IEEE J. **QE-11**, 838–843 (1975)

3.28 R.L. Aggarwal, B.Lax: "Optical Mixing of CO_2 Lasers in the Far-Infrared", in *Nonlinear Infrared Generation,* ed. by Y.-R. Shen (Springer, Berlin, Heidelberg 1977)

3.29 M. Inguscio, G. Moruzzi, K.M. Evenson, D.A. Jennings: J. Appl. Phys. **60**, R161–R192 (1986)

3.30 D.A. Jennings, K.M. Evenson, D.J.E. Knight: Proc. IEEE **74**, 168–179 (1986)

3.31 P.J. Epton, W.L. Wilson, Jr., F.K. Tittel, T.A. Rabson: Appl. Opt. **18**, 1704–1705 (1979)

3.32 L.O. Hocker, A. Javan: Appl. Phys. Lett. **12**, 124–125 (1968)

3.33 T. Kachi, S. Kon: Infrared Phys. **22**, 337–341 (1982)

3.34 M.S. Tobin: Proc. IEEE **73**, 61–85 (1985)

3.35 D.J.E. Knight, N.P.L.: private communication (1988)

4. Power Measurements

The lowest intensities of (sub)mm laser emission usually considered to be useful are at the microwatt level, and such emission can easily be detected with a Golay cell [4.1] or pyroelectric detector [4.2], especially when chopping and synchronous detection is used. These devices were not designed for such long wavelengths, however, and in particular they have very small entrance apertures through which (sub)mm radiation couples in a fairly unpredictable way. The Golay cell is furthermore totally saturated by radiation at the level of a few tens of microwatts, if properly coupled. This means that such detectors are unsuitable for absolute power measurement, unless considerable effort is made to calibrate them. The majority of reports of new (sub)mm laser lines therefore either give no indication of strengths, or else give relative indications such as "very strong", "medium", "very weak", etc.

Moreover, the power level achieved on a given line depends not only on fixed molecular properties, such the pump laser offset and the rotational constants of the laser molecule, but also on factors which vary between experiments, such as the particular design of the laser concerned. The most obvious example of this is that a hole-coupled laser working well at short wavelengths may give very little output at long wavelengths owing to the cutting off of radiation by the hole. Unless an attempt is made to optimize the laser, it seems futile — even counterproductive — to quote the power levels obtained.

It should therefore not seem surprising that no estimate of the strength of the laser emission is given for the laser lines in the second part of this book. Nevertheless there are many users who are mainly interested in "sure-fire" laser lines for particular purposes. They want to know what lines are available in various parts of the (sub)mm spectrum, and what results they might expect in a few cases that have been particularly well researched. For such purposes I have compiled Table 4.1. This is based partly on review articles, such as the excellent one by *Tobin* [4.3]. I have supplemented these with data from the literature on which Part II of this book is based. I have corrected where necessary to "true" power levels, using only papers in which enough information was given about the measurements to enable this to be done reliably. I also left out very weak lines, because for any conceivable purpose in which low power levels are adequate, the user can consult directly the compilation in Part II. The criterion for inclusion in Table 4.1 is that the power in milliwatts should be greater than 100 divided by the wavelength in μm; this represents a fixed value of the quantum efficiency and thus avoids discriminating against longer wavelengths.

Table 4.1. Calibrated Power Measurements

λ [μm]^a	molecule^b	pump	pump power [W]	(sub)mm power [mW]^c	reference(s)	footnotes
11 - 13	NH3			> 100	[4.4]	1
41.0	CD3OD	10R18	33	60	[4.5]	
42.159	CH3OH	9P32		50	[4.6]	16, 26
67.495	CH3OH	9R18	39	22	[4.5]	2
69.680	CH3OH	10R16	42	10	[4.5]	
70.512	CH3OH	9P34	30	100	[4.7]	
78.5	15NH3	13CO2	14	2	[4.8]	
81.497	14NH3	N2O	3	40	[4.9]	16
86.7	CH3OD	13CO2	6	12	[4.8]	
96.522	CH3OH	9R10		300	[4.6]	16, 24
109.296	CH2F2	9P24	22	5	[4.10]	
110.0	15NH3	13CO2	16	8	[4.8]	
111.9	15NH3	CO2S	8	11	[4.11]	
112.3	15NH3	13CO2	16	12	[4.8]	
112.6	D2O	CO2S	3	6	[4.12]	6
117.727	CH2F2	9R20	23	65	[4.10]	
118.834	CH3OH	9P36	125	1250	[4.13]	18
122.15	CD3OH	10R38		1.5	[4.42]	21
122.30	CD3OD	10R28		10	[4.42]	21
122.466	CH2F2	9R22	12	2	[4.10]	22
124.6	NH3	13CO2	15	3	[4.8]	
127.021	13CD3OH	10P08	8	33	[4.48]	
135.269	CH2F2	9P24		5	[4.14]	7, 17
139.6	SO2	9R14	16	5	[4.15]	
142.1	SO2	9R18	16	4	[4.15]	
143.4	CD3OH	13CO2	14	3	[4.8]	
147.845	CH3NH2	9P24		10	[4.16]	18
152.7	NH3	13CO2	19	180	[4.5]	
158.513	CH2F2	9P10		15	[4.10], [4.14]	17
160.0	SO2	9R28	11	1.3	[4.15]	
163.034	CH3OH	10R38	31	18	[4.5]	3
166.631	CH2F2	9R20	21	3	[4.10]	
167.5	CD3OD	13CO2	9	3	[4.8]	
171.8	CH2CHF	10P22		1	[4.17]	18
182.0	CD3OD	13CO2	6	5	[4.8]	
184.306	CH2F2	9R32	140	833	[4.18]	9
184.766	CD3OD	10R24	37	2	[4.19]	
185.	CHFCHF	9R08	7	1.5	[4.20]	
189.832	CD2F2	10R34	1.5	1.3	[4.21]	
189.949	DCN	ELEC		221	[4.46]	18
190.008	DCN	ELEC		149	[4.46]	18
194.701	DCN	ELEC		252	[4.46]	18
194.764	DCN	ELEC		147	[4.46]	18
196.1	CH2F2	9R32	37	44	[4.47]	18, 23

λ [μm]ᵃ	moleculeᵇ	pump	pump power [W]	(sub)mm power [mW]ᶜ	reference(s)	footnotes
214.579	CH2F2	9R34		25	[4.14]	17
218.6	15NH3	CO2S	7	23	[4.11]	
221.0	CD3OH	13CO2	6	0.5	[4.8]	
234.7	CD3OH	13CO2	11	1	[4.8]	
235.5	CH2F2	9R06		6	[4.14]	10, 17
235.654	CH2F2	9R32	18	5	[4.10],[4.14]	17
240.98	CH3CL	10P10	90	1.3	[4.22]	
255.0	CD3OD	10R36	25	19	[4.5]	
256.027	CH2F2	9P24		3	[4.14]	8, 17
260.1	CHFCHF	10R20	7	0.8	[4.20]	
261.729	CH2F2	9P38	19	0.3	[4.10]	
263.4	NH3	N20	3	0.5	[4.11]	
264.536	CH3OH	9P34	33	4	[4.30]	
272.339	CH2F2	9P10		0.5	[4.14]	17
287.667	CH2F2	9R34		9	[4.14]	17
289.5	CH2F2	9P04	16	3	[4.10],[4.41]	
298.736	CD3OD	10R24	37	0.4	[4.19]	
305.726	CH3OD	9R08	20	3	[4.30]	
307.5	CHFCHF	10R30	< 40	17	[4.3]	
309.8	CD3OH	13CO2	10	1.3	[4.8]	
310.	CHFCHF	10R34	5	1.6	[4.20]	
310.0	CD3OD	13CO2	13	3	[4.8]	
310.8	CHFCHF	9P34	< 40	12	[4.3]	
325.9	13CD3F	9R22		9	[4.3]	
326.6	CHFCHF	10P06	< 40	15	[4.3]	
336.558	HCN	ELEC		600	[4.23]	11
344.778	CD3OD	10R04	27	0.6	[4.19]	
349.387	CH3CL	10R18	89	7	[4.22]	
355.5	CD3OD	10R16	37	13	[4.8]	
369.968	HCOOD	10R28		19	[4.3]	
372.814	CH3CN	10P20	40	1	[4.30]	
373.4	15NH3	10R42	16	23	[4.11]	
380.565	DCOOD	10R12		12	[4.24]	
382.639	CH2F2	9P10	< 40	10	[4.3]	
384.319	CHD2F	10P28	< 30	12	[4.3]	
388.273	CDF3	10R32	1.2	1.1	[4.15]	
388.5	15NH3	13CO2	14	8	[4.8]	
393.631	HCOOH	9R18	150	330	[4.18],[4.25]	
405.585	HCOOH	9R18		6	[4.26]	
410.712	CD3OD	10R12	37	0.5	[4.19]	
418.3	HCOOH	9R18		6	[4.26]	
418.613	HCOOH	9R22	33	65	[4.27]	
432.109	HCOOH	9R22		10	[4.26]	
432.631	HCOOH	9R20	33	60	[4.27]	19
432.667	HCOOH	9R20		84	[4.3]	
432.6	HCOOH	9R20	150	222	[4.18]	20
435.427	CHD2F	10R38	< 30	13	[4.3]	
443.265	CD3CL	9P10		4	[4.28]	16

λ [μm]ᵃ	moleculeᵇ	pump	pump power [W]	(sub)mm power [mW]ᶜ	reference(s)	footnotes
444.4	CH2CHF	10R20		3	[4.17]	18
445.9	HCOOH	9R20		14	[4.26]	
446.505	HCOOH	9R22		0.9	[4.26]	
447.142	CH3I	10P18	< 90	100	[4.3]	4
459.6	CDF3	10R26	1.6	1.1	[4.21]	
460.0	HCOOH	9R20		1.2	[4.26]	
460.562	CD3I	9R12		18	[4.3]	
461.261	HCOOD	10P16		10	[4.24]	
470.065	13CD3F	10P34		14	[4.3]	
472.4	CH2CHF	10P20		0.4	[4.17]	18
478.072	COF2	10P24	19	21	[4.3]	
488.528	CDF3	10R38		28	[4.3]	
496.1	CH3F	9P20	135	230	[4.18], [4.25]	13
500.577	CD2F2	10R24	3	4	[4.21]	
505.829	COF2	10P22	19	6	[4.3]	
509.791	DCOOD	10P08		1	[4.24]	
513.002	HCOOH	9R28	65	10	[4.3]	
513.016	HCOOH	9R28		35	[4.3]	
513.0	HCOOH	9R28	33	37	[4.27]	12
514.951	DCOOD	10P34		2	[4.24]	
516.382	COF2	10R08	9	3	[4.3]	
530.4	CD3OH	13CO2	11	0.9	[4.8]	
538.415	COF2	10P16	17	1.4	[4.3]	
548.7	CD2F2	10R28	2	1.4	[4.21]	
554.365	CH2CF2	10P14		4.5	[4.3]	25
557.7	CH2CF2	10P22	1.4	3	[4.21]	
567.868	DCOOD	10R26		4	[4.24]	
570.569	CH3OH	9P16	37	38	[4.5]	
581.984	CDF3	10R28	2	1	[4.21]	
582.1	CDF3	10R12	2	2	[4.21]	
605.6	CDF3	10R20	2	0.6	[4.21]	
657.989	CDF3	10R10		5	[4.21]	
662.816	CH2CF2	10P24		7	[4.3]	
672.1	CH2CHF	10P36		0.6	[4.17]	
691.250	CHD2F	10R26	10	6	[4.3]	
699.423	CH3OH	9P34	33	1	[4.30]	
699.636	COF2	10R22	13	9	[4.3]	
742.573	HCOOH	9R40	90	24	[4.18]	
792.	CD3CL	9P28		2	[4.28]	16
858.300	CD3OH	10R20		6	[4.3]	
877.548	DCOOD	10P26		2	[4.24]	
888.862	CH2CF2	10P22		4	[4.3]	
917.0	CH3OD	13CO2	11	0.3	[4.8]	
919.936	HCOOD	10R32		10	[4.24]	
926.209	HCOOD	10R14		8	[4.24]	
944.019	CH3CL	9R12	45	13	[4.22]	
954.467	CH2CF2	10P30		1	[4.3]	
1008.558	CDF3	10R12	2	1	[4.21]	

λ [μm][a]	molecule[b]	pump	pump power [W]	(sub)mm power [mW][c]	reference(s)	footnotes
1135.070	COF2	10R14	13	6	[4.3]	
1221.893	13CH3F	9P32	140	50	[4.18], [4.25]	15
1253.738	CH3I	10P32		4	[4.3]	
1260.561	CDF3	10R16	3	1	[4.21]	
1900.	COF2	10R08	9	0.5	[4.3]	
1990.757	CD3CL	9P14		0.5	[4.28]	16
2070.188	CH2CF2	10P14		> 1	[4.29]	14

a – I have given the best wavelength available (see tables in Part II). The measurement accuracy is indicated by the number of significant figures.

b – Most of the molecule codes are self-evident. They are listed in Chap. 10.

c – I have corrected all values for the calorimeter response, according to the curves of *Foote* et al ([4.31]; see text) except where noted otherwise. Beyond 500 μm wavelength the (constant) correction factor was simply taken to be 0.5, owing to the lack of better data. All values greater than 2 mW were rounded to the nearest mW.

footnotes

1. Most of the power levels reported for these lines are greater than 100 mW. Under special conditions 3W [4.43] and 8W [4.44] can be obtained on some lines. The lines, which are pumped by various CO_2 laser lines, are separately listed in Part II of this book.
2. Reported as 65 μm wavelength.
3. On the basis of precise wavelength measurements [4.32] and later frequency measurements, the pump is assumed to be 10*R*38 not 10*R*34 as stated in this reference.
4. [4.18] reports 110 mW with 130 W pump power.
5. The line is of 15NH₃ but the reported power levels were in NH_3.
6. Estimated by comparison between a Golay detector and calorimeter.
7. [4.10] reports 3 mW with 22 W pump power.
8. [4.10] reports 2 mW with 22 W pump power.
9. [4.25] report a value of 660 mW (corrected) for this line, while [4.6] cite a value of 150 mW (without reference).
10. Probably the same line as that reported at 236.592 μm (Chap. 7).
11. [4.33] mentions a figure of 4100 mW but do not give further details.
12. Not stated which component is involved; presumed [4.45] to be that at 513.016 μm.

13 Not stated which component is involved; [4.7] reported 22 mW with 30 W pump power using a buffer gas.

14 Despite the lack of experimental details and the absence of this line in the reference cited by [4.29], it is included here as a dubious measurement because of its interestingly long wavelength.

15 [4.5] reported 14 mW from 35W pump.

16 Although the details of the power measurement were not reported the line is important enough to be included.

17 A non-standard meter correction was used.

18 This power measurement is presumably uncorrected for meter response.

19 Of the two lines near 432.6 μm the authors clearly identify the one of *shorter* wavelength as the stronger and the other as "approximately 50% less". The corresponding statements in the abstract and summary are incorrect [4.45].

20 Not stated which component is involved.

21 Obtained with a waveguide pump laser, which improved the absorption coincidence.

22 Using a waveguide pump laser, [4.42] obtained 16 mW submm output.

23 This is a cascade transition from the 184.306 μm line. The power on that line was 146 mW.

24 [4.30] reported 25 mW with 21 W pump power

25 [4.30] reported 1 mW with 40 W pump power

26 For the line at 41.7 μm, which could be the same as this one, [4.30] reported 55 mW with 33 W pump power

---

Please keep in mind my admonition that there are many potentially strong lines in the compilation which are not in Table 4.1 just because the measurements have not yet been made, or have not yet been reported, or because no-one has attempted to optimise on those lines.

Absolute power measurements are fairly difficult to perform in the (sub)mm part of the spectrum, because of uncertainties in the beam propagation, because of diffraction effects, and because of poor matching to detectors, which results in standing-wave effects and unterminated reflectances. Only a handful of researchers have addressed this problem, so their efforts can be summarized briefly.

Power measurements are very often carried out with the aid of a bolometric calorimeter. This consists of an absorbing disc attached to a thermopile and, usually, a housing to isolate the sensitive element thermally from the surroundings. Several different types are available commercially [4.34, 35], and the sensitivity of one of these (the Scientech 1 in. disc calorimeter) in the (sub)mm has been investigated in a much-quoted paper by *Foote* et al. [4.31]. Many researchers who have used this type of calorimeter, or even ones like it, have corrected their readings by a calibration factor derived from this paper (see Fig. 4.1). The calibration factor is derived by calculating the absorptance of the thermopile disc

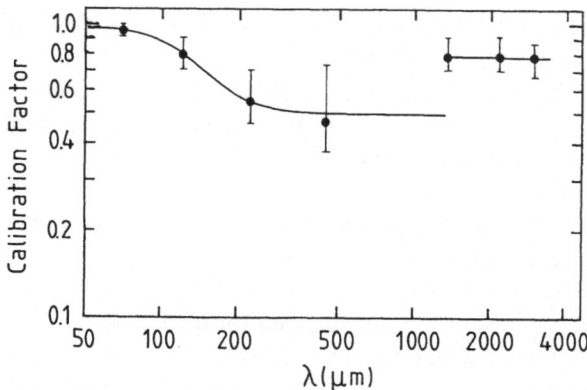

Fig. 4.1. Calibration curve of the Scientech calorimeter (see text) which is widely used in (sub)milli-meter power measurements. The calibration wavelengths are at 70, 119, 220, 447, 1400, 2150 and 3000 μm. The curve drawn through the data represents my choice for a "standard" fit – see text. Adapted from [4.31]

from reflectance and transmission measurements, but exactly how this was done is unclear from the data presented. At any rate, reference to Fig. 4.1 shows that the calibration factor at for example 447 μm wavelength is 0.45 (with considerable uncertainty). This means that a given calorimeter reading has to be divided by 0.45 to obtain the true power. The reading itself would either be the voltage delivered by the thermopile, interpreted according to some nominal sensitivity, or the value given by a meter which can be used with the calorimeter and which uses a dc substitution method to improve the accuracy.

There are several reasons why power measurements made on this basis may still suffer from systematic errors. For example, the assumption has been made that the beam does not diffract significantly within the isothermal enclosure, and that the reflectance of the disc is indeed small so that no standing waves are set up. It is also evident from Fig. 4.1 that there is a systematic difference between the microwave data and the (sub)mm data at wavelengths up to 447 μm, these sets of data having been obtained using different procedures. Finally, one has to assume that individual calorimeters all have the same behaviour, which might not be true if, for example, the precise thickness of various layers in the thermopile is crucial.

Nevertheless the use of this calibration curve is very general and *Vowinkel* and *Röser* [4.36], using a calorimeter of a quite different construction, obtained good agreement with the corrected sensitivity of the Scientech calorimeter at submm wavelengths. Their calorimeter used a waveguide input and a dc substitution calibration method. A calorimeter of somewhat similar design is commercially available; with appropriate compensation for waveguide losses it can be used at frequencies of up to 690 GHz [4.38]. A related design is described in [4.39]. An accuracy of 10% is claimed over a sensitive range of 0.4 μm to over 500 μm.

The National Physical Laboratory in the United Kingdom made measurements on a similar device, the Laser Instrumentation 17S radiometer. They assumed that reflectance was the major loss factor. Their measurements at 337 μm used a path-length modulator to eliminate both standing-wave effects and feedback to the HCN laser source. They obtained a calibration factor of 0.79, but reflection measurements showed that this would be highly variable with wavelength. However, a slight increase in the thickness of the absorbing paint on the disc gave an improvement to 0.94, remaining constant from a few tens of μm out to about 500 μm [4.37].

For the purposes of this book, including the preparation of Table 4.1, I took calibration factors for Scientech-type calorimeters from the curve drawn through the data in Fig. 4.1, choosing not to guess where the curve is supposed to turn up between 500 μm and 1 mm. Given the uncertainties involved, this curve is as good as any — except perhaps near the discontinuity — and the use of such a "standard" would enable results from different laboratories to be more easily compared.

To end this section on a necessarily sober note, I draw attention to a recent comparison between a thermopile laser calorimeter and a totally new calorimeter design [4.40]. Their responsivities happily varied linearly with power over a large range, but absolute readings differed by more than a factor ten! The devices were intended for high-power CO_2 lasers, so the details need not concern us here, but the result is an indication that we may be very far from eliminating systematic effects in absolute power measurements.

References

4.1 The Golay cell is made by Unicam Instruments Ltd., England; the device was first described by M.J.E. Golay: Rev. Sci. Instrum. **18**, 357 (1947)
4.2 A major supplier is Molectron Corporation, U.S.A.; a discussion is given by C.B. Roundy and R.L. Byer: J. Appl. Phys. **44**, 929 (1973)
4.3 M.S. Tobin: Proc. IEEE **73**, 61–85 (1985)
4.4 K.J. Siemsen, J. Reid, D.J. Danagher: Appl. Opt. **25**, 86–91 (1986)
4.5 R.A. Wood, A. Vass, C.R. Pidgeon, M.J. Colles, B. Norris: Opt. Commun. **33**, 89–90 (1980)
4.6 F. Strumia, N. Ioli, A. Moretti: *Physics of New Laser Sources,* ed. by N.B. Abraham, F.T. Arecchi, A. Mooradian, A. Sona, NATO ASI Series Vol. 132 (Plenum, New York 1985)
4.7 D.T. Hodges, F.B. Foote, R.D. Reel: Appl. Phys. Lett. **29**, 662–664 (1976)
4.8 B.W. Davis, A. Vass, C.R. Pidgeon, G.R. Allan: Opt. Commun. **37**, 303–305 (1981)
4.9 A. Tanaka, A. Tanimoto, N. Murata, M. Yamanaka, H. Yoshinaga: Japan. J. Appl. Phys. **13**, 1491–1492 (1974)
4.10 E.J. Danielewicz: "The Optically Pumped Difluoromethane Far-Infrared Laser", in *Reviews of Infrared and Millimeter Waves,* Vol. 2, ed. by K.J. Button, M. Inguscio, F. Strumia (Plenum, New York 1984)

4.11 C.O. Weiss, M. Fourrier, C. Gastaud, M. Redon: "Optically Pumped Far-Infrared Ammonia Lasers", in *Reviews of Infrared and Millimeter Waves*, Vol. 2, ed. by K.J. Button, M. Inguscio, F. Strumia (Plenum, New York 1984)

4.12 E.J. Danielewicz, C.O. Weiss: Opt. Commun. **27**, 98–100 (1978)

4.13 J. Farhoomand, H.M. Pickett: Int. J. Infrared Mmwaves **8**, 441–447 (1987)

4.14 E.J. Danielewicz, T.A. Galantowicz, F.B. Foote, R.D. Reel, D.T. Hodges: Opt. Lett. **4**, 280–282 (1979)

4.15 A.R. Calloway, E.J. Danielewicz: IEEE J. **QE-17**, 579–581 (1981)

4.16 S.F. Dyubko, V.A. Svich, L.D. Fesenko: JETP Lett. **16**, No. 11, 418–419 (1972)

4.17 A.R. Calloway, E.J. Danielewicz: Int. J. Infrared Mmwaves **2**, 933–942 (1981)

4.18 T. Lehecka: U.C.L.A.: private communication (1988)

4.19 E.C.C. Vasconcellos, A. Scalabrin, F.R. Petersen, K.M. Evenson: Int. J. Infrared Mmwaves **2**, 533–539 (1981)

4.20 K.B. Amos, J.A. Davis: IEEE J. **QE-16**, 574–575 (1980)

4.21 M.S. Tobin, J.P. Sattler, T.W. Daley: IEEE J. **QE-18**, 79–86(1982)

4.22 J-C. Deroche, G. Graner: "FIR Laser Lines Optically Pumped in Methyl Chloride, $CH_3\,^{35}Cl$ and $CH_3\,^{37}Cl$", in *Reviews of Infrared and Millimeter Waves*, Vol. 2, ed. by K.J. Button, M. Inguscio, F. Strumia (Plenum, New York 1984)

4.23 F.K. Kneubühl, Ch. Sturzenegger: "Electrically Excited Submillimeter-Wave Lasers", in *Infrared and Millimeter Waves* **3**, (Academic, New York 1980)

4.24 S.F. Dyubko, V.A. Svich, L.D. Fesenko: Sov. Phys. Tech. Phys. **20**, 1536–1538 (1976)

4.25 T. Lehecka, R. Savage, R. Dworak, W.A. Peebles, N.C. Luhmann, Jr., A. Semet: Rev. Sci. Instrum. **57**, 1986–1988 (1986)

4.26 P.A. Stimson, B.W. James, I.S. Falconer, L.B. Whitbourn, J.C. Macfarlane: Appl. Phys. Lett. **50**, 786–788 (1987)

4.27 L.B. Whitbourn, J.C. Macfarlane, P.A. Stimson, B.W. James, I.S. Falconer: Infrared Physics **28**, 7–20 (1988) (see note 19 in Table 4.1)

4.28 G. Graner, J-C. Deroche: "Far-Infrared Laser Lines Obtained by Optical Pumping of the CD_3Cl molecule", in *Reviews of Infrared and Millimeter Waves*, Vol. 2, ed. by K.J. Button, M. Inguscio, F. Strumia (Plenum, New York 1984)

4.29 G. Duxbury: "Submillimeter Laser Lines in 1,1 Difluoroethylene, CF_2CH_2", in *Reviews of Infrared and Millimeter Waves*, Vol. 2, ed. by K.J. Button, M. Inguscio, F. Strumia (Plenum, New York 1984)

4.30 D.T. Hodges, F.B. Foote, R.D. Reel: IEEE J. **QE-13**, 491–494 (1977)

4.31 F.B. Foote, D.T. Hodges, H.B. Dyson: Int. J. Infrared Mmwaves **2**, 773–782 (1981)

4.32 R.J. Wagner, A.J. Zelano, L.H. Ngai: Opt. Commun. **8**, 46–47 (1973)

4.33 H. Herman, B.E. Prewer: "Laser Sources in the 140 to 7500 GHz Frequency

Range", in *Proceedings, Military Microwaves '82,* London, (1982)

4.34 Scientech, 5649 Arapahoe Avenue, Boulder Colorado 80303, U.S.A.

4.35 Laser Instrumentation Ltd., Unit 4 Bear Court, Basingstoke Hamps. RG24 0QT, England

4.36 B. Vowinkel, H.P. Röser: Int. J. Infrared Mmwaves **3,** 471–487 (1982)

4.37 T.G. Blaney, D.G. Moss: NPL Report DES 68 (1980). National Physical Laboratory, Teddington, England

4.38 K + C Engineering, Lenaustrasse 38, D-6000 Frankfurt 1, West Germany; Application Note 834 describes calibration to submm wavelengths

4.39 K.M. Evenson, D.A. Jennings, F.R. Petersen, J.A. Mucha, J.J. Jiménez, R.M. Charlton, C.J. Howard: IEEE J. **QE-13,** 442–444 (1977)

4.40 H.J.J. Seguin, V.A. Seguin, A.K. Nath, J. Radzion: Rev. Sci. Instrum. **57,** 185–190 (1986)

4.41 A. Scalabrin, K.M. Evenson: Opt. Lett. **4,** 277–279 (1979)

4.42 G. Carelli, N. Ioli, A. Moretti, D. Pereira, F. Strumia, R. Densing: Appl. Phys. **B 45,** 97–100 (1988)

4.43 P. Wazen, J.-M. Lourtioz: Appl. Phys. **B 32,** 105–111 (1983)

4.44 K. Benzerhouni, F. Meyer, J.M. Lourtioz: Infrared Physics **26,** 377–380 (1986)

4.45 B.W. James, University of Sydney: private communication (1988), and L.B. Whitbourn, J.C. Macfarlane, P.A. Stimson, B.W. James, I.S. Falconer: Infrared Physics **28** (5), I (1988)

4.46 P. Belland, D. Véron: IEEE J. **QE-16,** 885–890 (1980)

4.47 B.W. Davis, A. Vass: Int. J. Infrared Mmwaves **9,** 279–293 (1988)

4.48 N. Ioli, A. Moretti, F. Strumia: Appl. Phys. **B 48,** 305–309 (1989). Note error in frequency given for this line.

5. Pump Lasers

In this section I give a brief overview of the aspects of pump lasers commonly encountered in the literature on (sub)mm lasers. In keeping with the preceding sections, the main intention is to acquaint the reader with various techniques and with the relevant terminology.

Apart from a handful of electrically excited lasers, the (sub)mm lasers reported in this book all use CO_2 lasers or, less frequently, N_2O lasers to pump them. It should be clear that, since we are dealing here with cw (sub)mm lasers, we are also dealing with cw pump lasers. These are electrically excited lasers generally operating at low pressure (10–25 torr) and current (20–100 mW).

Use is sometimes made of the *optoacoustic effect* in determining the efficiency of the pumping. This entails the use of a microphone in a vessel containing some of the (sub)mm laser vapour. This vessel might of course be the laser itself, as seen in Fig. 3.3 . The technique is used as an aid in stabilizing the CO_2 laser frequency on the absorption in the vapour or in the search for new absorption lines [5.1].

As far as *detection* of the pump laser radiation itself is concerned, the types of detectors discussed in Chap. 4 may all be used as infrared detectors as well. In addition, HgCdTe [5.2] detectors are also frequently employed.

The *wavelength* of the pump radiation can be readily determined with commercial grating spectrometers — they are usually equipped with a fluorescent screen which shows the diffracted spot under ultraviolet illumination. Experience shows that errors do arise in assigning pump lines — many examples can be found in the notes in Chap. 7. This can be avoided by calibrating the spectrometer with a high-order diffraction spot from a He-Ne laser, by checking the operation of the spectrometer with a well-established pump line/molecular gas combination, or by using heterodyne techniques.

The remainder of this section is devoted to a brief discussion of the types of infrared lasers encountered in the optical pumping of (sub)mm lasers. The most commonly used is the CO_2 laser operating on "regular" bands. A partial vibrational energy level diagram for CO_2 (Fig. 5.1) shows their location. The upper vibrational state is designated 00^01 — this nomenclature was explained in Chap. 2. The lower levels are designated 10^01 and 02^00. As was also mentioned in Chap. 2, one would expect the energy of the (020) levels to be at around $2\omega_2$, and this is indeed true for the 02^20 state. However the symmetries of the

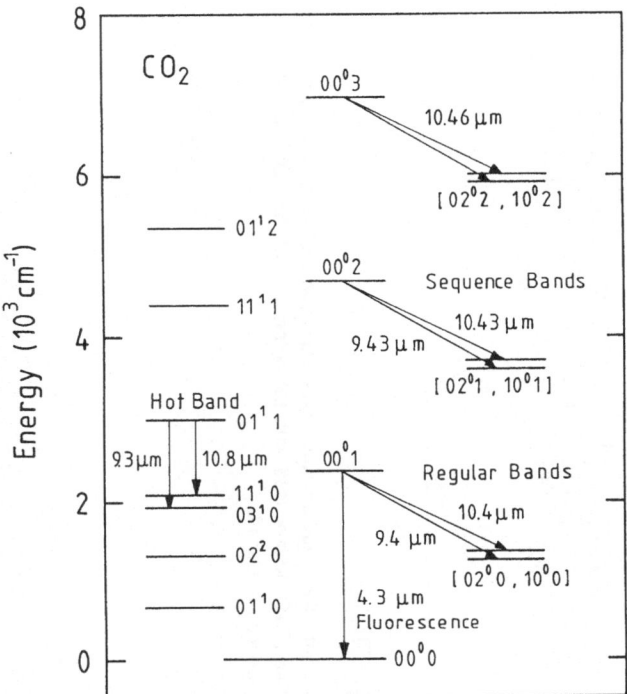

Fig. 5.1. Simplified vibrational energy level diagram for CO_2 showing hot, sequence, and regular bands. All the bands for which a wavelength is shown have been observed. Adapted from [5.11] and [5.12]. Reproduced with the permission of the National Research Council of Canada

02^00 and 10^01 states, and their close proximity to one another in energy, cause a type of resonant mixing referred to as *Fermi resonance* [Ref. 2.2, Sect. II.5]. The resulting mixed states are designated I and II, for the higher and lower energy levels respectively, and they are further apart in energy than would have been the case without the mixing.

The vibrational transitions in question are centred at 10.4 μm and 9.4 μm and are usually referred to as the "10 μm" and "9 μm" bands or I and II, in keeping with the above. Within each of these are two allowed possibilities for the change in the rotational quantum number J, namely $\Delta J = -1$ (P branch) or $\Delta J = +1$ (R branch). Conventionally then, a transition is designated by (for example) 9P34, indicating the 9 μm band P-branch transition down to the $J = 34$ level. Owing to the symmetry of the molecule, only transitions terminating on even values of J are allowed in these bands.

The frequency distribution of the lines in the four branches of the CO_2 laser is shown in Fig. 5.2 (upper half). We are dealing here with the usual isotopic form, $^{12}C^{16}O_2$. Gaps between the lines average about 56 GHz in the P-branch, and 34 GHz in the R-branch. Each line is slightly tunable — up to a few hundred megahertz for a waveguide laser — but obviously not enough to cover the gaps between lines. There are also gaps between the branches.

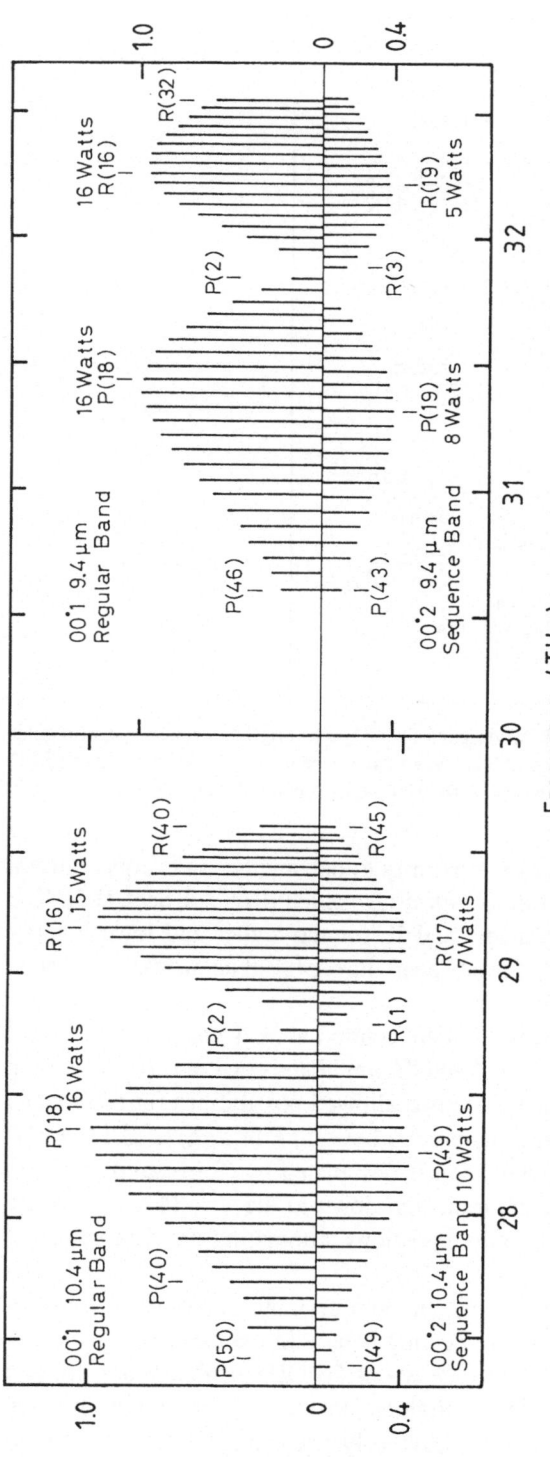

Fig. 5.2. Distribution of lines in the branches of the two regular CO₂ laser bands (*upper half*) and sequence bands (*lower half*). All the lines shown have been observed, with typical peak powers as shown. For clarity the sequence band lines are drawn pointing downwards. After [5.16]. Reproduced with the permission of the National Research Council of Canada

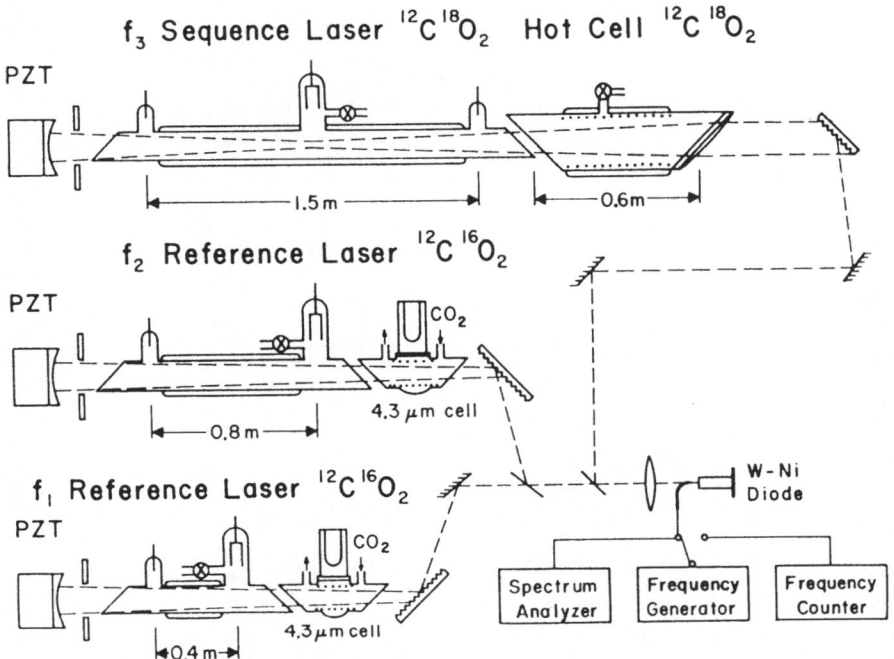

Fig. 5.3. Schematic diagram of the apparatus used to produce lines in the CO_2 laser sequence band and to measure their frequencies. The top laser oscillates on sequence or hot band transitions because of the presence of the hot cell. In this particular case the $C^{18}O_2$ isotope is in use. The other two lasers operate on distinct regular CO_2 lines. Other experimental details are discussed in the text. From [5.9]. Reproduced with the permission of the National Research Council of Canada

An important technique to improve frequency coverage is the use of a *hot cell* in the cavity. This is a cell containing CO_2 at about 400°C, which causes absorption at frequencies in the regular bands and suppresses laser action. This allows laser action on certain other transitions of lower gain, notably the so-called *sequence bands* whose relationship to the regular bands is shown in Fig. 5.1. Each transition in the sequence bands is slightly displaced in frequency with respect to the corresponding one in the regular bands, as shown in Fig. 5.2 (lower half).

Figure 5.3 shows some of these techniques in an experimental arrangement designed to produce sequence band lines and to measure their frequencies. The hot cell is clearly seen in the top laser; the two reference lasers operate on the regular bands and are frequency-stabilized by means of the flourescence cell technique [5.3] which I will not discuss further. Output from all three lasers combines on a non-linear device to produce mixing products — in this experiment neither of the known frequencies f_1 and f_2 needs to be particularly close to the unknown frequency f_3 because of the measuring scheme chosen ($f_3 \approx 2f_2 - f_1$).

The other bands shown in Fig. 5.1 are of the *hot band* family. The centres of the 10.8 μm and 9.3 μm hot bands are displaced rather a lot from those of

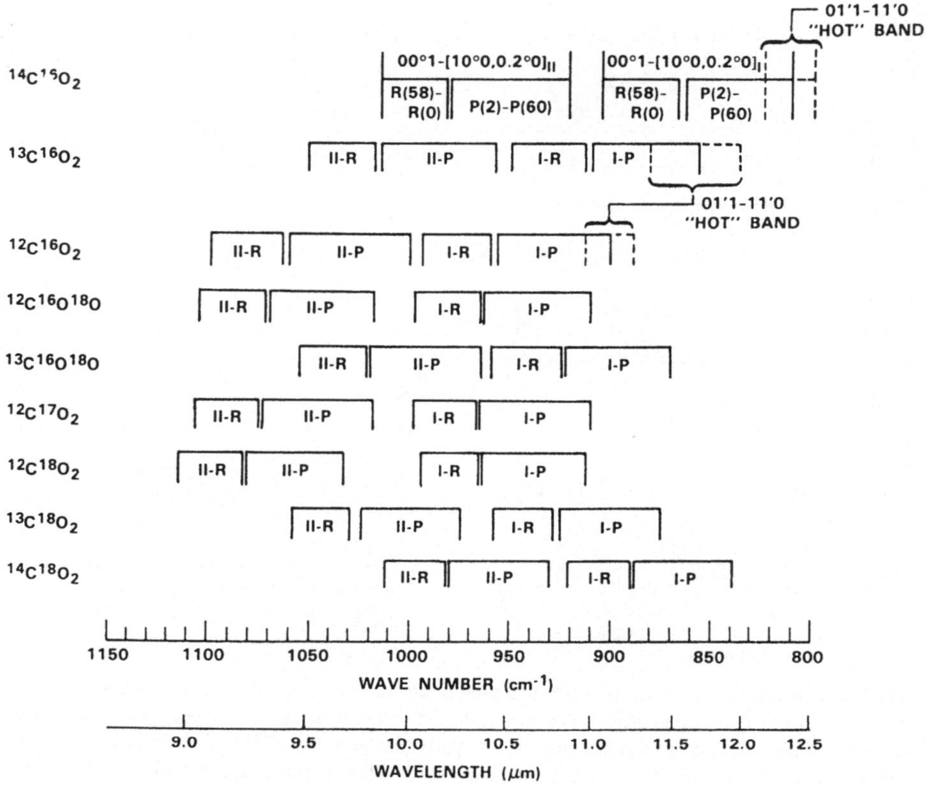

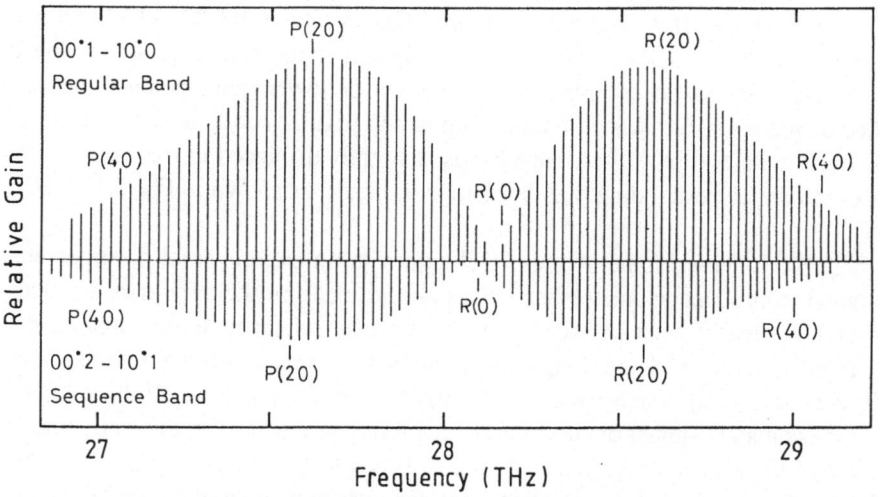

Fig. 5.4. Frequency and wavelength domain of nine CO_2 isotopic species, from [5.15]. Reprinted with permission of Lincoln Laboratory, Massachusetts Institute of Technology, Lexington, Massachusetts.

Fig. 5.5. Distribution of lines in the regular (*upper*) and sequence (*lower*) bands of the N_2O laser. Shown is the calculated relative gain, pointing downwards for the sequence band lines for the sake of clarity. From [5.5]

the corresponding regular and sequence bands, so there are transitions in the (P-branch) which do not overlap with lines of the regular bands. Many of these give rise to laser action. Part of the band does overlap however, and laser action has not been achieved on these lines even with the use of a hot cell. Evidently the proximity of the lower levels in the hot band and regular band prevents sufficient discrimination against the latter [5.4].

Substituting various *isotopic forms* of CO_2, for example $^{13}CO_2$ or $C^{18}O_2$, is another way of causing systematic shifts in the band centres. The frequency domains so obtained are shown in Fig. 5.4.

Another possibility is the use of the N_2O laser, whose emission band lies around 10.65 μm and has the same designation (00^01–10^00) as the 10 μm CO_2 laser band. The positions and relative strengths of the lines are shown in Fig. 5.5. Although in cw operation it has considerably less power than the CO_2 laser, the N_2O laser has been extremely successful in pumping the ammonia molecule NH_3, for which there are close coincidences with absorbing transitions. Operation of the N_2O laser has also been reported on the sequence band [5.5]. Frequency measurements have been made on both the hot band and the other regular band at 9.6 μm but without obtaining laser emission [5.4], although *Djeu* and *Wolga* [5.6] obtained laser action at 9.6 μm with a laser of 3.8 m length.

Table 5.1. References to the frequencies of pump laser frequencies

	CO_2	Isotopic	N_2O
Regular	[5.7]	[5.7]	[5.8]
Sequence	[5.11]	[5.9],[5.10]	[5.5]
Hot	[5.12],[5.14]	[5.13]	

The frequencies obtainable with all these techniques have been accurately calculated (and in some cases measured) and can be obtained from the references in Table 5.1. For convenience, the frequencies of the standard CO_2 laser lines are reproduced in Table 5.2. Most of the lines, sorted by category and wavelength, can also be found in [5.17].

Table 5.2. Frequencies of the lines of the regular CO_2 laser [5.7], together with calculated wavenumbers and wavelengths (in vacuum). Most of the frequencies are measured values. Wavelengths in air, which are sometimes used to denote pump transitions, can be found from $(n - 1) = 2.73 \times 10^{-4}$, which is true at 15°C over this range of wavelengths.

| \multicolumn{4}{c}{10 μm band (I)} | | | | \multicolumn{4}{c}{9 μm band (II)} | | | |
Line	Freq. [MHz]	σ [cm⁻¹]	λ [μm]	Line	Freq. [MHz]	σ [cm⁻¹]	λ [μm]
P(60)	27077607.5077	903.2118	11.072	P(60)	30143456.0742	1005.4775	9.946
P(58)	27146404.4578	905.5066	11.044	P(58)	30212223.6949	1007.7713	9.923
P(56)	27214396.1809	907.7745	11.016	P(56)	30280322.1201	1010.0428	9.901
P(54)	27281588.8741	910.0159	10.989	P(54)	30347743.7465	1012.2918	9.879
P(52)	27347988.4259	912.2307	10.962	P(52)	30414481.1364	1014.5179	9.857
P(50)	27413600.4235	914.4193	10.936	P(50)	30480527.0251	1016.7209	9.836
P(48)	27478430.1601	916.5818	10.910	P(48)	30545874.3277	1018.9007	9.814
P(46)	27542482.6413	918.7183	10.885	P(46)	30610516.1462	1021.0569	9.794
P(44)	27605762.5914	920.8291	10.860	P(44)	30674445.7759	1023.1894	9.773
P(42)	27668274.4599	922.9143	10.835	P(42)	30737656.7119	1025.2979	9.753
P(40)	27730022.4271	924.9740	10.811	P(40)	30800142.6555	1027.3822	9.733
P(38)	27791010.4094	927.0083	10.787	P(38)	30861897.5199	1029.4421	9.714
P(36)	27851242.0651	929.0174	10.764	P(36)	30922915.4360	1031.4774	9.695
P(34)	27910720.7986	931.0014	10.741	P(34)	30983190.7583	1033.4880	9.676
P(32)	27969449.7656	932.9604	10.719	P(32)	31042718.0700	1035.4736	9.657
P(30)	28027431.8776	934.8945	10.696	P(30)	31101492.1877	1037.4341	9.639
P(28)	28084669.8055	936.8037	10.675	P(28)	31159508.1671	1039.3693	9.621
P(26)	28141165.9839	938.6883	10.653	P(26)	31216761.3064	1041.2791	9.604
P(24)	28196922.6147	940.5481	10.632	P(24)	31273247.1518	1043.1632	9.586
P(22)	28251941.6703	942.3833	10.611	P(22)	31328961.5006	1045.0217	9.569
P(20)	28306224.8967	944.1940	10.591	P(20)	31383900.4054	1046.8542	9.552
P(18)	28359773.8165	945.9802	10.571	P(18)	31438060.1774	1048.6608	9.536
P(16)	28412589.7314	947.7420	10.551	P(16)	31491437.3897	1050.4413	9.520
P(14)	28464673.7246	949.4793	10.532	P(14)	31544028.8804	1052.1955	9.504
P(12)	28516026.6628	951.1923	10.513	P(12)	31595831.7547	1053.9235	9.488
P(10)	28566649.1983	952.8808	10.494	P(10)	31646843.3878	1055.6251	9.473
P(8)	28616541.7701	954.5451	10.476	P(8)	31697061.4264	1057.3002	9.458
P(6)	28665704.6061	956.1850	10.458	P(6)	31746483.7910	1058.9487	9.443
P(4)	28714137.7235	957.8005	10.441	P(4)	31795108.6771	1060.5707	9.429
P(2)	28761840.9300	959.3917	10.423	P(2)	31842934.5560	1062.1660	9.415
R(2)	28877902.4412	963.2631	10.381	R(2)	31958996.0672	1066.0374	9.381
R(4)	28923046.4336	964.7690	10.365	R(4)	32004017.3872	1067.5391	9.367
R(6)	28967457.0695	966.2504	10.349	R(6)	32048236.2544	1069.0141	9.354
R(8)	29011133.0097	967.7072	10.334	R(8)	32091652.6660	1070.4623	9.342
R(10)	29054072.7058	969.1395	10.318	R(10)	32134266.8953	1071.8838	9.329
R(12)	29096274.3988	970.5472	10.303	R(12)	32176079.4907	1073.2785	9.317
R(14)	29137736.1185	971.9303	10.289	R(14)	32217091.2743	1074.6465	9.305
R(16)	29178455.6817	973.2885	10.274	R(16)	32257303.3400	1075.9878	9.294
R(18)	29218430.6909	974.6219	10.260	R(18)	32296717.0518	1077.3025	9.282
R(20)	29257658.5324	975.9304	10.247	R(20)	32335334.0411	1078.5906	9.271
R(22)	29296136.3740	977.2139	10.233	R(22)	32373156.2043	1079.8523	9.261
R(24)	29333861.1629	978.4723	10.220	R(24)	32410185.7000	1081.0874	9.250
R(26)	29370829.6231	979.7054	10.207	R(26)	32446424.9456	1082.2962	9.240
R(28)	29407038.2525	980.9132	10.195	R(28)	32481876.6140	1083.4788	9.230
R(30)	29442483.3197	982.0955	10.182	R(30)	32516543.6298	1084.6351	9.220
R(32)	29477160.8609	983.2522	10.170	R(32)	32550429.1653	1085.7654	9.210
R(34)	29511066.6762	984.3832	10.159	R(34)	32583536.6360	1086.8698	9.201
R(36)	29544196.3256	985.4883	10.147	R(36)	32615869.6965	1087.9483	9.192
R(38)	29576545.1250	986.5674	10.136	R(38)	32647432.2354	1089.0011	9.183
R(40)	29608108.1417	987.6202	10.125	R(40)	32678228.3702	1090.0284	9.174
R(42)	29638880.1900	988.6466	10.115	R(42)	32708262.4421	1091.0302	9.166
R(44)	29668855.8259	989.6465	10.105	R(44)	32737539.0104	1092.0068	9.157
R(46)	29698029.3420	990.6196	10.095	R(46)	32766062.8469	1092.9582	9.149
R(48)	29726394.7621	991.5658	10.085	R(48)	32793838.9297	1093.8847	9.142
R(50)	29753945.8353	992.4848	10.076	R(50)	32820872.4368	1094.7865	9.134
R(52)	29780676.0297	993.3764	10.067	R(52)	32847168.7402	1095.6636	9.127
R(54)	29806578.5263	994.2404	10.058	R(54)	32872733.3987	1096.5164	9.120
R(56)	29831646.2123	995.0766	10.049	R(56)	32897572.1515	1097.3449	9.113
R(58)	29855871.6737	995.8847	10.041	R(58)	32921690.9108	1098.1494	9.106

References

5.1 G. Busse, R. Thurmaier: Appl. Phys. Lett. **31**, 194–195 (1977)

5.2 Société Anonyme de Télécommunications, France

5.3 C. Freed, A. Javan: Appl. Phys. Lett. **17**, 53–56 (1970)

5.4 K. Siemsen: private communications (1988); unpublished material on the N_2O laser which supplements [5.5]

5.5 K.J. Siemsen, J. Reid: Opt. Commun. **20**, 284–288 (1977)

5.6 N. Djeu, G.J. Wolga: IEEE J. **QE-5**, 50 (1969)

5.7 L.C. Bradley, K.L. Soohoo, C. Freed: IEEE J. **QE-22**, 234–267 (1986)

5.8 B.G. Whitford, K.J. Siemsen, H.D. Riccius, G.R. Hanes: Opt. Commun. **14**, 70–74 (1975)

5.9 K.J. Siemsen: Opt. Lett. **6**, 114–116 (1981)

5.10 K.J. Siemsen: Opt. Commun. **34**, 447–450 (1980)

5.11 K.J. Siemsen, B.G. Whitford: Opt. Commun. **22**, 11–16 (1977)

5.12 B.G. Whitford, K.J. Siemsen, J. Reid: Opt. Commun. **22**, 261–264 (1977)

5.13 F.R. Petersen, J.S. Wells, A.G. Maki, K.J. Siemsen: Appl. Opt. **20**, 3635–3640 (1981)

5.14 F.R. Petersen, J.S. Wells, K.J. Siemsen, A.M. Robinson, A.G. Maki: J. Mol. Spec. **105**, 324–330 (1984)

5.15 C. Freed: SPIE Vol. 709, 36–45 (1986)

5.16 J. Reid, K. Siemsen: J. Appl. Phys. **48**, 2712–2717 (1977)

5.17 R. Beck, W. Englisch, K.Gürs: *Table of Laser Lines in Gases and Vapors*, Springer Series in Optical Sciences Vol. 2, (Springer, Berlin, Heidelberg, Third Edition 1980)

Part II

A Compilation of cw (Sub)millimetre Laser Lines

6. General Notes

In this part of the book the reader will find data pertaining to 4555 cw laser lines with wavelengths between 10.06 μm and 3.03 mm. It is an up-to-date, critical compilation of the (sub)millimetre lines available with standard techniques. It has been necessary to employ a number of criteria in selecting lines for inclusion but despite the unavoidably arbitrary nature of some of these I hope that the resulting compilation is of maximum use to most researchers.

What has been included? Atomic and molecular species are included. All lines longer than 10 μm in wavelength are included, except those of the well-known CO_2 and N_2O laser bands. A few lines belonging to miscellaneous CO_2 bands have been included. There is no long wavelength cutoff as such – in a time-honoured tradition I have included several millimetre lines, and called them "laser lines": there is no obvious confusion in the literature with "masers" at centimetre wavelengths. Included are:

  (i) true cw lines;
  (ii) quasi-cw lines;
  (iii) long-pulsed lines.

A line is considered true cw even if mechanical chopping of the pump beam was used as a detection technique because such chopping (usually below 30 Hz) does not significantly enhance peak pump power. True cw lines are listed without further comment.

  Quasi-cw are those lines which are pumped by pulses long compared with typical vibrational lifetimes in the FIR medium but where the peak pump power is somewhat enhanced by the chopping. This may be electrical chopping or Q-switching so long as the pulse length is 0.5 ms or longer, which gives a pump power enhancement of usually 2–3 times. There is a fair certainty that the line could be obtained in true cw mode. Quasi-cw lines are indicated by an appropriate comment, usually in the *references*, (Chap. 11).

  A third, rather problematical, group is formed by the long-pulsed lines. These are defined for present purposes as having pulse lengths (full-width at half-maximum) between 1.0 μs and 0.5 ms. At the long end of this range they merge with quasi-cw lines and at the short end they approach the domain of TEA-laser pumped lines. It is by no means likely that most of these lines can be obtained cw without special techniques. However the literature shows many cases where this was the case; moreover it is unfortunately true that cw observations of previously reported long-pulsed lines are often not published. It is therefore necessary to include them here: but in all cases where such a line has not been reported cw or quasi-cw a star (*) is printed next to the pump code.

Optically pumped and electrically excited lines are included. Cascade- and "refilling-"transition lines are of course included. Lines indicated as "Raman" rather than laser transitions are included. Lines pumped by waveguide lasers, which are slightly tunable, are also included. All papers available to the author in April 1989 were incorporated.

What has not been included? As mentioned, the lines of CO_2 and N_2O lasers themselves are not included. They are very numerous, and are already catalogued in a number of references (see Chapter 5). TEA-laser pulsed lines, or any others with pulse lengths less than 1 μs, are not included. Stark-shifted and multiple photon lines are excluded, as are molecular beam masers and lasers relying on chemical reactions for their excitation.

Sources of Data. All of the quoted references have been viewed. They have been exhaustively compared with later references (such as review articles and articles including improved measurements). In a few cases it was necessary to take data from such secondary sources, usually because the reviewer's own unpublished results had been incorporated. Such cases are clearly noted.

Elimination of Double Listings. Some published results became superceded. I followed these criteria:

(i) I gave preference to publications reporting frequencies to those reporting wavelengths;
(ii) journal papers were preferred to conference papers, to private communications, and to other journals judged to be obscure;
(iii) improved measurement accuracy was preferred to older references.

Historical precedence was given when all else was equal, but the above criteria show that my priority was for accuracy, rather than for giving credit for the discovery of lines. In *some* cases, mention is made of the elimination of older references in either the footnotes to Table A or in the references themselves. This is usually because the older reference contains useful information (such as polarization) not given in the later reference.

Under normal circumstances, lines were considered to be identical if they had the same molecule and pump codes and differed in wavelength by less than the experimental accuracy or 0.3% , whichever was smaller. Exceptions were provided by the following cases:

(i) the lines had the same reference—it is assumed that the author(s) had reason to believe that the lines were not identical;
(ii) the lines had differing proposed assignments;
(iii) the lines had differing polarizations with respect to the pump beam.

49

Cases in which either assignment or polarization information was used to establish that lines are distinct are footnoted in Table A; any other special cases are likewise footnoted. Note that this compilation does *not* itself list assignments or polarizations; for these the reader is referred to the references.

Elimination of Errors. It is possible that published data contain errors. Obvious errors have been corrected, with accompanying notes in the footnotes or references, as appropriate. A further check I employed was to program a computer to search for pairs of lines which were identical (see above) except for small changes in either pump or molecule code. Such cases indicate possible typographical errors, and these were *not* eliminated but appropriately footnoted.

The most likely remaining errors are:

(i) wavelength inaccuracies of more than 0.3% (and hence not detected as double listings even when better data are published);
(ii) pump transitions misidentified by more than one step.

It is also inevitable that I have introduced some errors or have made some omissions. In these and other cases I would be grateful to hear of any corrections which can be made.

7. Table A. **Wavelength-Sorted Data**

Column headings are:

1. Wavelength in μ**m.** Wavelengths are given to a maximum of three decimal places indicated. Where possible the wavelength was derived from the frequency using $c = 299\ 792\ 458$ m/s. Trailing zeros were not eliminated, but this does not lead to any ambiguity in practice. Unless noted, measured rather than calculated values were used. The distinction between air and vacuum wavelengths is too small to be relevant considering the experimental errors involved.

2. Molecule code. The codes used are chosen for ease of recognition and are usually self-explanatory. Explicit forms (e.g. CH2CHF) were preferred to more compressed ones (e.g. "C2H3F"). A full list is given in Chapter 10.

3. Pump code. These are usually self-explanatory. Only the "regular" CO_2 laser transitions are given explicitly (e.g. 9P34, 10P08). Occasionally the pump transition is poorly identified — this is signified by for example 9P?. Other codes are:

13CO2	–	isotopic pump laser $^{13}CO_2$
C18O2	–	isotopic pump laser $C^{18}O_2$
13C18O2	–	isotopic pump laser $^{13}C\ ^{18}O_2$
CO2S	–	sequence band CO_2 pump laser
CO2H	–	hot band CO_2 pump laser
N2O	–	N_2O pump laser

In these cases the individual transitions are not identified. This leads to no ambiguity in practice. Of course in checking for doubly-listed lines (see Chapter 6) the individual transitions were taken into account.

ELEC	–	electric discharge pumped

The symbol * indicates long-pulsed mode, i.e.
1.0 μs < pulse length (FWHM) < 0.5 ms.

4. Reference – see Chapter 11. Only one reference is given per line, this being chosen in accordance with the priorities outlined in Chapter 6 above. In many cases the footnotes are used to indicate other relevant references.

5. Measured frequency in MHz, if available. In cases where several measurements were available I have chosen to be consistent with reference IMEJ86 wherever possible; this means that I have taken either the most accurate measurement or a suitable average value. The resulting value may not be precisely equal to the particular measurement in the cited reference; such cases are always footnoted.

6. Footnote numbers, if applicable. The footnotes themselves can be found immediately after Table A. They tend to be specific to a particular line or a few lines – notes which apply to a whole paper are to be found along with the reference to that paper. Cases of possibly misidentified molecules or pump lines are indicated by the tags mm# or pp# respectively, where # is an integer. A definition of these cases is given alongside the appropriate footnotes on p. 163.

λ [μm]	Molecule	Pump	Reference	Freq. [MHz]	Footnotes
10.060	NE	ELEC	FMPG64	138	
10.198	HF	*ELEC	Deuts67B		
10.338	14NH3	9R30	KR86		
10.342	14NH3	9R30	KR86		
10.359	14NH3	9R30	KR86		
10.367	14NH3	9R30	KR86		
10.400	O	ELEC	Wille71	145	
10.458	HF	*ELEC	Deuts67B		
10.507	14NH3	9R30	KR86		
10.546	14NH3	9R30	KR86		
10.582	HF	*ELEC	Deuts67B		
10.718	14NH3	9R30	KR86		
10.732	14NH3	9R30	KR86		
10.737	14NH3	9R30	KR86		
10.744	14NH3	9R30	SRD86	127	
10.744	HF	*ELEC	Deuts67B		
10.754	14NH3	9R30	KR86		
10.762	14NH3	9R30	KR86		
10.767	14NH3	9R30	KR86		
10.784	14NH3	9R30	SRD86	127	
10.789	15NH3	10R42	SRD86		
10.812	HF	*ELEC	Deuts67B		
10.855	14NH3	9R30	KR86		
10.981	NE	ELEC	Patel68	140	
11.011	14NH3	9R30	SRD86	127	
11.057	HF	*ELEC	Deuts67B		
11.209	14NH3	9R30	SRD86	127	

λ [μm]	Molecule	Pump	Reference Freq. [MHz]		Footnotes
11.212	14NH3	9R30	KR86		
11.257	15NH3	10R42	SRD86		
11.260	14NH3	9R30	KR86		
11.263	14NH3	9R30	KR86		
11.297	XE	ELEC	FMPG64		138
11.403	HF	*ELEC	Deuts67B		
11.460	14NH3	9R30	KR86		
11.471	14NH3	9R30	KR86		
11.482	CS2	ELEC	Patel65		160
11.489	CS2	ELEC	Patel65		160
11.503	CS2	ELEC	Patel65		160
11.510	CS2	ELEC	Patel65		160
11.517	CS2	ELEC	Patel65		160
11.520	14NH3	*CO2S	ZRGB80		104
11.521	14NH3	9R30	KR86		104
11.521	14NH3	9R30	SRD86		104, 127
11.524	CS2	ELEC	Patel65		160
11.531	CS2	ELEC	Patel65		160
11.538	CS2	ELEC	Patel65		160
11.541	HF	*ELEC	Deuts67B		
11.545	CS2	ELEC	Patel65		160
11.586	15NH3	10R42	SRD86		
11.596	CS2	ELEC	Patel65		160
11.712	14NH3	9R30	SRD86		127
11.716	14NH3	9R30	KR86		
11.727	14NH3	9R30	KR86		
11.746	14NH3	9R30	SRD86		127
11.763	15NH3	10R42	SRD86		
11.785	HF	*ELEC	Deuts67B		
11.794	14NH3	9R30	KR86		
11.798	14NH3	9R30	SRD86		127
11.798	15NH3	10R42	SRD86		
11.830	H2O	*ELEC	TP68		
11.863	HF	*ELEC	AY70		
11.865	NE	ELEC	FMPG64		138
11.866	15NH3	10R42	SRD86		
11.960	H2O	*ELEC	TP68		
11.979	14NH3	9R30	KR86		
11.990	14NH3	9R30	KR86		
12.010	14NH3	9R30	SRD86		104, 127
12.039	14NH3	9R30	KR86		

λ [μm]	Molecule	Pump	Reference	Freq. [MHz]	Footnotes
12.063	15NH3	10R42	SRD86		
12.079	14NH3	9R30	SRD86		127
12.080	14NH3	9R30	KR86		
12.100	14NH3	9R30	KR86		
12.140	AR	ELEC	FMPG64		138
12.147	AR	ELEC	Patel68		138
12.148	15NH3	10R42	SRD86		
12.208	HF	*ELEC	Deuts67B		
12.245	14NH3	9R30	SRD86		104, 127
12.249	14NH3	9R30	KR86		104
12.261	14NH3	9R30	KR86		
12.262	HF	*ELEC	Deuts67B		
12.266	XE	ELEC	Patel68		135
12.281	14NH3	9R30	SRD86		127
12.299	15NH3	10R42	SRD86		
12.311	14NH3	9R30	KR86		
12.336	15NH3	10R42	SRD86		
12.350	14NH3	9R30	KR86		
12.384	14NH3	9R30	SRD86		127
12.403	14NH3	9R30	KR86		
12.528	14NH3	9R30	KR86		105
12.540	14NH3	9R30	KR86		
12.561	14NH3	9R30	SRD86		127
12.591	14NH3	9R30	KR86		
12.616	15NH3	10R42	SRD86		
12.631	14NH3	9R30	KR86		
12.678	HF	*ELEC	Deuts67B		
12.682	14NH3	9R30	SRD86		127
12.697	14NH3	9R30	KR86		
12.701	HF	*ELEC	Deuts67B		
12.720	14NH3	9R30	KR86		
12.739	15NH3	10R42	SRD86		
12.811	14NH3	9R30	SRD86		127
12.820	NE	ELEC	FMPG64		138
12.849	14NH3	9R30	SRD86		127
12.850	HCN	*ELEC	TP68		
12.905	15NH3	10R42	SRD86		
12.917	XE	ELEC	Patel68		135
12.967	15NH3	10R42	SRD86		
12.972	14NH3	9R30	SRD86		127
12.977	15NH3	CO2S	SRD86		

λ [μm]	Molecule	Pump	Reference	Freq. [MHz]	Footnotes
13.024	14NH3	9R30	KR86		
13.030	15NH3	10R42	SRD86		
13.051	14NH3	9R30	KR86		
13.146	14NH3	9R30	SRD86		127
13.151	CO2	*ELEC	HK66		161
13.188	HF	*ELEC	Deuts67B		
13.201	HF	*ELEC	Deuts67B		
13.221	HF	*ELEC	Deuts67B		
13.270	14NH3	9R30	SRD86		127
13.415	14NH3	9R30	KR86		
13.453	14NH3	9R30	KR86		
13.537	CO2	*ELEC	HK66		161
13.578	14NH3	9R30	KR86		
13.728	HF	*ELEC	Deuts67B		
13.735	NE	ELEC	FMPG64		138
13.757	NE	ELEC	FMPG64		138
13.784	HF	*ELEC	Deuts67B		
13.826	14NH3	9R30	KR86		
13.872	HCL	*ELEC	Deuts67C		
14.099	HCL	*ELEC	Deuts67C		
14.288	HF	*ELEC	Deuts67B		
14.343	HCL	*ELEC	Deuts67C		
14.441	HF	*ELEC	Deuts67B		
14.780	NH3	*ELEC	AW69		
14.930	NE	ELEC	FMPG64		
15.016	HF	*ELEC	Deuts67B		
15.037	AR	ELEC	Patel68		140
15.040	NH3	*ELEC	AW69		
15.042	AR	ELEC	Patel68		140
15.080	NH3	*ELEC	AW69		
15.410	NH3	*ELEC	AW69		
15.470	NH3	*ELEC	AW69		
16.000	CF4	9R12	Telle83		
16.022	HF	*ELEC	Deuts67B		
16.213	HCL	*ELEC	Deuts67C		
16.444	HF	*ELEC	AY70		
16.580	CO2	*ELEC	HK66		161
16.609	HCL	*ELEC	Deuts67C		
16.630	NE	ELEC	FMPG64		138
16.644	HCL	*ELEC	AY70		
16.655	HF	*ELEC	AY70		

λ [μm]	Molecule	Pump	Reference	Freq. [MHz]	Footnotes
16.668	NE	ELEC	Patel68		140
16.765	HCL	*ELEC	AY70		
16.893	NE	ELEC	Patel68		140
16.931	H2O	*ELEC	MC64		
16.947	NE	ELEC	Patel68		140
16.975	HF	*ELEC	AY70		
17.024	CO2	*ELEC	HK66		161
17.034	HCL	*ELEC	Deuts67C		
17.095	HF	*ELEC	AY70		
17.125	HCL	*ELEC	AY70		
17.156	NE	ELEC	FMPG64		138
17.189	NE	ELEC	FMPG64		138
17.325	HF	*ELEC	AY70		
17.372	CO2	*ELEC	HK66		161
17.492	HCL	*ELEC	Deuts67C		
17.575	HCL	*ELEC	AY70		
17.645	HF	*ELEC	AY70		
17.802	NE	ELEC	FMPG64		138
17.841	NE	ELEC	Patel68		140
17.888	NE	ELEC	Patel68		140
17.987	HCL	*ELEC	Deuts67C		
17.997	HCL	*ELEC	AY70		
18.035	HCL	*ELEC	AY70		
18.085	HF	*ELEC	AY70		
18.210	NH3	*ELEC	AW69		
18.300	BCL3	ELEC	KKPPS68		
18.396	NE	ELEC	Patel68		140
18.514	XE	ELEC	FMPG64		138
18.522	HCL	*ELEC	Deuts67C		
18.555	HCL	*ELEC	AY70		
18.593	HCL	*ELEC	AY70		
18.800	BCL3	ELEC	KKPPS68		
18.801	HF	*ELEC	Deuts67B		
19.100	BCL3	ELEC	KKPPS68		
19.113	HF	*ELEC	Deuts67B		
19.122	HCL	*ELEC	AY70		
19.145	HCL	*ELEC	AY70		
19.183	HCL	*ELEC	AY70		
19.399	HBR	*ELEC	AY70		
19.400	BCL3	ELEC	KKPPS68		
19.520	CH3OH	13CO2	PD82A		

λ [μm]	Molecule	Pump	Reference	Freq. [MHz]	Footnotes
19.700	HCL	*ELEC	Deuts67C		
19.783	HCL	*ELEC	AY70		
19.821	HCL	*ELEC	AY70		
19.988	HBR	*ELEC	AY70		
20.134	HF	*ELEC	Deuts67B		
20.200	BCL3	ELEC	KKPPS68		
20.346	HCL	*ELEC	Deuts67C		
20.351	HF	*ELEC	Deuts67B		
20.360	HBR	*ELEC	AY70		
20.411	HCL	*ELEC	Deuts67C		
20.480	NE	ELEC	Patel68	140	
20.600	BCL3	ELEC	KKPPS68		
20.896	HBR	*ELEC	AY70		
20.939	HF	*ELEC	Deuts67B		
20.949	HBR	*ELEC	AY70		
20.999	HCL	*ELEC	Deuts67C		
21.047	HCL	*ELEC	Deuts67C		
21.156	HCL	*ELEC	Deuts67C		
21.460	NH3	*ELEC	AW69		
21.501	HBR	*ELEC	AY70		
21.546	HBR	*ELEC	AY70		
21.699	HF	*ELEC	Deuts67B		
21.752	NE	ELEC	Patel68	140	
21.789	HF	*ELEC	Deuts67B		
21.813	HCL	*ELEC	Deuts67C		
21.971	HCL	*ELEC	Deuts67C		
22.136	HBR	*ELEC	AY70		
22.226	HBR	*ELEC	AY70		
22.400	BCL3	ELEC	KKPPS68		
22.540	NH3	*ELEC	AW69		
22.651	HCL	*ELEC	Deuts67C		
22.710	NH3	*ELEC	AW69		
22.836	NE	ELEC	Patel68	140	
22.855	HBR	*ELEC	AY70		
22.864	HCL	*ELEC	Deuts67C		
23.000	BCL3	ELEC	KKPPS68		
23.359	H2O	*ELEC	MC64	129	
23.436	HBR	*ELEC	AY70		
23.571	HCL	*ELEC	Deuts67C		
23.680	NH3	*ELEC	AW69		
23.849	HCL	*ELEC	Deuts67C		

λ [μm]	Molecule	Pump	Reference	Freq. [MHz]	Footnotes
23.860	NH3	*ELEC	AW69		
24.318	HCL	*ELEC	Deuts67C		
24.583	HCL	*ELEC	Deuts67C		
24.618	HCL	*ELEC	Deuts67C		
24.920	NH3	*ELEC	AW69		
24.937	HCL	*ELEC	Deuts67C		
25.120	NH3	*ELEC	AW69		
25.270	CH3OH	*9P26	Henni86		
25.423	NE	ELEC	Patel68		140
25.704	HCL	*ELEC	Deuts67C		
26.146	HCL	*ELEC	Deuts67C		
26.247	HCL	*ELEC	AY70		
26.270	NH3	*ELEC	AW69		
26.666	H2O	*ELEC	MC64		
26.944	AR	ELEC	Patel68		140
27.508	HCL	*ELEC	AY70		
27.700	CD3OH	9P34	SBW84		
27.971	H2O	ELEC	BBEK73	10718068.7	37, 38
28.054	H2O	*ELEC	MC64		
28.064	NE	ELEC	FMPG64		138
28.273	H2O	*ELEC	MC64		
28.356	H2O	*ELEC	MC64		
29.786	HBR	*ELEC	AY70		
30.445	HBR ·	*ELEC	AY70		
30.500	CH3OH	10R32	SBW84		
30.690	NH3	*ELEC	AW69		
30.700	CD3OH	9P30	SBW84		
30.948	HBR	*ELEC	AY70		
31.100	CD3OH	9R20	SBW84		
31.368	HBR	*ELEC	AY70		
31.470	NH3	*ELEC	AW69		
31.550	NE	ELEC	FMPG64		138
31.849	HBR	*ELEC	AY70		
31.920	NH3	*ELEC	AW69		
31.928	NE	ELEC	PFMG64B		135
32.020	NE	ELEC	FMPG64		138
32.130	NH3	*ELEC	AW69		
32.469	HBR	*ELEC	AY70		
32.520	NE	ELEC	FMPG64		138
32.799	HBR	*ELEC	AY70		
32.830	NE	ELEC	FMPG64		138, 139

λ [μm]	Molecule	Pump	Reference	Freq. [MHz]	Footnotes
32.929	H2O	ELEC	MC64		141
33.033	H2O	*ELEC	MC64		
33.400	CH3OH	9R16	SMNM83		
33.409	HBR	*ELEC	AY70		
33.470	H2S	*ELEC	HC69		
33.522	CH3OH	*9R16	Henni86		pp11
33.540	CH3OH	*9R18	Henni83		pp11
33.640	H2S	*ELEC	HC69		
33.896	D2O	*ELEC	MC64		
34.100	CD3OH	10R34	SBW84		
34.200	CD3OH	10R38	SBW84		
34.550	NE	ELEC	FMPG64		138
34.600	CH318OH	9P30	IMPSG89		
34.679	NE	ELEC	PFMG64B		135
34.790	13CH3OH	10R22	HP78		
34.800	CD3OH	10P22	DW78B		
35.000	H2O	*ELEC	MC64		
35.000	CD3OD	10R28	HP79		
35.000	CH318OH	10R36	IMPSG89		
35.090	D2O	*ELEC	MC64		
35.500	CD3OH	9R06	SEJZS87		
35.602	NE	ELEC	PFMG64B		128, 135
35.700	CD3OH	10P26	SBW84		
35.841	H2O	*ELEC	MC64		
35.860	CH3OH	*9P16	Henni83		
35.968	CH3OH	*9P16	Henni86		
36.319	D2O	*ELEC	MC64		
36.524	D2O	*ELEC	MC64		
36.619	H2O	*ELEC	MC64		
36.666	CH3OH	9R16	Henni86		pp12, 66, 115
36.690	CH3OH	*9R18	Henni83		pp12
37.000	CH3OH	9P14	Knigh81		
37.100	CD3OH	9P24	SBW84		
37.231	NE	ELEC	PFMG64B		135
37.500	CH3OH	9P32	HRB73		
37.600	CD3OH	10R34	DW78B		
37.791	D2O	*ELEC	MC64		
37.854	CH3OH	9P32	PEJS80	7919660.2	
37.859	H2O	*ELEC	MC64		
38.094	H2O	*ELEC	MC64		
39.698	H2O	*ELEC	MC64		

λ [μm]	Molecule	Pump	Reference	Freq. [MHz]	Footnotes
39.785	CH3OH	*9R18	TOH88		
39.924	CH3OH	9P34	PEJS80	7509036.2	
40.000	CD3OH	9R28	SEJZS87		
40.000	CH318OH	10P26	IMPSG89		
40.030	CH3OH	*10R34	TOH88		
40.100	CD3OH	10P22	DW78B		
40.200	CH3OH	9P34	HRB73		
40.526	HBR	*ELEC	AY70		
40.629	H2O	*ELEC	MC64		
40.994	D2O	*ELEC	MC64		
41.034	CH3OH	*9P16	Henni86		
41.060	CH3OH	C18O2	PD82A		
41.171	CH3OH	*9P26	Henni86		
41.250	CD3OH	10R08	Knigh81		
41.330	CH3OH	13C18O2	PD84		
41.355	CD3OH	10R18	SEJZS87	7249266.0	74
41.460	CD3OH	10R34	CIMPS87		
41.500	CH3OH	*9P30	TOH88		
41.500	CD3OH	10R08	DW78B		62
41.700	CH3OH	9P32	HRB73		
41.741	NE	ELEC	PFMG64B		135
41.800	CD3OH	10R18	DW78B		
41.800	CHD2OH	10R26	FPVSF89		
41.871	CH3OH	*9P16	Henni86		
41.900	13CH3OH	10P16	HP78		
41.910	CH3OH	9R16	VHHR86		
42.159	CH3OH	9P32	PEJS80	7110981.4	
42.300	CH3OH	9P34	Henni82		
42.400	CH3OH	*10P12	Henni86		
42.500	CD3OH	9P38	SBW84		
42.500	CD3OH	10R34	SBW84		
42.500	CH2DOH	9P38	SPEJ80		
42.600	14NH3	N2O	Wille81		
42.600	CD3OH	9R28	SBW84		
42.920	CD3OH	9P38	CIMPS87		
42.953	CH3OH	*9P26	Henni86		
43.100	CH3OH	10R36	IMS80		
43.400	CH3OH	9P34	HRB73		
43.450	CH3OH	9P26	MH84B		
43.470	CH3OH	10R34	Henni78		
43.697	CD3OH	10R18	SEJZS87	6860664.2	39

λ [μm]	Molecule	Pump	Reference	Freq. [MHz]	Footnotes
43.700	CD3OH	10R48	CIMPS87		
43.700	CH318OH	9P30	IMPSG89		
43.970	CH3OH	*9P24	Henni83		
44.000	CH2DOH	9P30	SPEJ80		
44.240	CH3OH	9P16	Henni86		119
44.300	CD3OH	9R28	SBW84		
44.307	CH3OH	9P34	TOH88		150
44.550	CD3OH	10R32	CIMPS87		
44.700	CD3OH	9P08	CIMPS87		
44.800	CD3OH	10R08	SBW84		
45.000	CD3OH	10R08	CIMPS87		
45.100	CHD2OH	10R26	FPVSF89		
45.523	H2O	*ELEC	MC64		
45.660	CD3OH	10P08	CIMPS87		
46.164	CH3OH	*9P10	Henni86		
46.450	CH3OH	9P12	MH84B		
46.600	CH3OD	9P52	FK86		
46.700	CH3OD	9R08	BP77		
47.100	CD3OH	9P28	SBW84		
47.251	H2O	ELEC	MC64		141
47.463	H2O	ELEC	DHJRS69		43, 130
47.650	CH3OD	9R08	VHHR86		
47.693	H2O	ELEC	MC64		141
47.800	CH3OH	9P22	VHHR86		
47.800	CH3OH	*9P24	Henni83		
48.100	CH3OH	9P22	SBW84		
48.363	CH3OH	*9P16	Henni86		
48.400	CH318OH	10R38	IMPSG89		
48.600	CD3OH	9R34	SBW84		
48.630	CH3OH	9R22	Henni86		81, 119
48.677	H2O	*ELEC	MC64		
48.700	CH3OH	9P16	MH84B		
48.700	CD3OH	9R06	SBW84		
48.740	CH3OH	9R22	SH84		pp20, 81
48.760	CH3OH	10R34	TOH88		153
48.784	CH3OH	*9R18	TOH88		pp20
49.070	CD3OH	10R32	CIMPS87		
49.500	CH318OH	9P16	IMPSG89		
49.620	H2S	*ELEC	HC69		
49.780	CD3OH	10R20	CIMPS87		
49.800	CD3OH	9R28	GW78		

λ [μm]	Molecule	Pump	Reference	Freq. [MHz]	Footnotes
49.980	CH3OH	13CO2	PD84		
50.000	CD3OH	9R26	SEJZS87		
50.000	CD3OH	10R38	SEJZS87		
50.015	CH3OH	*9P34	TH86		
50.100	CD3OH	10R34	SBW84		
50.224	CH3OH	*9P10	Henni86		
50.300	CD3OH	10R24	CIMPS87		
50.699	CH3OH	*9P04	TOH88		
50.700	NE	ELEC	PFMG64A		136
51.207	CH3OH	9P20	TOH88		150
51.240	CH3OH	9P34	TOH88		150
51.370	15NH3	13CO2	DP84		
51.850	CH3OH	13CO2	PD82A		
52.000	NH3(D)	10R34	GDEFJ83		
52.100	13CD3OH	9R32	IEPSV84		
52.171	CH3OH	*9P30	TH86		
52.200	13CD3OH	9R18	IEPSV84		
52.270	CH3OH	13C18O2	PD84		
52.390	NE	ELEC	PFMG64A		136
52.400	H2S	*ELEC	HC69		
52.400	CD3OD	10R16	VSPE81		
52.480	CH3OH	13CO2	PD82A		
52.700	CH318OH	10R36	IMPSG89		
52.800	CD3OH	10R20	SBW84		62
52.900	CD3OH	9R34	GW78		
53.000	CH3OH	10R46	IMMS81B		
53.100	CD3OH	9P44	CIMPS87		
53.300	CD3OH	9R34	YYSF82		
53.486	NE	ELEC	PFMG64B		128, 135
53.500	CH3OH	10R36	IMS80		
53.600	CD3OD	9R34	FK86		
53.600	CH318OH	10R30	IMPSG89		
53.820	CD3OH	9R34	Knigh81		
53.831	CH3OH	9P20	TOH88		150
53.861	CH3OH	10R36	PEJS80	5566052.7	
53.906	H2O	*ELEC	MC64		
53.988	CH3OH	*9R06	TOH88		
54.019	NE	ELEC	PFMG64B		128, 135
54.100	CD3OH	9P28	SBW84		
54.117	NE	ELEC	PFMG64B		128, 135
54.700	CD3OH	10R18	PFS86		

λ [μm]	Molecule	Pump	Reference	Freq. [MHz]	Footnotes
55.077	H2O	ELEC	MC64		141
55.370	CH3OH	9P40	PEJS80	5414344.1	
55.400	CD3OH	10R20	YYSF82		113
55.560	CD3OH	9R28	Knigh81		80
55.590	CH3OD	C18O2	PD85		
55.600	CHD2OH	10R18	FPVSF89		
55.680	NE	ELEC	PFMG64A		136
55.800	13CD3OH	9R32	IEPSV84		
56.150	CH3OH	9R24	SH84		
56.230	CH3OH	13CO2	PD82A		
56.500	CD3OH	9R06	SBW84		62
56.700	CD3OH	10R30	SBW84		
56.730	CH3OH	9R16	Henni86		119
56.840	H2S	*ELEC	HC69		
56.845	D2O	*ELEC	MC64		
56.870	CD3OH	9P26	CIMPS87		
57.240	CH3OD	9R08	PD85		68
57.340	CH3OD	9R16	PD85		
57.355	NE	ELEC	PFMG64B		128, 135
57.660	H2O	*ELEC	MC64		
57.900	CHD2OH	10P04	FPVSF89		
58.000	NH3(D)	10R36	GDEFJ83		
59.600	CD3OH	10R40	CIMPS87		
60.000	13CH3OH	10R14	IMSD85		
60.086	CH3OH	*9R08	TOH88		
60.100	CD3OH	9R34	Knigh81		
60.173	CH3OH	9P40	PEJS80	4982153.1	
60.290	H2S	*ELEC	HC69		
60.350	CH3OH	13CO2	PD82A		
60.580	CH3OH	C18O2	PD82A		
60.600	CD3OD	10P04	FK86		
60.800	CD3OH	9R34	GW78		
61.400	CD3OH	10P08	CIMPS87		
61.500	H2S	*ELEC	HC69		
61.613	CH3OH	9R18	PEJS80	4865709.8	
61.700	CD3OH	10R24	SBW84		
62.171	CH3OH	*9R08	TOH88		
62.500	C2H4O2H2	*10R16	PCD73		
62.700	CD3OD	9P28	YKYSF81		
62.966	CH3OH	10R16	PEJS80	4761182.4	
63.006	CH3OH	10R34	TOH88		153

λ [μm]	Molecule	Pump	Reference	Freq. [MHz]	Footnotes
63.096	13CH3OH	9P12	HP78	4751340.9	19
63.370	CH3OH	9P34	PEJS80	4730860.6	147
63.681	CH3OH	*9R18	TOH88		
63.880	CH3OH	13C18O2	PD84		
63.948	CH3OH	*9P30	TOH88		151
64.397	CH3OH	*9P36	TOH88		
64.400	CD3OD	9R38	VSPE81		
65.000	CH3OH	10R46	IMMS81B		
65.300	CH3OH	10R28	MH84B		
65.400	13CD3OH	9R10	IEPSV84		
65.544	CH3OH	*9P08	TOH88		
65.550	CH318OH	9P22	IMPSG89		
65.600	CH3OH	9P34	HRB73		
65.870	CD3OH	10R32	CIMPS87		
66.210	CH3OH	13C18O2	PD84		
66.249	CH3OH	9R16	Henni86		119
66.400	CD3OH	10R34	CIMPS87		
66.780	CD3OD	C18O2	PD85		
66.800	CD3OH	9R18	CIMPS87		
66.870	14NH3	13CO2	DP84		
67.177	H2O	*ELEC	MC64		
67.200	14NH3	9R30	RGF79		51
67.224	CH3OH	*9R06	TOH88		
67.430	CH3OH	*9R22	Henni86		
67.479	CD3OH	10R30	SEJZS87	4442724.8	75
67.495	CH3OH	9R18	PEJS80	4441675.2	
67.800	13CD3OH	10R12	IEPSV84		
68.000	CH3NH2	9P14	DFSY81B		
68.100	CD3OH	10R14	YYSF82		
68.329	NE	ELEC	Patel68		135
68.450	CD3OH	9R06	CIMPS87		
68.698	CH3OH	*9P10	Henni86		
68.700	CD3OH	10R14	CIMPS87		63
68.800	CD3OH	10R46	SEJZS87		64
68.930	CD3OD	13CO2	PD85		
69.100	C2H4O2H2	*10R16	PCD73		
69.180	CD3OH	10R36	CIMPS87		
69.400	CH3OD	9R16	YYSF82		
69.500	CH3OD	9R06	LD79		
69.680	CH3OH	10R16	PEJS80	4302444.9	
69.900	CH318OH	10R12	IMPSG89		

λ [μm]	Molecule	Pump	Reference	Freq. [MHz]	Footnotes
69.950	CH3OH	13C18O2	PD84		
70.000	13CH3OH	9P42	IMSD85		
70.100	C2H4O2H2	*9P34	PCD73		
70.300	CH3OD	9R16	LD79		
70.512	CH3OH	9P34	PEJS80	4251674.0	1
70.600	CD3OH	10R24	YYSF82		
70.989	CD3OH	10R08	SEJZS87	4223062.0	82, 84
71.400	CD3OH	10R08	YYSF82		113
71.500	CD3OH	10R40	CIMPS87		
71.700	CD3OH	10R10	CIMPS87		
71.700	13CH3OH	9P32	IMSD85		
71.899	HCN	*ELEC	MCW68		133, 143
71.965	D2O	ELEC	MC64		154
72.000	13CH3OH	10R40	IMSD85		
72.150	NE	ELEC	PFMG64A		136
72.429	D2O	*ELEC	MC64		
72.748	D2O	ELEC	PEJWG75	4120984.3	130
72.900	13CD3OH	9P10	IEPSV84		
73.101	HCN	*ELEC	MCW68		133, 143
73.200	CH3OH	C18O2	PD82A		
73.306	CH3OH	9P40	PEJS80	4089579.6	
73.337	D2O	*ELEC	MC64		
73.402	H2O	*ELEC	MC64		
73.467	13CD3OH	10R20	IEPSV84	4080637.2	
73.520	H2S	*ELEC	HC69		
73.800	CD3OD	10R28	VSPE81		
74.100	CHD2OH	10R04	FPVSF89		
74.380	CH3OD	13CO2	PD85		
74.384	CH3OH	*9P10	Henni86		
74.545	D2O	*ELEC	MC64		
74.800	CHD2OH	10P28	FPVSF89		
75.060	CH3OH	C18O2	PD82A		
75.200	C2H4O2H2	*9P32	PCD73		
75.275	13CD3OD	9R24	VE85	3982631.1	
75.500	13CD3OD	9R10	VE85		
75.578	XE	ELEC	PP65		
75.821	CH3OH	*10P16	Henni86		
75.932	CH3OH	*10P18	Henni86		
76.000	CD3OH	10R22	CIMPS87		
76.093	HCN	*ELEC	MCW68		133, 143
76.100	CD3OH	10P32	DW78B		

λ [μm]	Molecule	Pump	Reference	Freq. [MHz]	Footnotes
76.300	CD3OD	10P38	FK86		
76.300	CD3OH	10P42	SEJZS87		
76.305	D2O	*ELEC	MC64		
76.900	CD3OH	10R34	CIMPS87		
76.930	CD3OH	10P48	CIMPS87		
77.000	14NH2D	10R30	Lands80D		51, 57
77.000	14NH2D	10R14	Lands80D		51, 57
77.001	HCN	*ELEC	MCW68		133, 143
77.341	CH3OH	*9P34	TH86		
77.400	C2H4O2H2	*10R16	PCD73		
77.406	CH3OH	9R08	PEJS80	3873005.1	
77.487	CH3OH	*9P08	TOH88		
77.489	13CH3OH	10R26	HP78	3868819.0	
77.650	CH318OH	9P18	IMPSG89		
77.840	CH3OH	13CO2	PD84		
77.905	CH3OH	10R16	PEJS80	3848185.5	
77.930	CH3OD	10R22	PD85		
78.000	CD3OD	10P10	HP79		
78.200	CH318OH	10R24	IMPSG89		
78.390	CH3OH	13C18O2	PD84		
78.443	H2O	ELEC	EWME70	3821775.0	37
78.500	15NH3	13CO2	DVPA81		
78.600	CD3OH	9R24	CIMPS87		
78.780	CD3OH	9R42	CIMPS87		
79.050	CD3OD	C18O2	PD85		
79.091	H2O	ELEC	PEJWG75	3790474.5	37
79.600	CD3OD	10R42	PVSEP85		
79.980	CH3OH	C18O2	PD82A		
80.000	CHD2OH	10R40	FPVSF89		
80.100	CD3OD	9R32	PVSEP85		
80.180	CH3OD	9P32	VHHR86		70
80.300	13CH3OH	10P34	PS87		
80.300	CH3OH	9P34	HRB73		
80.440	CD3OH	10R28	CIMPS87		mm16
80.500	CD3OD	10R20	VSPE81		
80.500	CD3OD	10R28	VSPE81		mm16
80.500	H2S	*ELEC	HC69		
80.600	CH3OH	CO2S	WGS77		
80.843	CH3OH	*9P14	Henni86		
80.900	CD3OH	10R16	YYSF82		
81.010	13CH3OH	9P20	IMSD85		

λ [μm]	Molecule	Pump	Reference	Freq. [MHz]	Footnotes
81.497	14NH3	N2O	KMNT79	3678577.0	5
81.554	HCN	*ELEC	MCW68		133, 143
81.557	CD3OH	10R16	SEJZS87	3675859.9	73, 75
81.900	CH3OD	9P30	NH80		
81.903	CH3OH	*9P26	Henni86		
82.100	13CD3OD	9R34	VE85		
82.200	CD3OD	10R16	VSPE81		
82.400	13CD3OD	9R14	VE85		
82.600	CD3OD	9R18	PVSEP85		
82.700	CD3OH	9P18	PFS86		
82.700	CH3OH	9P26	MH84B		
83.430	H2S	*ELEC	HC69		
83.600	CD3OD	10P30	PVSEP85		
83.600	CD3OH	10R32	YYSF82		
83.700	CD3OH	10R14	CIMPS87		mm20
83.700	CHD2OH	10R14	FPVSF89		mm20
83.770	PH3	10R34	SCLB81		
83.900	CD3OH	10R32	PFS86		
83.900	CHD2OH	10R32	FPVSF89		
84.005	CH3OH	*10P10	Henni86		
84.111	D2O	*ELEC	MC64		
84.279	D2O	ELEC	HSJ69	3557143.0	130, 142
84.400	13CD3OD	10P12	VE85		
84.406	13CD3OH	10R22	IEPSV84	3551805.8	
84.500	CD3OH	10P06	PFS86		
84.908	CH3OH	*9R22	TOH88		
84.913	CH3OH	*10P16	Henni86		
85.047	NE	ELEC	Patel68		135
85.317	13CH3OH	9P22	HP78	3513853.0	
85.500	CD3OD	10P38	PVSEP85		
85.601	CH3OH	9P40	PEJS80	3502210.2	
85.790	13CH3OH	10R28	HP78		
85.825	CH3OH	*9P34	TH86		
85.830	CH3OH	13CO2	PD82A		
86.000	14NH2D	10P40	Lands80D		51, 57
86.112	13CH3OH	9P10	HP78	3481433.0	
86.239	CH3OH	9R08	PEJS80	3476282.5	
86.300	CD3OH	10P56	CIMPS87		
86.400	CD3OH	10R16	DW78B		
86.500	CD3OD	10R10	VSPE81		
86.700	CH3OD	13CO2	DVPA81		

λ [μm]	Molecule	Pump	Reference	Freq. [MHz]	Footnotes
86.741	CD3OH	10R34	SEJZS87	3456161.2	
86.900	NE	ELEC	PFMG64A		136
87.000	CH3NH2	9P34	DFSY81B		
87.000	14ND3	9R40	Lands80D		51
87.100	CD3OD	10R44	PVSEP85		
87.100	CH2DOH	9P18	SPEJ80		
87.300	CD3OD	10R16	VSPE81		
87.400	14NH3	CO2S	DW78D		
87.470	H2S	*ELEC	HC69		
87.650	CH318OH	9R06	IMPSG89		
87.800	CD3OH	10P20	PFS86		
87.900	CD3OH	9P28	PFS86		
87.900	CH2DOH	9P40	SPEJ80		
87.900	13CH3OH	9P08	HP78		
88.000	NH3(D)	9P40	GDEFJ83		
88.460	NE	ELEC	PFMG64A		136
88.720	CH3OD	9P32	VHHR86		
88.819	CH3OH	*10P16	Henni86		
89.000	NH3	9R30	GDEFJ83		
89.000	13CH3OH	9P44	IMSD85		
89.310	CH3OH	13C18O2	PD84		
89.600	CH3OD	9P30	NH80		
89.775	H2O	*ELEC	MC64		
89.930	NE	ELEC	PFMG64A		136
89.980	CH3OH	13C18O2	PD84		
90.000	NH3(D)	10P04	GDEFJ83		
90.160	CD3OH	9P22	CIMPS87		
90.303	CH3OH	*9R06	TOH88		
90.400	CH2DOH	10P30	SPEJ80		
90.800	C2H4O2H2	*9P32	PCD73		
90.970	CH318OH	10P42	IMPSG89		
91.080	CH3OD	13CO2	PD85		
92.000	CH3NH2	9R20	DFSY81B		
92.000	H2S	*ELEC	HC69		
92.544	CH3OH	9P24	PEJS80	3239461.6	
92.600	CH318OH	9P34	IMPSG89		
92.664	CH3OH	10R34	PEJS80	3235253.6	
93.000	CHD2OH	10R32	FPVSF89		
93.020	NE	ELEC	PFMG64A		
93.400	CH318OH	9P22	IMPSG89		
93.600	13CD3OD	10R16	VE85		

λ [μm]	Molecule	Pump	Reference	Freq. [MHz]	Footnotes
93.600	15NH3	CO2S	FWSGW82		
93.690	15NH3	C18O2	DP84		
93.880	CD3OH	10R32	CIMPS87		
93.960	CH3OH	C18O2	PD84		
94.300	CD3OD	9P06	FK86		
94.500	D2O	CO2S	DW78D		
94.700	CH3OH	9R02	SSKYM81		
94.900	CD3OH	10R32	CIMPS87		
95.100	CH3OH	13C18O2	PD84		
95.240	13CH3OH	9P32	IMSD85		
95.250	CH3OH	C18O2	PD82A		
95.551	CH2F2	9R12	PSE80	3137510.6	100
95.763	HE	ELEC	TM76		125
95.800	C2H4O2H2	*9R10	PCD73		
96.000	CH3OH	9R10	Radfo75		
96.380	H2S	*ELEC	HC69		
96.401	HCN	*ELEC	MCW68		133, 143
96.522	CH3OH	9R10	PEJWG75	3105936.8	121
96.812	CH3OH	*9P08	TOH88		
97.290	CH3OH	13CO2	PD84		
97.500	CD3OD	10R26	VSPE81		
97.519	CH3OH	10R40	PEJS80	3074210.0	
97.800	CH3OH	*10R48	Henni86		
98.000	13CH3OH	9P28	PS87		
98.000	CH3OH	10R40	TYY75		
98.500	13CD3OH	9R34	IEPSV84		
98.650	CH318OH	10R10	IMPSG89		
98.693	HCN	*ELEC	MCW68		133, 143
98.862	CH3OH	*9P40	TOH88		
99.140	CH318OH	9P28	IMPSG89		
99.280	CH3OH	9P36	IIMMS81		
99.500	CH3NH2	*9R14	PCD73		
99.861	CH3OH	10P16	VWE87	3002087.5	
100.000	CH3NH2	9R14	Lands80D		
100.000	CH2DOH	9P18	SPEJ80		
100.010	CH3OH	*10P18	Henni86		
100.166	CH3OH	10R32	VWE87	2992957.0	
100.800	CH3OD	9P26	LD79		
100.806	CH3OH	9R14	PEJS80	2973940.6	
101.257	HCN	*ELEC	MCW68		133, 143
101.300	13CH3OH	9R40	PS87		

λ [μm]	Molecule	Pump	Reference	Freq. [MHz]	Footnotes
101.440	CH3OD	9P26	VHHR86		69
101.900	H2CO	*ELEC	HM76		
102.000	CH3NH2	9P18	DFSY81B		
102.023	CH2DOH	9P16	SPEJ80	2938465.1	
102.200	CD3OD	9R24	PVSEP85		
102.300	CH3OH	C18O2	PD82A		
102.600	CH3OH	13C18O2	PD84		
102.600	CD3OH	10R34	DW78B		
102.900	14NH3	C18O2	DP84		
103.000	13CH3OH	9P24	IMSD85		
103.000	CD3OH	10P14	PFS86		
103.000	CHD2OH	10P16	FPVSF89		
103.000	CHD2OH	10P14	FPVSF89		
103.125	CH3OD	9P30	RPJM77	2907088.9	
103.300	H2S	*ELEC	HC69		
103.481	13CH3OH	9P22	HP78	2897083.0	
103.586	13CH3OH	10R26	HP78	2894132.0	
104.000	CH3OD	9P30	KYHH75		
104.000	CH3NH2	*9P28	PCD73		
104.000	PH3	9R14	SCLB81		
104.300	CD3OD	10R12	VSPE81		
104.600	CH318OH	9P22	IMPSG89		
104.600	CHD2OH	9P18	FPVSF89		
104.850	CH3OH	*9P36	TH86		
105.000	CHD2OH	10R06	FPVSF89		
105.000	CH3NH2	9P28	DFSY81B		
105.070	13CH3OH	9P10	IMSD85		
105.100	CH3OH	9R02	SSKYM81		
105.147	13CH3OH	10R18	HP78	2851169.0	
105.400	CD3OD	10P46	PVSEP85		
105.518	CH2F2	9P16	PSE80	2841142.9	
106.000	CH3OD	9R02	Lands80C		
106.000	CH3OD	9R26	Lands80C		
106.020	NE	ELEC	PFMG64A		136
106.400	13CH2F2	9R14	STPVE85		
107.200	CD3OH	10R14	CIMPS87		
107.400	CHD2OH	10R24	FPVSF89		
107.538	CD3OD	10R12	VSPE81	2787789.4	
107.720	D2O	ELEC	PEJWG75	2783066.6	130, 142
107.800	13CH3OH	9R20	PS87		
108.000	14NH2D	10R26	Lands80D		51, 57

λ [μm]	Molecule	Pump	Reference	Freq. [MHz]	Footnotes
108.600	CH3OH	C18O2	PD82A		
108.668	CD3OH	10P10	SEJZS87	2758781.7	75
108.700	CD3OD	10R12	VSPE81		
108.720	CH3OD	9P32	VHHR86		67
108.800	H2S	*ELEC	HC69		
108.818	CH2DOH	9P12	SPEJ80	2754995.7	
108.941	CH2DOH	9P32	SPEJ80	2751872.9	
109.000	CH3NH2	9P40	DFSY81B		
109.100	CD3OH	9P22	PFS86		
109.100	C2H4O2H2	*9R16	PCD73		
109.296	CH2F2	9P24	PSE80	2742946.0	21
109.300	CH318OH	10R04	IMPSG89		
109.300	CHD2OH	9R38	FPVSF89		
109.400	CH3OH	C18O2	PD82A		
109.926	13CD3OD	10R16	VE85	2727211.7	
109.938	13CD3OD	10R16	VE85	2726923.5	
110.000	13CD3OH	10R22	IEPSV84		
110.000	CH3OD	9P32	YYSF82		
110.000	CH3OD	10R44	KYHH75		
110.000	15NH3	13CO2	DVPA81		
110.240	HCN	*ELEC	MCW68		134, 144
110.432	13CH3OH	10R18	HP78	2714715.0	
110.700	CH3OD	9P32	NH80		
110.716	CH3OH	9P36	VWE87	2707749.3	
110.900	13CH3OH	10R46	PS87		
111.300	CD3OD	9R38	PVSEP85		
111.400	CHD2OH	9R38	FPVSF89		
111.400	CD3OH	10R28	CIMPS87		
111.600	CHD2OH	10R40	FPVSF89		
111.600	CH318OH	10P06	IMPSG89		
111.900	15NH3	CO2S	DW78A		
111.900	CHD2OH	10R34	FPVSF89		
112.000	13CH2F2	9P44	STPVE85		
112.066	HCN	*ELEC	MCW68		133, 143
112.100	CD3OH	9R32	CIMPS87		
112.300	15NH3	13CO2	DVPA81		35
112.300	CD3OH	10R34	DW78B		
112.300	14NH3	CO2S	FWSGW82		
112.532	CH2DOH	9P12	SPEJ80	2664058.3	
112.600	D2O	CO2S	DW78D		
112.946	CH3OH	9P44	VWE87	2654310.7	

λ [μm]	Molecule	Pump	Reference	Freq. [MHz]	Footnotes
113.000	CD3OD	9P24	YKYSF81		
113.000	14NH2D	10P40	Lands80D		51
113.311	HCN	*ELEC	MCW68		134, 144
113.400	13CH3OH	9P06	PS87		
113.450	CH3OH	*9P16	Henni86		
113.500	CH3OD	CO2S	FK86		
113.600	13CH3OH	9P30	IMSD85		
113.732	CH3OH	9R08	PEJS80	2635958.0	
113.800	CH3OD	9P32	NH80		
114.000	CD3OD	9P24	YKYSF81		
114.200	CH318OH	9P32	IMPSG89		
114.400	CD3OH	10R32	CIMPS87		
114.960	CH3OD	9R02	GSRF80		
115.000	ND2ND2	10P16	SDEF85		
115.000	13CH3OH	9P22	IMSD85		
115.420	H2O	ELEC	MC64		141
115.500	CH3NH2	*9P44	PCD73		
115.700	CH318OH	9P36	IMPSG89		
115.800	CH318OH	9P40	IMPSG89		
115.823	13CH3OH	10R16	HP78	2588363.0	
115.935	CH2F2	C18O2	GCK86	2585856.8	
116.000	CH3NH2	9P44	Lands80D		
116.000	CH3OH	13C18O2	PD84		
116.000	CH3SH	9R34	Lands80B		
116.132	HCN	*ELEC	MCW68		133, 143
116.500	CD3OH	9P48	PFS86		
116.800	H2S	*ELEC	HC69		
117.000	CH3SH	10R34	Lands80B		
117.000	CH3OH	*9R24	Henni86		
117.000	CH2F2	9R20	DW78C		
117.085	CH2DOH	9P32	SPEJ80	2560467.0	
117.100	C2H4O2H2	*9P14	PCD73		
117.227	CH3OD	9P26	RPJM77	2557365.4	112
117.300	CHD2OH	9R38	FPVSF89		
117.400	CH3OH	C18O2	PD82A		
117.620	CD3OH	10R08	CIMPS87		
117.630	CH3OH	*9P06	Henni86		114
117.727	CH2F2	9R20	PSE80	2546495.0	
117.920	13CH3OH	9P26	IMSD85		
117.960	CH3OH	9P14	PEJS80	2541485.6	
118.000	CH3NH2	9P06	DFSY81B		pp13

λ [μm]	Molecule	Pump	Reference	Freq. [MHz]	Footnotes
118.000	C2H4O2H2	*9P36	PCD73		
118.000	CH3NH2	*9P08	PCD73		pp13
118.000	C2H4O2H2	*10P30	PCD73		
118.013	13CH3OH	9P22	HP78	2540332.0	
118.553	13CD3OD	9R14	VE85	2528772.8	
118.591	H2O	ELEC	FSPB67	2527952.8	33, 130
118.800	CD3OH	9P38	PFS86		
118.834	CH3OH	9P36	PEJS80	2522781.6	2
118.900	C2H4O2H2	*9P34	PCD73		
119.000	CH3NH2	9P08	Lands80D		
119.000	CD3OH	9R14	YYSF82		
119.057	CD3OD	10R26	VSPE81	2518067.7	102
119.100	13CD3OH	10P42	IEPSV84		
119.400	13CD3OH	9R14	IEPSV84		
119.600	H2CO	*ELEC	HM76		
119.700	CH3OH	13CO2	PD82A		
119.800	CH3OH	*9R28	Henni86		
119.840	CH318OH	9P26	IMPSG89		
119.900	CD3OH	10R34	CIMPS87		
120.000	HFCO	9R08	JDL83		
120.000	HFCO	9P24	JDL83		
120.000	CH3NH2	10P24	Lands80D		
120.080	H2O	*ELEC	MC64		
120.300	CD3OH	9R14	PFS86		
120.450	CD3OH	9R40	CIMPS87		
120.469	CD2F2	10R36	VPE81	2488553.4	
120.661	CD3OH	9R14	SEJZS87	2484584.9	
120.900	CHD2OH	9R30	FPVSF89		
120.902	CH3OH	10R44	VWE87	2479622.2	
121.000	O3	*9P14	WZN73		
121.000	CH2CHF	N2O	GRBF84		
121.200	13CH3OH	10R28	HP78		
121.270	CH3OH	*10P10	Henni86		
121.700	CH2F2	9R22	DW78C		
122.000	13CH3OH	10R02	IMSD85		
122.154	CD3OH	10R38	SEJZS87	2454225.9	75
122.304	CD3OD	10R28	VSPE81	2451203.1	
122.402	13CH3OH	9P20	CIMPS88	2449245.9	83
122.466	CH2F2	9R22	PSE80	2447968.5	
122.466	CH2F2	9P08	PSE80	2447974.6	
122.800	H2CO	*ELEC	HM76		

λ [μm]	Molecule	Pump	Reference	Freq. [MHz]	Footnotes
122.885	13CH3OH	10R30	CIMPS88	2439623.2	83
123.000	OCS	*ELEC	HC69		
123.260	13CH3OH	10P16	HP78		
123.550	CD3OH	10R12	CIMPS87		
123.640	CH3OH	*10P16	Henni86		
123.800	CHD2OH	10P08	FPVSF89		
123.850	CH318OH	9P34	IMPSG89		
123.900	CHD2OH	9P38	FPVSF89		
123.900	CH318OH	9P30	IMPSG89		
123.900	CH3OH	C18O2	PD82A		
124.000	14NH2D	10R14	Lands80D		51, 57
124.000	DFCO	9R12	JDL83		
124.000	CH3SH	10R24	Lands80B		
124.253	13CD3OD	10P24	VE85	2412757.9	
124.300	13CD3OH	10P22	IEPSV84		
124.400	NE	ELEC	PFMG64A		136, 137
124.400	CHD2OH	10P28	FPVSF89		
124.432	CH2DOH	10P34	SPEJ80	2409293.3	
124.500	CD3OD	9R04	VSPE81		
124.600	NH3	13CO2	DVPA81		
124.700	CH3OH	13CO2	PD82A		
124.798	CD3OD	10R26	VSPE81	2402224.0	
124.930	CD3OH	10P10	CIMPS87		
125.000	CH2DOH	10P34	ZD78		
125.100	CH3OH	10R10	MH84B		
125.400	CHD2OH	10P40	FPVSF89		
125.800	C2H4O2H2	*9P34	PCD73		
125.900	H2CO	*ELEC	HM76		
126.000	CH3NH2	*10R06	PCD73		
126.100	NE	ELEC	PFMG64A		
126.100	13CD3OH	9P10	IEPSV84		
126.164	HCN	ELEC	Maki68		89, 126
126.200	13CD3OD	9P12	VE85		
126.200	H2S	*ELEC	HC69		
126.545	CH2F2	C18O2	GCK86	2369056.7	
126.600	CH3OH	13CO2	PD82A		
127.000	CH3SH	9P18	Lands80B		
127.021	13CD3OH	10P08	IEPSV84	2360174.8	
127.300	CH2F2	9P10	IMMSD85		
127.300	CD3OH	10P04	PFS86		
127.400	CHD2OH	10R18	FPVSF89		

λ [μm]	Molecule	Pump	Reference	Freq. [MHz]	Footnotes
127.656	13CD3OH	10R22	IEPSV84	2348438.4	
127.770	CH318OH	10R30	IMPSG89		
127.800	CH2F2	9P10	MH84A		
128.000	CH3SH	9P24	Lands80B		
128.000	HFCO	9R22	JD81		
128.000	CH3OD	9P22	DSF74A		
128.000	CH3NH2	10R12	Lands80D		
128.034	CD3OH	10R34	SEJZS87	2341508.9	75
128.100	13CD3OD	10R16	VE85	2340291.8	
128.100	SO2	9R18	SL84		
128.100	CH2F2	9P20	MH84A		
128.500	CHD2OH	9P26	FPVSF89		
128.629	HCN	ELEC	MCW68		131, 132
128.700	CD3OH	10R34	DW78B		62
129.000	CD3OD	9P16	YKYSF81		
129.100	CH2F2	9P20	IMMSD85		
129.100	H2S	*ELEC	HC69		
129.200	13CD3OD	9P38	VE85		
129.550	CH3OH	10R34	PEJS80	2314111.3	
129.600	CD3OD	9P42	PVSEP85		
130.000	CH3OH	10R34	KYHH75		
130.000	CH3NH2	9P20	DFSY81B		
130.100	CH3OH	*9R14	Henni86		
130.500	CD3OD	10R06	VSPE81		
130.800	H2S	*ELEC	HC69		
130.839	HCN	ELEC	Maki68		89, 126
131.200	CH3OD	10R36	FK86		
131.500	CD3OD	10R32	PVSEP85		mm17
131.560	CH3OH	*9R32	TOH88		
131.563	CD3OH	10R32	SEJZS87	2278703.0	mm17
131.600	CH3OH	C18O2	PD82A		
131.690	CH318OH	9R30	IMPSG89		
132.000	OCS	*ELEC	HC69		
132.000	C2H4O2H2	*9P24	PCD73		
132.000	C2H4O2H2	*9P36	PCD73		
132.100	CD3OH	10P22	CIMPS87		
132.200	CHD2OH	9R30	FPVSF89		
132.800	NE	ELEC	PFMG64A		
132.900	CH3OH	13CO2	PD82A		
133.000	CH3OD	9P10	YYSF82		
133.120	CH3OH	9P24	PEJS80	2252054.2	

λ [μm]	Molecule	Pump	Reference	Freq. [MHz]	Footnotes
133.700	CD3OH	9P14	PFS86		
133.700	13CH3OH	10P12	PS87		
133.900	HCOOH	9R22	Lands80A		
133.998	CH2F2	9P22	PSE80	2237296.4	
134.000	CH3NH2	*9R18	PCD73		
134.000	CH3OD	10P10	HP79		
134.000	CH3NH2	*9R14	PCD73		
134.000	CH3OD	9P10	LD79		
134.000	ND2ND2	10P18	SDEF85		
134.200	SO2(I)	9P10	BS84		164
134.600	CH318OH	9P30	IMPSG89		
134.700	CH3OD	9P06	BP77		
134.900	CH2F2	C18O2	PD82B		
134.933	HCN	ELEC	Maki68		89, 126
135.000	CH3OH	*9R28	Henni86		
135.000	CHD2OH	9R26	FPVSF89		
135.000	C2H4O2H2	*9P36	PCD73		
135.172	CH2DOH	10R32	SPEJ80	2217863.3	
135.173	CH2DOH	10R32	SPEJ80	2217849.9	
135.269	CH2F2	9P24	PSE80	2216263.5	21
135.400	CD3OH	10R22	PFS86		
135.500	H2S	*ELEC	HC69		
135.523	13CH2F2	9R44	STPVE85	2212111.0	
135.710	CH3OH	9P36	IIMMS81		
135.834	CH2DOH	9R08	SPEJ80	2207058.3	
135.900	CD3OD	9P52	FK86		
135.940	PH3	9P12	SCLB81		
136.000	CH3OD	10R22	YYSF82		
136.000	CH3OD	10P24	HP79		
136.120	CH3OH	*9R10	Henni86		
136.200	CHD2OH	10R26	FPVSF89		
136.500	CD3OH	9R06	PFS86		
136.627	CD3OH	10R14	SEJZS87	2194236.9	
136.800	CH3OH	9R34	PS87		
136.959	CD3OD	10R46	IMEJ86	2188929.0	155
137.000	CD3OD	9P26	YKYSF81		
137.000	CHD2OH	9P34	FPVSF89		
137.000	CH3NH2	9P18	DFSY81B		
137.170	CH3OD	10R22	GSRF80		66
138.281	13CH2F2	9R22	STPVE85	2168000.0	
138.400	CD3OH	10R34	CIMPS87		

λ [μm]	Molecule	Pump	Reference	Freq. [MHz]	Footnotes
138.768	HCN	*ELEC	MCW68		134, 144
139.000	CH3NH2	*9R14	PCD73		
139.266	CD2F2	10R20	VPE81	2152662.4	
139.500	CD3OD	9P26	PVSEP85		
139.600	SO2	9R14	CD81A		20
139.800	SO2	*ELEC	HHCS71		
139.850	CH3OD	N2O	GSRF80		
140.000	CD3OH	9R10	PFS86		
140.300	CH2DOH	9P20	SPEJ80		
140.400	CH3OD	CO2H	FK86		
140.405	13CH2F2	9R38	STPVE85	2135193.0	
140.600	H2S	*ELEC	HC69		
140.780	SO2	ELEC	KS80		32
140.880	SO2	ELEC	KS80		32
140.900	13CH3OH	10P30	PS87		
140.950	CD3OH	10R08	CIMPS87		
141.000	CH3OD	10R42	Lands80C		
141.000	CH3OD	9P32	YYSF82		
141.000	CH3NH2	*10R22	PCD73		
141.300	CD3OD	9R08	VSPE81		
141.700	CD3OD	9R12	PVSEP85		
142.000	CH3NH2	10R22	DFSY81B		pp1
142.000	SO2	*ELEC	HHCS71		
142.000	CH3NH2	10R20	Lands80D		pp1
142.100	SO2	9R28	SL84		
142.100	SO2	9R18	CD81A		20
142.430	CH318OH	9P32	IMPSG89		
142.600	CH2CHF	10P54	RGF84		
142.800	CH318OH	10R26	IMPSG89		
142.900	CHD2OH	10P40	FPVSF89		
143.000	CH3NH2	*9R14	PCD73		
143.186	CH2F2	C18O2	GCK86	2093728.7	
143.400	CD3OH	13CO2	DVPA81		mm1
143.640	CH318OH	9P40	IMPSG89		
143.800	CD3OD	13CO2	PD85		mm1
143.800	CD3OH	10R12	CIMPS87		
143.800	CD3OH	9P12	PFS86		
143.800	CD3OD	C18O2	PD85		
144.000	DFCO	9R16	JDL83		
144.118	CD3OH	10P18	SEJZS87	2080189.3	73, 112
144.180	CH318OH	10R16	IMPSG89		

λ [μm]	Molecule	Pump	Reference	Freq. [MHz]	Footnotes
144.400	CD3OH	10R26	CIMPS87		
144.800	CHD2OH	9R26	FPVSF89		
145.000	CH3NH2	9R14	Lands80D		
145.000	CH3OH	13CO2	PD82A		
145.050	CH3OH	10R20	Strum84		
145.081	CH2F2	C18O2	GCK86	2066379.1	
145.252	CH3OH	10R32	VWE87	2063941.1	114
145.300	CHD2OH	9R32	FPVSF89		
145.563	13CD3OH	10R24	IEPSV84	2059531.6	
145.600	CH3OD	9P32	NH80		pp2
145.662	CH3OD	9P30	BKM78	2058141.8	pp2, 112
145.700	CD3OH	9P08	PFS86		
146.000	CH3NH2	9R08	Lands80D		
146.097	13CH3OH	9P10	HP78	2052004.0	
146.200	SO2	9R14	SL84		
146.326	13CD3OH	10R26	IEPSV84	2048803.6	
147.000	CH3SH	10R34	Lands80B		
147.000	CH3NH2	10R36	Lands80D		
147.000	CH3NH2	*9P24	PCD73		
147.040	14NH3	9R30	WZN73		124
147.280	CD3OH	9R26	CIMPS87		
147.349	CD3OH	10P32	SEJZS87	2034573.6	
147.650	CD3OH	10P12	CIMPS87		
147.845	CH3NH2	9P24	RPJM77	2027752.6	
147.970	13CH3OH	9P30	HP78		
148.000	NH3	9R30	GDEFJ83		
148.000	CD3OH	10P32	PFS86		
148.000	CH3OH	13CO2	PD82A		
148.000	CD3OD	C18O2	PD85		
148.000	HCCCHO	10P18	DJ80		
148.200	CH2CHF	10R50	RGF84		
148.200	SO2(I)	9P16	CD81B		
148.300	13CD3OH	10P42	IEPSV84		
148.500	CH3NH2	9P24	DSF72		
148.590	13CH3OH	10R16	HPPJE79	2017576.1	
148.617	13CD3OD	10R30	VE85	2017218.5	mm13
148.940	CD3OH	9R32	CIMPS87		
149.000	SIHF3	C18O2	DFS85		
149.000	CH318OH	9P30	IMPSG89		
149.200	O3	9P06	DDM83		
149.272	13CH3OH	9P22	HP78	2008360.0	

λ [μm]	Molecule	Pump	Reference	Freq. [MHz]	Footnotes
149.388	CH2DOH	10P36	SPEJ80	2006805.2	
149.613	CH2DOH	10R32	SPEJ80	2003788.3	
149.700	SO2	9R18	SL84		
149.800	CH3OH	C18O2	PD82A		
149.800	CH3OH	*10R48	Henni86		
150.000	CH3NH2?	9P10	DFSY81B		
150.000	SO2	*ELEC	HHCS71		
150.000	CD3OD	10R30	HP79		mm13
150.200	13CD3OH	9R30	IEPSV84		
150.300	13CD3OD	9R32	VE85		
150.438	CD2F2	10R18	VPE81	1992795.2	
150.500	CD3OD	9R16	PVSEP85		
150.572	CH2DOH	10P26	SPEJ80	1991028.3	
150.800	CD3OH	10R16	PFS86		
150.816	CH2DOH	10R34	SPEJ80	1987798.9	
151.000	13CD3OD	9P24	VE85		
151.000	CH3OH	10R08	DFSY81B		
151.000	13CD3OH	9R38	IEPSV84		
151.190	SO2	ELEC	KS80		111
151.254	CH3OH	9R26	PEJS80	1982050.6	121
151.300	CD3OH	9R32	PFS86		62
151.300	CD3OD	10P10	PVSEP85		88
151.310	SO2	ELEC	KS80		111
151.560	CH3OD	9R34	GSRF80		
151.650	CH318OH	9P22	IMPSG89		
151.800	CD3OH	10R20	CIMPS87		
151.800	13CD3OD	9P28	VE85		
152.000	HDCO	9R22	DDG80		
152.000	CH3OH	9R02	Lands80C		
152.076	13CH3OH	10R16	HPPJE79	1971337.2	
152.300	CD3OD	9P44	PVSEP85		
152.420	CH3OH	*9P30	TH86		
152.500	CD3OD	9R04	VSPE81		
152.600	CHD2OH	10P08	FPVSF89		
152.670	CH3OH	*9P12	TOH88		
152.700	15NH3	13CO2	WDVP80		35, 51
152.700	CH2DOH	9R24	SPEJ80		
153.000	CH3NH2	*9P08	PCD73		
153.195	CH2F2	C18O2	GCK86	1956935.8	
153.540	CH318OH	9P36	IMPSG89		
153.694	13CD3OH	9R28	IEPSV84	1950581.6	

λ [μm]	Molecule	Pump	Reference	Freq. [MHz]	Footnotes
153.700	CD3OH	10P46	CIMPS87		
154.160	CH2F2	C18O2	GCK86	1944679.6	
154.200	CD3OD	C18O2	PD85		
155.000	13CH3OH	10P08	PS87		
155.000	HDCO	10R34	DDG80		
155.100	H2CO	*ELEC	HM76		
155.600	CD3F	10P10	TSW79B		
156.000	HCCCHO	10P22	DJ80		
156.000	13CD3OH	9P38	IMSD85		
156.510	CH3OH	*9P10	TOH88		
157.000	CH2CHCL	10P22	LSB81		
157.600	H2CO	*ELEC	HM76		
157.600	CH3OH	9R28	Henni86		119
157.929	13CH3OH	9P12	HP78	1898280.0	
158.000	CD3OH	9R28	GW78		
158.513	CH2F2	9P10	PSE80	1891274.3	21
158.900	CD3OH	9R18	PFS86		
158.960	CH2F2	9P20	PSE80	1885959.3	
159.000	CH3NH2	*9P24	PCD73		
159.200	CH3OH	CO2S	WGS77		
159.218	CH2DOH	10R34	SPEJ80	1882906.3	
159.400	CD3OH	9R28	SEJZS87	1880754.6	75, 112
159.500	H2CO	*ELEC	HM76		
159.500	ND2ND2	9P36	SDEF85		
159.500	SO2	9R28	BS81		
159.676	CH3OH	9R26	PEJS80	1877508.5	
160.000	SO2	9R28	CD81A		
161.000	CH3SH	9R30	Lands80B		
161.100	CD3OH	10R38	CIMPS87		
161.300	CD3OD	10P08	PVSEP85		
161.530	CH3OH	*10P10	Henni86		
161.800	CHFCHF	10P10	BH82		
162.218	CH3OH	9P36	VWE87	1848083.8	
162.400	H2S	*ELEC	HC69		
162.670	CH3OH	10R44	Henni86		114
162.700	CH2DOH	10P30	SPEJ80		
162.850	CD3OH	10P24	CIMPS87		
163.010	CH3OH	*10R34	WZN73		pp14
163.034	CH3OH	10R38	PEJWG75	1838839.3	pp14, 114
163.120	CH2F2	C18O2	GCK86	1837861.1	
163.300	CHD2F	9P34	Tobin84		

λ [μm]	Molecule	Pump	Reference	Freq. [MHz]	Footnotes
163.574	CH3OH	9P12	VWE87	1832768.6	
163.610	O3	9P40	WZN73		116
163.800	H2CO	*ELEC	HM76		
164.000	C2H4O2H2	*9R10	PCD73		
164.000	C2H4O2H2	*9P14	PCD73		
164.000	CH3NH2	*9R18	PCD73		
164.000	CH3OH	10R48	IMS80		
164.000	C2H4O2H2	*9P16	PCD73		
164.000	CH2DOH	9R18	ZD78		
164.000	DFCO	9R12	JDL83		
164.000	CH3OH	10R38	DTS74		
164.090	CH3OH	*9P34	TH86		
164.200	CH3OH	9P12	IIMMS81		
164.200	CH3OD	10P38	FK86		
164.400	CH3OH	13C18O2	PD84		
164.400	CHD2OH	9R24	FPVSF89		
164.508	CH3OH	9P14	PEJS80	1822362.7	
164.564	CH3OH	9P14	Knigh81	1821735.5	
164.600	CH3OH	9P16	PEJS80	1821335.2	
164.656	13CH2F2	9P12	STPVE85	1820715.0	
164.697	CH3OH	9P24	PEJS80	1820261.5	
164.746	CH2DOH	9R08	SPEJ80	1819720.3	
164.783	CH3OH	9R10	PEJS80	1819314.0	
164.815	13CH2F2	9R04	STPVE85	1818964.0	
165.000	CD3OH	10R32	SEJZS87		
165.000	CH3NH2	9R18	Lands80D		
165.000	CH3NH2	10P24	Lands80D		
165.000	CHD2OH	9R18	ZD78		
165.100	CHD2OH	9R04	FPVSF89		
165.100	CH318OH	9P10	IMPSG89		
165.150	HCN	*ELEC	MCW68		134, 144
165.200	SO2	9R26	SL84		
165.300	CD3OD	10P14	PVSEP85		
165.600	CH3OH	9P24	DSF74A		
165.604	CD3OD	10R22	VSPE81	1810294.3	102
165.800	CH2F2	9R22	DW78C		
165.900	CH2F2	9R20	DW78C		
166.000	CH3NH2?	10P24	DFSY81B		
166.000	CH3NH2	9P32	PCD73		115
166.280	13CH3OH	9P12	IMSD85		
166.631	CH2F2	9R20	PSE80	1799139.3	

λ [μm]	Molecule	Pump	Reference	Freq. [MHz]	Footnotes
166.677	CH2F2	9R22	PSE80	1798647.0	
166.760	CD3OH	10R32	CIMPS87		
166.800	CH2F2	C18O2	PD82B		
166.800	SO2(l)	9R18	BS84		162
166.879	CD2F2	9R10	VPE81	1796461.7	
167.000	CH2DOH	9P18	ZD78		
167.000	CH3OD	9R22	YYSF82		
167.100	CH3OH	C18O2	PD82A		
167.352	CH2DOH	9P32	SPEJ80	1791384.9	
167.500	CD3OD	13CO2	DVPA81		94
167.541	CH2DOH	9P18	SPEJ80	1789365.9	
167.587	CH3OH	10R40	PEJS80	1788876.6	
167.810	CH3OH	*9P06	Henni86		114
168.000	CH3NH2	*9R22	PCD73		
168.000	CHD2OH	10R38	ZD78		
168.083	CD3OH	10R34	SEJZS87	1783601.1	
168.100	CH3OD	9P30	DSF74A		
168.840	13CH3OH	9P40	HP78		
169.000	CH3NH2	9R22	Lands80D		
169.000	SIH2F2	C18O2	DFS85		
169.000	C2H4O2H2	*9P36	PCD73		
169.250	CH3OD	9R22	VHHR86		67
169.600	SO2	9R28	BS84		
170.100	CH318OH	10P24	IMPSG89		
170.180	CH318OH	9R26	IMPSG89		
170.200	H2CO	*ELEC	HM76		
170.576	CH3OH	9P36	RPJM77	1757526.3	
171.000	CH2DOH	9P12	ZD78		
171.000	C2H4O2H2	*9R08	PCD73		
171.100	CHD2OH	10P12	FPVSF89		
171.300	CH3OH	CO2S	WGS77		
171.400	SO2	9R28	BS84		
171.500	O3	9P30	WZN73		116
171.670	D2O	ELEC	BPT69		142
171.700	CD2CL2	10R16	VHHR86		
171.758	13CH3OH	10R18	HP78	1745439.0	
171.800	CH2CHF	10P22	CD81B		
171.800	CH2DOH	9R22	SPEJ80		
172.000	CD3OD	9P46	PVSEP85		
172.000	CH3OH	*9R20	Henni86		
172.020	CH3OH	*9P40	TOH88		

λ [μm]	Molecule	Pump	Reference	Freq. [MHz]	Footnotes
172.100	CHD2OH	10R20	FPVSF89		
172.400	CHD2OH	9R38	FPVSF89		
172.600	CH3OD	10R14	PD85		
172.620	CD3OH	10P12	CIMPS87		
172.700	CH3OD	C18O2	PD85		
172.800	CH2CHCL	N2O	GRBF84		
172.800	CD3F	10R06	TSW79B		
172.846	CH2DOH	9P12	SPEJ80	1734446.4	
173.500	CH3OD	10P40	FK86		
173.637	13CD3OD	10R20	VE85	1726548.5	
174.000	CD3OH	10R32	CIMPS87		
174.300	CH3OH	9R20	PS87		
174.300	CD3OD	C18O2	PD85		
174.700	SO2(I)	9R24	BS84		
175.000	CH3NH2	*10R06	PCD73		
175.100	13CD3OD	9R22	VE85		
175.260	13CD3OH	10P08	IMS89		
175.500	SIH2F2	9P20	DFS85		
176.000	CH3OH	*9R24	Henni86		
176.000	CH3OH	9R02	Lands80C		
176.000	CLO2	C18O2	DJ80		
176.000	CH3NH2	*10R32	PCD73		
176.000	CH2CLF	9P12	ADF84		
176.000	CH3CHO	9R40	LSB81		
176.450	CH318OH	10R04	IMPSG89		
176.500	CH3OH	C18O2	PD82A		
176.800	CD3OH	9R32	CIMPS87		
177.000	CH3OD	9R26	YYSF82		
177.000	CH3NH2	*9R12	PCD73		
177.000	CH3OH	13CO2	PD82A		
177.000	CD3OH	10R12	CIMPS87		
177.400	CD3OH	9P30	PFS86		
177.510	CH3OH	*9P34	TH86		
177.600	13CD3OH	9R40	IEPSV84		
178.000	CH3OH	10R02	Lands80C		
178.000	CH3NH2	10R32	Lands80D		
178.000	CH3OD	9P32	YYSF82		
178.000	CH3NH2	9R12	Lands80D		
178.000	CH3NH2?	9P12	DFSY81B		
178.000	CD2CL2	10R20	SDFY86		
178.410	CH3OH	*9R04	TOH88		

λ [μm]	Molecule	Pump	Reference	Freq. [MHz]	Footnotes
178.600	CD3OD	9R02	PVSEP85		
178.820	CH3OD	9P32	VHHR86		70
179.000	CH3NH2	9P22	Lands80D		
179.000	CHD2OH	10R16	ZD78		
179.000	CD3OH	9R14	GW78		
179.728	CH3OH	10R04	PEJS80	1668035.0	
179.800	CHD2OH	9R32	FPVSF89		
179.800	CH318OH	9P18	IMPSG89		
180.000	CH3NH2	9P46	Radfo75		
180.000	SO2	9R40	SL84		
180.600	CH3OH	*9R20	Henni86		
180.600	13CH2F2	9P34	STPVE85		
180.676	CH3OH	9P34	PEJS80	1659278.6	
180.741	CD3OH	10R34	SEJZS87	1658689.9	mm2, 73, 75
180.750	CD3OH	10R08	CIMPS87		
181.000	CD3OH	9R28	YYSF82		
181.100	CH3OH	*9P08	HL82		
181.100	CH3OH	10R06	PS87		mm21
181.100	CH318OH	10R30	IMPSG89		
181.200	CH318OH	9P32	IMPSG89		
181.200	13CH3OH	10P16	PS87		
181.500	CD3OD	10R34	DVPA81		mm2
181.580	CH3OH	*9R30	TOH88		
181.600	CH318OH	10R06	IMPSG89		mm21
181.711	CD3OH	9R28	SEJZS87	1649830.3	75
181.788	DCN	ELEC	Maki68		106, 126
181.926	NH2NH2	10P06	RPJM77	1647877.4	
182.000	CH3OD	9R26	Lands80C		
182.000	SO2	9R18	BS81		
182.000	CD3OD	13CO2	DVPA81		
182.100	CH2DOH	9R22	SPEJ80		
182.100	CH3OD	9P06	NH80		
182.190	CH318OH	9P14	IMPSG89		
182.200	CH2F2	9P10	MH84A		
182.200	CH3OD	10P46	FK86		
182.381	13CH2F2	10R20	STPVE85	1643769.0	
182.566	CD3OH	9R14	SEJZS87	1642101.9	75
183.000	CH3NH2	*9R14	PCD73		
183.289	13CH2F2	9P28	STPVE85	1635628.0	
183.360	CH318OH	10P26	IMPSG89		
183.400	13CD3F	9P36	TF81		

λ [μm]	Molecule	Pump	Reference	Freq. [MHz]	Footnotes
183.621	CH2DOH	9P10	SPEJ80	1632666.9	
183.680	CH3OH	*9R32	TOH88		
183.900	CH3OH	*9P06	HL82		
184.000	CD3OH	9R08	DSF75A		
184.100	SO2(I)	9R22	BS84		164
184.200	CD3OD	9P10	PVSEP85		
184.306	CH2F2	9R32	PSE80	1626602.6	21
184.400	H2CO	*ELEC	HM76		
184.400	CD2CL2	10P02	BF86		
184.500	SIH2F2	13CO2	DFS85		
184.766	CD3OD	10R24	VSPE81	1622555.2	103
184.800	CH318OH	10P24	IMPSG89		
185.000	CH3SH	9R18	Lands80B		
185.000	CH3NH2	9R14	Lands80D		
185.000	C2H4O2H2	*9P34	PCD73		
185.000	C2H4O2H2	*9R18	PCD73		
185.000	CD3OH	10R14	PFS86		
185.000	CHFCHF	9R08	AD80		
185.100	SO2(I)	9R20	BS84		164
185.500	CH3OH	9P34	PEJS80	1616128.4	
186.000	CH2CHCL	9P18	LSB81		
186.000	CH3OD	9R28	Lands80C		
186.042	CH3OH	9R18	PEJS80	1611421.9	121
186.043	13CH2F2	9P38	STPVE85	1611414.0	
186.300	CH3OH	*9R20	HL82		
186.319	CH3OH	9P34	PEJS80	1609026.7	
186.400	CH3OD	9R02	FK86		
186.828	CH3F	C18O2	XKP84	1604647.7	14
187.000	SIH3F	10R10	DS82		
187.050	CD3OH	10P24	CIMPS87		
187.200	CH3OH	13C18O2	PD84		
187.500	CHD2OH	9P38	FPVSF89		
187.819	CD2F2	10R14	VPE81	1596174.9	
188.411	CH2DOH	10P26	SPEJ80	1591161.2	
188.424	CD3OH	10P42	SEJZS87	1591053.2	75
188.900	CD3OH	10P28	PFS86		
188.900	CH3OH	*9R16	Henni83		
188.960	13CH3OH	9P26	IMSD85		
189.000	C2H4O2H2	*9P34	PCD73		
189.000	C2H4O2H2	*9P36	PCD73		
189.190	CH3OH	*9R12	Henni86		

λ [μm]	Molecule	Pump	Reference	Freq. [MHz]	Footnotes
189.200	CD3OD	10P42	FK86		
189.300	CH2DOH	10P28	SPEJ80		
189.730	CD3OH	9P36	SEJZS87	1580101.8	75
189.832	CD2F2	10R34	VPE81	1579250.3	
189.900	CD3OD	C18O2	PD85		
189.942	CH3OH	9R10	IIMSD86	1578339.2	
189.949	DCN	ELEC	HJ68A	1578278.7	106
190.000	CHFCHF	9R04	AD80		
190.000	CD3OH	10P28	SEJZS87		
190.008	DCN	ELEC	HJ68B	1577789.0	
190.270	CH3OH	*9R06	TOH88		
190.300	13CH3OH	9P04	PS87		
190.300	CH2F2	C18O2	PD82B		
190.500	SIH2F2	C18O2	DFS85		
190.650	CH3OH	9R10	HIMS82		
190.726	CH3OH	9P34	PEJS80	1571849.7	
191.040	CH318OH	9P42	IMPSG89		
191.200	CH3OH	*10R04	WZN73		
191.356	CD3OH	10R34	SEJZS87	1566672.8	73
191.500	CD2CL2	10P18	SDFY86		
191.620	CH3OH	10R10	PEJS80	1564518.7	121
191.683	CH3OH	9R16	Henni86		121, 119
191.848	CH2F2	9P22	PSE80	1562655.9	
192.000	C2H4O2H2	*9P38	PCD73		
192.000	CD2F2	10R42	VPE81		
192.000	SIH2F2	13CO2	DFS85		
192.000	SO2(I)	9P10	BS84		166
192.500	CD3OD	10R30	VSPE81		
192.710	SO2	ELEC	KS80		32
192.780	CH3F	*10R34	CM71		
192.790	CD2F2	9P08	VPE81	1555020.1	
192.900	H2S	*ELEC	HC69		
192.907	NH2NH2	10P24	RPJM77	1554076.0	
193.000	SIH2F2	C18O2	DFS85		
193.100	SO2	9R40	SL84		
193.142	CH3OH	9P38	PEJS80	1552190.1	
193.173	CH2F2	C18O2	GCK86	1551938.8	
193.250	CH318OH	9P16	IMPSG89		
193.400	CHD2F	9P42	Tobin84		
193.497	13CH2F2	9P22	STPVE85	1549340.0	
193.500	NH2NH2	10P24	DSF74C		

λ [μm]	Molecule	Pump	Reference	Freq. [MHz]	Footnotes
193.500	CH3OH	C18O2	PD82A		
193.500	CH2F2	9R12	DW78C		
193.550	CH318OH	10R18	IMPSG89		
193.904	CH2F2	9R22	PSE80	1546083.4	
194.000	PH3	10P42	SCLB81		
194.000	CH3NH2	*9R08	PCD73		
194.063	CH3OH	9R14	PEJS80	1544818.7	
194.200	CH2CHF	10P54	RGF84		
194.260	CH3OH	*9R30	TOH88		pp21
194.300	CH3OH	*9R28	Henni86		pp21
194.352	HDCO	9P08	DDG80	1542524.6	
194.448	CH2F2	9R12	PSE80	1541764.7	
194.500	SO2(l)	9R20	BS84		163
194.703	DCN	ELEC	HJ68B	1539745.0	40, 106
194.764	DCN	ELEC	HJ68B	1539257.0	
195.000	CH3OH	9R28	PS87		
195.000	CH2CL2	10P12	HW82		
195.000	HDCO	9R26	DDSB77A		
195.000	NH2NH2	13C18O2	JTT82		
195.158	13CH2F2	9R20	STPVE85	1536153.0	
195.496	CH2DOH	9P36	SPEJ80	1533499.9	
195.500	SIH2F2	10R28	DFS85		
196.000	CLO2	C18O2	DJ80		
196.000	HDCO	9P08	DDSB77A		
196.000	HFCO	9R20	JD81		
196.000	HFCO	9R06	JDL83		
196.100	CH2F2	9R32	DV88		
196.100	CH2DOH	10P28	SPEJ80		
196.200	13CD3OH	9R08	IEPSV84		
196.300	CHFCHF	10R04	BH82		
196.500	HCOOH	9R22	Lands80A		
196.564	CH3OH	9P44	VWE87	1525164.1	
196.600	CD3OH	10P08	PFS86		
196.800	CHD2OH	9P08	FPVSF89		
196.950	CD3OH	9R36	CIMPS87		
197.000	C2H4O2H2	*9P38	PCD73		
197.046	13CD3OH	10R26	IEPSV84	1521430.9	
197.388	13CH2F2	9P20	STPVE85	1518795.0	
197.940	CH3NH2	9P24	IMEJ86	1514562.6	155
198.000	CH3NH2	*9R20	PCD73		
198.000	CHFCHF	10R04	AD80		

λ [μm]	Molecule	Pump	Reference	Freq. [MHz]	Footnotes
198.000	DFCO	C18O2	JDL83		
198.600	CD3OH	9P22	CIMPS87		
198.664	CH3OH	9P38	PEJS80	1509040.2	
198.682	CD3OH	9P40	SEJZS87	1508908.6	75
198.790	13CH3OH	10R30	IMSD85		
199.000	CH3NH2	9R20	DFSY81B		
199.140	CH3F	*10R34	CM71		
199.500	CD3OH	9P36	CIMPS87		
199.810	CD3OH	10R44	CIMPS87		
199.900	CH318OH	10R26	IMPSG89		
200.000	CD3F	9P28	TSW79B		
200.000	CH2DOH	9P38	SPEJ80		
200.000	C2H4O2H2	*9P36	PCD73		
200.000	NH2NH2	N2O	JTT82		
200.210	CH3OH	*9P30	Henni86		
200.295	13CH2F2	9P16	STPVE85	1496757.0	
200.870	CD3OH	9R36	CIMPS87		
200.900	CH3OH	13CO2	PD84		
201.000	CD3OH	9P40	GW78		
201.000	CH3NH2	*9R12	PCD73		
201.059	HCN	*ELEC	MCW68		133, 143
201.500	CD3F	10P12	TSW79B		
201.800	CH2F2	9R06	DW78C		
201.900	CH2CHF	CO2S	RGF84		
202.000	CD3OD	10R02	PVSEP85		
202.300	CD2F2	10R16	TSD82		
202.400	CH3OH	9P36	CBB70B		
202.465	CH2F2	9R06	PSE80	1480712.9	
202.600	CHD2OH	9R26	FPVSF89		
203.000	CH3NH2?	9P34	DFSY81B		
203.000	CH2CHF	10R20	TW82		
203.100	CHD2OH	9P24	FPVSF89		
203.300	CD3OH	10R08	SEJZS87		
203.300	CD2F2	9P28	VPE81		
203.500	CD3OH	10P16	PFS86		
203.636	13CH3OH	10R16	HP78	1472199.0	
203.800	CH318OH	10P06	IMPSG89		
203.960	13CH3OH	9P20	IMSD85		
204.000	CHD2F	10P48	Tobin84		
204.000	CLO2	C18O2	DJ80		
204.387	DCN	ELEC	HJ68B	1466787.0	106

λ [μm]	Molecule	Pump	Reference	Freq. [MHz]	Footnotes
204.700	CHD2OH	9P26	FPVSF89		
204.800	CD3OD	13CO2	PD85		
205.000	13CH3OH	9P06	IMSD85		
205.000	CH3SH	9R18	Lands80B		
205.100	CH3OH	C18O2	PD82A		
205.200	CH3OH	*9P18	HL82		
205.300	SO2	9R14	SL84		
205.600	CH3OH	9P34	IIMMS81		
205.800	CD3OH	10P36	PFS86		
205.981	CH2F2	C18O2	GCK86	1455434.8	
206.000	CD3F	9P16	TSW79B		
206.043	13CH2F2	10R18	STPVE85	1455000.0	
206.400	SO2	*ELEC	HHCS71		
206.600	CH3CH2F	*10P36	WZN73		
206.600	CH318OH	9P20	IMPSG89		
206.687	CH2DOH	9P14	SPEJ80	1450463.1	
206.785	CH3OH	9P12	PEJS80	1449778.0	
207.000	CLO2	9R20	DSF75B		
207.200	CH2F2	9P34	MH84A		
207.835	CD2F2	10R38	VPE81	1442454.3	
208.000	SO2	9R14	BS81		
208.000	CH3NH2	9R12	PCD73		115
208.300	CH3OH	9P34	IIMMS81		
208.300	CD3OD	10R08	PVSEP85		
208.300	CH2CL2	10P18	HW82		
208.400	CH2F2	C18O2	PD82B		
208.412	13CH3OH	9P10	HP78	1438460.0	
208.800	SO2(I)	9R18	BS84		162
208.950	CH3OH	*9P10	TOH88		
209.000	13CD3OH	10R26	IEPSV84		
209.030	CH3OH	10R20	Strum84		
209.100	13CD3F	9P24	TF81		
209.233	13CD3OD	10R12	VE85	1432817.8	
209.300	CH3OCH3	9R22	BSKK82		
209.600	CH3OD	C18O2	PD85		
209.930	CH3OH	9R14	PEJS80	1428057.6	
210.000	CH3OH	9P30	DFSY81B		
210.100	CH2F2	9P20	IMMSD85		
210.500	CD3OD	10R40	VSPE81		
210.500	CD3OD	9P16	PVSEP85		
211.000	CH3OH	C18O2	PD82A		

λ [μm]	Molecule	Pump	Reference	Freq. [MHz]	Footnotes
211.001	HCN	ELEC	MCW68		131, 132, 143
211.263	CH3OH	10R04	PEJS80	1419049.3	
211.315	CH3OH	9P12	PEJS80	1418701.0	
211.400	CH2F2	9P20	MH84A		
212.200	CD2CL2	C18O2	PD82B		
212.200	CH3OH	13CO2	PD82A		
212.500	CH2DOH	10R16	SPEJ80		
212.900	CHD2OH	10P18	FPVSF89		
212.900	CH3OD	9R04	PD85		91
213.300	CHFCHF	9P36	BH82		
213.351	13CH2F2	9R44	STPVE85	1405163.0	
213.462	CH3OH	9P22	PEJS80	1404427.0	
213.900	CH3OH	*10R38	HL82		
214.200	CH318OH	9P14	IMPSG89		
214.300	13CH3OH	10P28	PS87		
214.300	CH3OH	9P04	PS87		
214.350	CH3OH	9P10	Henni78		
214.579	CH2F2	9R34	PSE80	1397118.6	21
214.597	13CH2F2	9R34	IMEJ86	1397005.0	
214.714	CD2F2	10R36	VPE81	1396238.8	
214.800	CH3OH	9P22	DSF74A		
215.000	CLO2	10R32	DSF75B		
215.081	CD3OH	10P32	SEJZS87	1393856.9	75
215.250	CD3OH	10R20	CIMPS87		
215.300	SO2(I)	9P16	BS84		165
215.330	SO2	ELEC	KS80		111
215.372	CH3OD	9R14	BKM78	1391972.1	112
215.600	CD3OH	10P36	PFS86		
215.800	CH318OH	9P06	IMPSG89		
216.000	CLO2	C18O2	DJ80		
216.100	CH3OH	*9R18	TOH88		
216.120	HE	ELEC	TM76		125
216.300	HE	ELEC	KS80		
216.356	13CD3OD	10P24	VE85	1385646.1	
216.500	13CH3OH	9R32	PS87		
216.500	CH3OH	C18O2	PD82A		
216.800	CH2DOH	9R16	SPEJ80		
216.800	CH3OH	*9R14	HL82		
216.900	CD3OD	10P20	PVSEP85		
217.000	ND2ND2	10R14	SDEF85		
217.100	CH3CH2F	*9R14	WZN73		

λ [μm]	Molecule	Pump	Reference	Freq. [MHz]	Footnotes
217.200	CD3OH	10R28	CIMPS87		
217.830	O3	9P30	DDM83	1376271.1	
217.900	CHD2OH	9R16	FPVSF89		
217.900	CH3OD	13CO2	PD85		
218.000	DCOOD	10R20	DSF76		
218.000	CH3NH2	9P24	DSF72		
218.000	CH2CLF	9P22	ADF84		
218.000	CH2DOH	9R22	SPEJ80		
218.200	SO2(l)	9P32	BS84		
218.220	CH3OH	9P10	Henni78		
218.267	CD2F2	10R38	VPE81	1373513.3	
218.500	CH3OH	*9R14	Henni86		
218.600	15NH3	CO2S	FWSGW82		55
218.700	CH318OH	9P30	IMPSG89		
218.749	CH3NH2	9P24	IMEJ86	1370485.0	155
219.000	CH3NH2	*9P32	PCD73		
219.000	CD3OH	10P12	PFS86		
219.000	NH2NH2	13CO2	JTT82		
219.096	CH2DOH	9R24	SPEJ80	1368315.4	
219.300	CHFCHF	10P08	BH82		
219.400	CD2CL2	N2O	BF86		
219.500	CHFCHF	10R18	BH82		
219.600	CD2F2	9P28	VPE81		
219.700	CD3OH	9R40	CIMPS87		
219.800	CH318OH	10R20	IMPSG89		
219.900	CH318OH	10R20	IMPSG89		
219.900	CD3OH	10R18	DW78B		
220.000	CH3NH2	9P32	Lands80D		
220.000	HFCO	9R36	JD81		
220.100	CH3OD	9R06	PD85		90
220.230	H2O	ELEC	PFS68	1361282.6	142
220.270	CH318OH	10R18	IMPSG89		
220.300	CH3OCH3	9R32	BSKK82		
220.700	CHFCHF	10P26	BH82		
221.000	CH3NH2	10R20	Lands80D		
221.000	CD3OH	13CO2	DVPA81		
221.000	13CD3OH	9R08	IEPSV84		
221.000	CH3OD	9R06	Lands80C		
221.000	CH2CLF	9R22	ADF84		
221.000	SIH3F	10R36	DS82		
221.200	SO2(l)	9R30	BS84		164

λ [μm]	Molecule	Pump	Reference	Freq. [MHz]	Footnotes
221.200	CHD2OH	9R10	FPVSF89		
221.500	CH3OD	CO2H	FK86		
221.860	CH318OH	9P30	IMPSG89		
221.880	CD3OH	9P40	CIMPS87		
221.900	CD3OH	9P08	PFS86		59
222.000	CD3OD	10R26	PVSEP85		
222.000	CD3OH	9P06	DSF75A		59
222.217	CD3OH	10R34	SEJZS87	1349100.1	
222.300	CH2CHF	10R50	RGF84		
222.500	CH318OH	9P26	IMPSG89		
222.700	CD3OH	10P48	CIMPS87		
222.800	13CH3OH	10R02	PS87		
222.949	HCN	*ELEC	MCW68		133, 143
223.000	CD3OH	9P08	GW78		
223.500	CH3OH	9P16	CBB70B		
223.570	CH2F2	9R14	IMMSD85		
223.600	CH2F2	C18O2	PD82B		
223.700	CH3OD	10R34	FK86		mm18
223.800	CH3OH	*10R34	TOH88		mm18
223.840	CH3OH	*9P28	Henni86		
224.000	CD3CL	9R28	DFBS75		26
224.000	CH3SH	9P44	Lands80B		
224.226	CH2DOH	10P36	SPEJ80	1337012.5	
224.530	CH3OH	9R10	HIMS82		
224.700	CH3OH	C18O2	PD82A		
225.000	CD3OH	9R46	PFS86		
225.000	CD3OH	9P20	PFS86		
225.200	CH3OD	10R22	PD85		93
225.300	H2S	*ELEC	HC69		
225.516	CH3OH	9R08	PEJS80	1329362.9	
225.800	CD3OH	9P06	PFS86		
226.000	CH3NH2?	9P14	DFSY81B		
226.297	CH2DOH	9P46	SPEJ80	1324771.9	
226.300	CH3OD	CO2H	FK86		
226.800	CHD2OH	9P16	FPVSF89		
226.900	CD3OH	9R24	PFS86		
226.900	CH3CH2F	*10P40	WZN73		
227.000	13CD3OD	9R28	VE85		
227.000	CH318OH	10P10	IMPSG89		
227.150	CH3CL	*9P48	CM76		
227.500	CHD2OH	10R28	FPVSF89		

λ [μm]	Molecule	Pump	Reference	Freq. [MHz]	Footnotes
227.657	CH2F2	9P18	PSE80	1316860.5	pp6
227.660	CH2F2	9R18	IMMSD85		pp6, 88
227.661	CD3OD	10R10	VSPE81	1316838.7	103
228.100	CHFCHF	9P20	BH82		
228.300	CD3OH	10R34	CIMPS87		
228.700	CHD2OH	10R10	FPVSF89		
229.067	CD2F2	10R20	VPE81	1308755.5	
229.100	CD3OH	9P44	CIMPS87		
229.100	CH3OD	9P06	BP77		
229.300	CH2CHF	N2O	GRBF84		
229.390	HCOOH	*9R32	WZN73		
229.400	CH318OH	9P44	IMPSG89		
230.106	CH2F2	9R42	PSE80	1302845.8	
230.200	CH2F2	9R34	LBG85		58
230.700	CH318OH	9P10	IMPSG89		
231.000	C2H4O2H2	*9R10	PCD73		
231.000	CH2CL2	10P22	HW82		
231.000	CHD2F	9P10	Tobin84		
231.100	CD3OH	10P40	PFS86		
231.100	CHFCHF	10R38	BH82		
231.300	CH3OH	*10R44	HL82		
232.000	CHD2F	9P16	Tobin84		
232.000	CH3OH	9R10	Radfo75		
232.000	CD2CL2	10P24	SDFY86		
232.080	CH3OH	9P10	Henni86		115
232.100	CD3OH	9R44	PFS86		
232.400	CD3OD	10R30	VSPE81		
232.650	CH318OH	10P06	IMPSG89		
232.788	CH3OH	9R22	PEJS80	1287832.2	
232.800	CHFCHF	10R26	BH82		
232.900	SO2(I)	9R30	BS84		
232.930	CH3OH	*9R08	WZN73		pp15
232.939	CH3OH	9R10	PEJWG75	1286999.5	pp15
233.000	CH3OH	10R36	DFSY81B		
233.000	CLO2	10R08	DSF75B		
233.126	D2CO	9R14	DDG80	1285968.5	
233.400	CH3OD	9R14	VHHR86		81, 67, 112
233.400	CH3OH	C18O2	PD82A		
233.685	CD2F2	9P10	VPE81	1282892.0	
233.916	NH2NH2	10R08	RPJM77	1281625.8	
234.000	NH2NH2	10R34	DSF74C		

λ [μm]	Molecule	Pump	Reference	Freq. [MHz]	Footnotes
234.000	CH3SH	10R34	Lands80B		
234.000	CH3OD	9R14	NH80		81
234.400	14NH3	10P02	Wille81		51
234.610	CH3OH	*9R08	TOH88		
234.700	CD3OH	13CO2	DVPA81		
234.800	13CH2F2	9R18	STPVE85		
234.800	CD3OH	10P40	PFS86		
235.000	NH2NH2	10R08	DSF74C		
235.100	CH3OH	*9R04	HL82		
235.200	CD3OD	13CO2	PD85		
235.400	CD3OH	9P04	CIMPS87		
235.500	CH2F2	9R06	DW78C		
235.500	CH2CL2	10P24	HW82		
235.654	CH2F2	9R32	PSE80	1272171.4	21
235.700	CD3OD	10P06	PVSEP85		
235.800	CD3OH	10R30	CIMPS87		
235.800	CH3OH	*9R06	HL82		
235.900	CH3OD	CO2H	FK86		
236.000	CD3OH	9R14	GW78		
236.000	SIH3F	10R06	DS82		
236.000	CH3OD	10R44	Lands80C		
236.100	CD3OD	9P38	PVSEP85		
236.108	CD2F2	9R34	VPE81	1269723.6	
236.250	CH3CL	*9R02	CM76		
236.530	13CH3OH	9P10	Knigh81	1267459.0	
236.592	CH2F2	9R06	PSE80	1267131.0	
236.599	CH2F2	C18O2	GCK86	1267091.3	
236.601	CH2F2	9R06	PSE80	1267081.5	
237.000	NH2NH2	N2O	JTT82		
237.100	CD3OH	10P24	PFS86		
237.500	CD2CL2	10P26	SDFY86		
237.523	13CH3OH	9P12	HP78	1262162.0	
237.600	CH3OH	9P34	CBB70B		
237.758	CH2F2	C18O2	GCK86	1260914.2	
238.000	CH2DOH	10P18	ZD78		mm3
238.000	CH3OD	9R14	LD79		
238.000	CHD2OH	10P18	ZD78		mm3
238.000	CH3OD	10R34	NH80		
238.300	CD3OH	10P24	DW78B		
238.523	13CH3OH	9P12	HP78	1256872.0	
239.650	CD3OH	10R34	CIMPS87		

λ [μm]	Molecule	Pump	Reference	Freq. [MHz]	Footnotes
240.000	C2H4O2H2	*9R10	PCD73		
240.000	CH3OH	9P32	DFSY81B		
240.000	HCOOD	10R14	DSF76		
240.100	13CH3OH	10R06	PS87		
240.290	CH3OH	*9R24	Henni86		114
240.400	CH3OH	C18O2	PD82A		
240.980	CH3CL	10P10	CM76		22, 122
241.000	DCOOD	10R36	DSF76		
241.000	CH3OD	10R44	NH80		
241.100	CD2CL2	10R02	BF86		
241.500	CHFCHF	10P20	BH82		
241.500	CH318OH	10P10	IMPSG89		
241.600	13CD3OD	9R20	VE85		
241.750	CH318OH	10P06	IMPSG89		
242.200	CH3OH	*9R10	HL82		pp18
242.310	CH3OH	*9R06	TOH88		pp18, 157
242.470	CH318OH	9P22	IMPSG89		
242.473	CH3OH	10R34	PEJS80	1236396.8	
242.600	CHFCHF	10R14	DRH80		
242.847	CH3OH	10R32	VWE87	1234490.4	121
242.900	CD3OD	10P32	PVSEP85		
243.000	CH3NH2	*9P24	PCD73		
243.000	13CD3OD	9P32	VE85		
243.356	CH2F2	C18O2	GCK86	1231911.0	
243.500	CH3OD	CO2H	FK86		
243.848	D2CO	9R24	DDG80	1229421.8	
244.000	D2CO	10P16	Lands80A		
244.000	CH2CLF	9P26	ADF84		
244.000	CH3OH	10R40	DFSY81B		
244.000	ND2ND2	10P22	SDEF85		
244.100	CH2CHF	10P50	RGF84		
244.890	CH3NH2	9P24	DFSY81B		
245.000	D2CO	9R14	Lands80A		
245.000	CD3CL	9P32	DFBS75		26
245.000	CH3NH2?	9P34	DFSY81B		
245.000	D2CO	9R24	DDSB77A		
245.040	CH3BR	9P28	CM76		99
245.652	13CH2F2	9P28	STPVE85	1220395.0	
246.000	CH3NH2	10P12	Lands80D		
246.000	CD3CL	10R14	DFBS75		26
246.000	CH2CLF	9R26	ADF84		

λ [μm]	Molecule	Pump	Reference	Freq. [MHz]	Footnotes
246.000	CH3OH	10R38	DTS74		
246.100	CHD2OH	9P20	FPVSF89		
246.330	CH2F2	9R18	IMMSD85		
246.500	NH2NH2	10P06	DSF74C		
246.800	CHD2OH	9P04	FPVSF89		
247.000	CLO2	10R30	DSF75B		
247.000	13CD3OD	9P12	VE85		
247.300	CD3F	10P46	TSW79B		
247.400	CH3OH	C18O2	PD82A		
247.400	13CH3OH	10P10	PS87		
247.500	CD3OD	9P04	PVSEP85		
247.500	CD3F	9R10	TSW79B		
247.679	CH2F2	C18O2	GCK86	1210408.8	
248.108	CD2F2	10R16	VPE81	1208313.9	
248.122	CH2DOH	10P34	SPEJ80	1208246.0	
248.606	13CH2F2	9R26	STPVE85	1205896.0	
248.620	CH3OH	*9R20	TOH88		
248.800	CH2F2	9R34	LBG85		
248.900	CD2CL2	10R16	VHHR86		71
249.000	ND2ND2	9P22	SDEF85		41
249.000	CD3CL	9P38	DFBS75		26
249.000	CH3OH	10R34	Radfo75		
249.100	13CH3OH	9P26	PS87		
249.392	CD2F2	10R22	VPE81	1202093.2	
249.600	CHD2OH	9R20	FPVSF89		
249.700	CHD2OH	9P36	FPVSF89		
249.700	CD3OD	9R06	PVSEP85		
249.700	CH3OD	10R38	FK86		
249.720	CH2DOH	10P34	SPEJ80	1200512.7	
249.800	CD2F2	9P42	VPE81		
249.900	13CD3F	10R34	TF81		
250.000	CH3NC	10R22	LSB81		
250.000	C2H4O2H2	*9R18	PCD73		
250.000	CH3NH2	9P22	DFSY81B		pp19
250.100	CD3OH	9P04	PFS86		
250.138	CH3NH2	9P24	IMEJ86	1198510.1	pp19
250.500	NH2NH2	9R08	DSF74C		
250.700	CD2CL2	C18O2	PD82B		
250.781	CH3OH	10R34	PEJWG75	1195433.9	
250.970	CH2F2	9R26	IMMSD85		
251.000	CH3NH2	9R22	Lands80D		

λ [μm]	Molecule	Pump	Reference	Freq. [MHz]	Footnotes
251.000	CH3OH	10R44	TYY75		
251.140	CH3OH	10R38	PEJWG75	1193727.3	
251.180	CH3NH2	9P24	DFSY81B		
251.400	CH3OH	*10R38	HL82		
251.400	CD3OH	10P24	CIMPS87		
251.432	CH3OH	9R18	PEJS80	1192338.3	
251.900	CH318OH	9P32	IMPSG89		
251.910	CH3F	*10R34	CM71		
251.912	CH3OH	10R44	VWE87	1190069.1	
252.000	C2H4O2H2	*9P34	PCD73		
252.000	ND2ND2	10R36	SDEF85		
252.000	CD3OH	10R36	YYSF82		
252.300	CD3OH	9R24	PFS86		
252.336	CH2F2	C18O2	GCK86	1188068.7	
253.100	CD3OH	10P12	PFS86		
253.500	13CH3OH	10R20	IMSD85		
253.553	CH3OH	9P34	PEJS80	1182366.2	
253.720	CD3OH	10R36	SEJZS87	1181588.9	73, 79
253.800	CD3OH	10R38	PFS86		
254.000	CH3OH	10R38	Radfo75		
254.000	CD2CL2	10R36	ZD78		
254.000	CH3NH2	10R40	Lands80D		
254.041	CH3OH	9P34	PEJS80	1180092.5	
254.230	CH3OH	*9R08	TOH88		
254.300	CHD2OH	9P20	FPVSF89		
254.700	CH2CL2	10P26	HW82		22
254.700	CH3OH	13C18O2	PD84		
254.800	HCOOH	9P20	WZN73		116
254.802	13CH2F2	9P04	STPVE85	1176570.0	
255.000	CLO2	9R18	DSF75B		
255.000	H13COOH	9P14	DG82		
255.000	CD3OD	9P48	PVSEP85		
255.200	CD3OH	10R04	PFS86		
255.300	CD3OD	10R40	VSPE81		
255.300	CHD2OH	9P26	FPVSF89		
255.500	CH3OH	9P34	Strum84		
256.000	D2CO	9R24	Lands80A		
256.027	CH2F2	9P24	PSE80	1170941.0	21
256.400	CD3OD	9P08	PVSEP85		
257.000	CD3OH	10P22	YYSF82		
257.400	CH2CF2	10P08	HW84		

λ [μm]	Molecule	Pump	Reference	Freq. [MHz]	Footnotes
257.800	CH3OH	*9P12	HL82		
258.000	HFCO	C18O2	JD81		
258.000	CD3OH	9P20	YYSF82		
258.000	SO2	9R28	BS81		
258.300	CD3OH	9P46	CIMPS87		
258.425	H13COOH	9P16	DG82	1160071.8	
258.436	CD3OH	10P22	SEJZS87	1160027.8	75
259.000	CH3OH	9R08	DFSY81B		
259.000	NH3(D)	10R36	GDEFJ83		
259.500	CH2CF2	13CO2	MPD83		
259.900	CHD2OH	10R18	FPVSF89		pp22
260.000	CD3OH	10P22	DSF75A		
260.000	CH2F2	9R26	DW84		
260.000	CHD2OH	10R20	ZD78		pp22
260.000	CHD2F	9R16	Tobin84		
260.000	HFCO	9P16	JD81		
260.042	13CH2F2	10R20	STPVE85	1152860.0	
260.100	CHFCHF	10R20	DRH80		
261.000	CH3OH	9R02	Lands80C		
261.030	CH3CL	*10P34	CM76		
261.200	CH3OH	*9R04	TOH88		
261.500	SIH2F2	C18O2	DFS85		
261.500	CH3OH	9P12	IIMMS81		
261.700	CH3OH	10R38	HL82		118
261.729	CH2F2	9P38	PSE80	1145430.1	21
262.000	NH2NH2	10P28	DSF74C		
262.000	CH3OH	10R08	DFSY81B		
262.000	CHFCHF	10R24	AD80		
262.000	C2H4O2H2	*9P34	PCD73		
262.000	CH3SH	9P38	Lands80B		
262.100	CD3OD	10P48	PVSEP85		
262.248	CH2F2	C18O2	GCK86	1143163.2	
262.400	CH3OH	C18O2	PD82A		
262.400	CH318OH	9P26	IMPSG89		
262.700	CH3OH	*10R22	HL82		
263.000	SIH2F2	13CO2	DFS85		
263.200	CD3OD	9R42	FK86		
263.400	NH3	N2O	CBB70A		
263.500	CH2CHF	10R12	RGF84		
263.683	CH3OH	9P34	PEJS80	1136942.0	
264.000	CLO2	9R24	DSF75B		

λ [μm]	Molecule	Pump	Reference	Freq. [MHz]	Footnotes
264.000	NH2NH2	10R18	DSF74C		
264.000	CLO2	10R24	DSF75B		
264.050	CH3BR	*10R10	CM76		
264.350	CH3BR	10R20	VHHR86		
264.500	SIH3F	10R26	DS82		
264.536	CH3OH	9P34	PEJS80	1133277.0	
264.700	CH3CH2F	9P18	WZN73		116
264.700	CD3OH	10R04	CIMPS87		
264.759	CD3OH	10R34	SEJZS87	1132320.1	76
264.801	NH2NH2	10R20	RPJM77	1132140.6	
264.900	CH2NOH	13CO2	DP84		
265.000	CH3BR	10R20	VHHR86		
265.000	DCOOH	10R20	DSF76		
265.000	CD3OH	10R34	PFS86		76
265.000	NH2NH2	10R28	DSF74C		
265.000	DCOOD	10R32	DSF76		
265.000	CD3F	9P52	TSW79A		
265.300	CD3OH	9P14	PFS86		
265.600	CD2CL2	10P20	BF86		
265.600	CH3OH	9P44	PS87		
265.800	CH3BR	10R20	IMMS82		
266.000	CD3OH	9P20	PFS86		
266.000	CDF3	10P14	TF80		
266.500	CD2CL2	10P20	SDFY86		
266.735	CH2DOH	9P32	SPEJ80	1123932.7	
266.866	13CH2F2	9R24	STPVE85	1123382.0	
266.900	CDF3	9P44	TF80		17
267.000	CH3NH2	*9P40	PCD73		
267.000	NH2NH2	13C18O2	JTT82		
267.000	CD3OH	10R14	DSF75A		
267.200	CD3OH	10P22	CIMPS87		
267.443	CH3OH	10R34	PEJS80	1120957.7	
267.823	CD2F2	9P18	VPE81	1119368.0	
267.900	CD2CL2	N2O	BF86		
268.000	CD3OH	9P14	DSF75A		
268.000	CH3NH2	*9R12	PCD73		
268.062	CH2F2	C18O2	GCK86	1118369.3	
268.300	CH318OH	9R38	IMPSG89		
268.572	13CH3OH	10R16	HP78	1116245.0	pp7
268.600	CD3OH	10R14	PFS86		
268.600	13CH3OH	10R18	PS87		pp7

λ [μm]	Molecule	Pump	Reference	Freq. [MHz]	Footnotes
269.900	13CH3OH	10R14	PS87		
270.000	CH3NH2	9R18	Lands80D		
270.000	CHD2OH	10R10	FPVSF89		
270.000	CH3OH	9P32	IMS80		
270.005	CH2F2	9R22	PSE80	1110319.9	
270.600	CH2CHCN	10P26	DSF72		
270.700	CH3OH	*10P10	Henni86		
270.733	CD3OD	9R08	VSPE81	1107337.9	
271.000	CH3NH2	9R12	Lands80D		
271.290	CH3CL	*10P20	CM76		
271.500	CH3OH	C18O2	PD82A		
271.500	NH2NH2	10P18	DSF74C		
272.000	DCOOH	10P30	DSF76		
272.000	CD3I	9P12	DFBS75		
272.100	CHFCHF	10P14	BH82		
272.252	CH2DOH	9R24	SPEJ80	1101159.4	
272.300	CD3OH	10P12	CIMPS87		
272.339	CH2F2	9P10	PSE80	1100806.7	21
272.500	CD3OD	9P32	PVSEP85		
272.900	CH3OH	C18O2	PD82A		
272.958	13CD3OD	10R26	VE85	1098307.9	
273.000	CH3OH	13C18O2	PD84		
273.004	CH2DOH	9P06	SPEJ80	1098125.9	
273.400	15NH3	CO2S	FWSGW82		
273.400	CH2F2	C18O2	PD82B		
273.764	13CH2F2	9P26	STPVE85	1095077.0	
274.245	CH3OH	10R46	VWE87	1093154.7	
274.776	CD2F2	10R26	VPE81	1091044.7	
275.000	ND2ND2	10R12	SDEF85		
275.000	CH3CL	*9R14	CM76		
275.000	CH3OH	9P32	DFSY81B		
275.090	CH3CL	*9R36	CM76		
275.100	CH3OD	9R06	PD85		
275.500	CH2CHF	10P48	RGF84		
275.610	13CH3OH	10R26	IMSD85		
276.000	DCOOD	9P16	DSF76		
276.600	CD3OH	9R26	PFS86		
276.716	CD3OH	10P28	SEJZS87	1083395.1	75
276.900	CD3OH	10R24	PFS86		
277.000	NH2OH	9R24	TJD86		
277.000	C2H4O2H2	*9P38	PCD73		

λ [μm]	Molecule	Pump	Reference	Freq. [MHz]	Footnotes
277.000	CH3NC	10R24	LSB81		
277.000	CH318OH	9P44	IMPSG89		
277.200	CD2CL2	N2O	BF86		
278.000	ND2ND2	10P38	SDEF85		
278.000	CD3OH	10R24	DSF75A		
278.300	CH2NOH	13CO2	DP84		
278.399	D2CO	10P08	DF78	1076842.8	
278.400	CHD2OH	10R36	FPVSF89		
278.570	CH3CL	*9R16	CM76		
278.610	HCOOH	9P30	WZN73		116
278.805	CH3OH	9P38	PEJS80	1075277.1	
279.000	CHD2OH	9R34	FPVSF89		
279.000	TRIOX	9P20	DWC81		
279.014	13CH2F2	9P22	STPVE85	1074471.0	
279.400	CH3OD	9P32	DSF74A		
279.400	CH3OD	9P02	FK86		
279.800	CD2CL2	N2O	BF86		
279.800	CH3OH	13C18O2	PD84		
279.810	CH3BR	*10R52	CM76		
279.900	SO2(I)	9P10	BS84		166
280.000	CH3OD	10P18	NH80		
280.000	CH3NC	10R24	DJ80		
280.000	HFCO	9R28	JD81		
280.000	HFCO	C18O2	JD81		
280.218	13CH3OH	10R16	HPPJE79	1069853.4	
280.240	13CH3OH	10R16	HPPJE79	1069771.4	
280.500	SIH3F	10R38	DS82		
280.512	CD2F2	9P30	VPE81	1068733.7	
280.800	13CD3F	10P20	TF81		
280.800	CHD2OH	9R14	FPVSF89		
280.934	CH3OH	9R18	PEJS80	1067127.2	
281.000	13CH3OH	9P32	PS87		
281.000	CH3NH2	9R14	DFSY81B		
281.053	CH2F2	C18O2	GCK86	1066675.4	
281.180	CH3CN	*9P34	CM76		
281.200	CH2F2	9P32	MH84A		
281.500	CH3OH	9R?	DSF74A		
281.600	CH2CF2	10P38	HW84		
281.600	CH2CHF	10R44	RGF84		
281.670	CH3CL	*9R14	CM76		
281.800	CD2CL2	C18O2	PD82B		

λ [μm]	Molecule	Pump	Reference	Freq. [MHz]	Footnotes
281.980	CH3CN	*9P50	CM76		
282.000	CH3OD	9R06	PD85		
282.000	HFCO	C18O2	JD81		
282.100	SO2	9R28	BS84		
282.300	CH3CH2F	*9R12	WZN73		
282.800	CD3OH	10R32	CIMPS87		
282.900	CH2F2	C18O2	PD82B		
282.960	13CH3OH	10R20	IMSD85		
283.000	CH3NH2	9P46	Lands80D		
283.000	CH2CHBR	10R20	DESF76		
283.000	DCOOD	9R16	DSF76		
283.200	CH3OD	10P18	FK86		
283.750	CD3OH	9P42	CIMPS87		
283.783	CH2F2	C18O2	GCK86	1056414.8	
284.000	CH3NC	10R18	LSB81		
284.000	HCN	ELEC	HJ67		
284.000	CH2CLF	9R16	ADF84		
284.150	CH318OH	9R10	IMPSG89		mm22
284.300	CD3OH	9P40	PFS86		
284.330	CH3OH	*10R44	Henni86		114
284.354	CH2F2	C18O2	GCK86	1054291.8	
284.400	CD3OH	9R32	CIMPS87		
284.500	CH318OH	10R06	IMPSG89		
284.600	CHFCHF	10R18	BH82		
284.900	CH318OH	9P30	IMPSG89		
285.000	CLO2	10R20	DSF75B		
285.000	ND2ND2	10R38	SDEF85		
285.000	CH3OH	9R10	DFSY81B		mm22
285.000	CD3OH	10P24	YYSF82		
285.100	CHD2F	9R20	Tobin84		
285.250	CH318OH	10R04	IMPSG89		
285.300	CH3CH2OH	9P40	BS85		
285.500	ND2ND2	10P38	SDEF85		
286.000	ND2ND2	10P32	SDEF85		
286.000	CD3OH	10P18	YYSF82		
286.155	CH3OH	10R48	VWE87	1047657.6	
286.197	CD3OH	9P40	SEJZS87	1047502.3	
286.200	CD3OH	10R22	PFS86		
286.300	CDF3	10R48	TF80		
286.398	CD2F2	9P08	VPE81	1046768.2	
286.500	CHFCHF	9P28	DRH80		

λ [μm]	Molecule	Pump	Reference	Freq. [MHz]	Footnotes
286.724	CD3OH	10P24	SEJZS87	1045578.0	75
286.790	CH3CL	*10R34	CM76		
286.800	CDF3	10R10	TF80		
286.800	CDF3	10R04	TF80		
286.880	CH3CN	*9P50	CM76		
287.308	CD3OH	10P18	SEJZS87	1043454.5	75
287.667	CH2F2	9R34	PSE80	1042150.4	21
287.700	15NH3	13CO2	MPD83		
287.908	CH2F2	C18O2	GCK86	1041279.4	
287.950	CD3OH	9P50	CIMPS87		
288.000	CH3NH2	9R04	Radfo75		
288.000	C2H4O2H2	*9P12	PCD73		
288.000	CD3CL	10R18	DFBS75		26
288.000	CH3NC	10R18	DJ80		
288.000	CD3CL	9P16	DFBS75		26
288.300	CHD2OH	10R22	FPVSF89		
288.500	CH2CF2	10P12	DSF72		
289.000	NH2NH2	13CO2	JTT82		
289.000	CH3OH	9R28	DFSY81B		
289.139	CH2F2	C18O2	GCK86	1036844.9	
289.170	CH3OH	*9R24	Henni86		
289.400	CD2CL2	10R08	BF86		
289.500	CH2F2	9P04	PSE80	1035552.7	
289.500	CD3OD	10R14	PVSEP85		
289.600	CD3OD	9R34	PVSEP85		
289.600	CH3OD	9R08	PD85		
289.700	CH3OH	9P10	Henni82		
289.800	CH2CF2	10P24	HW84		
289.800	CHFCHF	10R24	BH82		
290.000	ND2ND2	10R08	SDEF85		
290.000	C2H4O2H2	*9P38	PCD73		
290.000	CD3OH	10P18	DW78B		49
290.000	NH2OH	9P12	TJD86		
290.000	CH3OH	9R26	DFSY81B		
290.000	CD3BR	10P18	Lands80A		
290.000	CH2CHF	10P36	TW82		
290.200	14NH3	CO2S	DW78D		
290.200	CHD2OH	9R34	FPVSF89		
290.620	CH3OH	9P12	Henni78		
290.670	CD3OD	10R14	IMEJ86	1031384.4	
290.812	CH2F2	C18O2	GCK86	1030879.8	

λ [μm]	Molecule	Pump	Reference	Freq. [MHz]	Footnotes
291.000	13CD3OH	10P22	IEPSV84		
291.000	14NH3	10R06	Wille81		
291.300	CH2CF2	10P10	HW84		
291.300	CHD2OH	10P20	FPVSF89		
291.300	CH2NOH	13CO2	DP84		
291.620	13CH3OH	9P36	HP78		
292.000	HCOOD	10P32	DSF76		
292.000	CH3OH	10R10	Radfo75		
292.000	NH2OH	13CO2	TJD86		
292.000	CH2CLF	9R20	ADF84		
292.141	CH3OH	9P38	PEJS80	1026189.3	
292.500	CH3OH	9P34	CBB70B		
292.700	CHD2F	9R30	Tobin84		
293.000	ND2ND2	9P14	SDEF85		
293.130	13CH3I	CO2S	GRF87		
293.400	CH2CHF	N2O	GRBF84		
293.648	CD3CL	9P24	DFBS75	1020924.7	26, 47
293.800	CH2CHCL	CO2H	FBM84		
293.800	CH2CF2	10R08	HW84		
293.822	CH3OH	10R10	PEJS80	1020321.1	
293.901	CH2F2	9P20	PSE80	1020044.0	
294.000	D2CO	10P08	Lands80A		
294.040	13CH3OH	10R20	IMSD85		
294.280	CH3BR	*10R28	CM76		
294.300	CH318OH	9P06	IMPSG89		
294.300	CH3OD	9R16	PD85		
294.600	CH2CL2	10P26	HW82		
294.600	CD3OD	9P24	PVSEP85		
294.700	CH3OD	CO2H	FK86		
294.811	CH3OD	9R08	BKM78	1016897.2	
295.397	CH2DOH	9P10	SPEJ80	1014881.0	
295.639	CH2DOH	10R34	SPEJ80	1014047.7	
296.000	CH2CLF	9R26	ADF84		
296.000	ND2ND2	9P20	SDEF85		
296.000	ND2ND2	10P32	SDEF85		
296.480	CH3OH	10P18	Henni86		115
297.000	CD3OD	10R34	FK86		mm11, 62
297.000	CD3OH	10R34	DSF75A		mm11
297.000	CD3BR	9P32	Lands80A		
297.090	COF2	10P40	GRF88		
297.100	CD3OH	9R20	PFS86		

λ [μm]	Molecule	Pump	Reference	Freq. [MHz]	Footnotes
297.700	CH3OH	13CO2	PD82A		
298.000	DCOOD	10R06	DSF76		
298.000	CH3SH	9P30	Lands80B		
298.000	CD3I	C18O2	MPD83		
298.000	CH2CHF	10P08	TW82		
298.049	CHCLF2	9R24	DFP78	1005850.0	
298.211	CH2F2	9R36	PSE80	1005303.3	
298.470	CH2F2	C18O2	GCK86	1004430.7	
298.500	CH2CL2	10P22	HW82		
298.736	CD3OD	10R24	VSPE81	1003536.6	
298.900	SO2(I)	9R20	BS84		163
299.000	CD3OH	9R06	DSF75A		
299.000	C2H4O2H2	*9P34	PCD73		
299.500	13CD3F	9P48	TF81		
299.900	CH2CF2	10P18	HW84		
300.000	CLO2	9R12	DSF75B		
300.100	CD3OD	10R22	FK86		
300.233	13CH2F2	9R36	STPVE85	998532.1	
300.246	13CH2F2	10R38	STPVE85	998487.9	
300.476	CH2F2	C18O2	GCK86	997725.4	
300.600	CH318OH	10R20	IMPSG89		
301.000	CD3I	9R26	DFBS75		
301.000	SIHF3	10R14	DFS85		
301.000	CHCLF2	9P28	DFP78		
301.000	ND2ND2	10R24	SDEF85		
301.000	CHD2F	9P44	Tobin84		
301.100	CD2CL2	N2O	BF86		
301.200	CH2NOH	10P10	DP84		
301.275	NH2NH2	10R12	RPJM77	995077.8	
301.370	COF2	10R54	GRF88		
301.654	13CH2F2	10R04	STPVE85	993829.9	
301.994	CH3OH	9P14	PEJS80	992708.9	
302.080	HCOOH	*9P08	WZN73		
302.278	HCOOH	9R04	RPJM77	991777.8	3
302.500	CH3I	N2O	GRF87		
303.000	CH3OH	9P34	DFSY81B		
303.540	CH3CN	*10P10	CM71		
303.800	CD2F2	10R48	VPE81		
304.000	HCOOD	9R06	DSF76		
304.083	DCOOD	10R24	DSF76	985889.7	
304.300	CH3OCH3	9R40	BSKK82		

λ [μm]	Molecule	Pump	Reference	Freq. [MHz]	Footnotes
304.350	COF2	10R52	GRF88		
304.800	CH3OH	9P18	Strum84		
305.240	COF2	10R50	GRF88		
305.600	CHD2OH	10R08	FPVSF89		
305.726	CH3OD	9R08	BKM78	980591.6	
306.000	HFCO	C18O2	JD81		
306.053	CHCLF2	9R24	DFP78	979544.5	
306.300	CH3OH	*9P16	HL82		
306.500	13CH3OH	9P20	PS87		
306.500	CH3OD	CO2S	FK86		
306.700	CH2CF2	10P16	HW84		
306.993	13CH2F2	9R40	STPVE85	976543.7	
307.000	CH3OD	9R08	KYHH75		
307.070	13CH3OH	10R26	IMSD85		
307.200	CH318OH	9P42	IMPSG89		
307.500	CHFCHF	10R30	DRH80		
307.500	CH3OD	CO2H	FK86		
307.650	CH3CL	*CO2S	CM76		
307.780	13CH3OH	9P22	HP78		
308.000	CH2CLF	9R22	ADF84		
308.000	FCN	C18O2	DJ80		
308.000	CD3OD	10R52	FK86		
308.040	CH2DOH	9P14	SPEJ80	973224.3	
308.296	CH2DOH	10R34	IMEJ86	972418.7	
308.500	CD3OH	10R28	PFS86		
308.600	14NH3	13CO2	DP84		
309.000	HCOOH	*10R22	PCD73		
309.000	CD3OH	10P20	DSF75A		
309.193	CH2F2	C18O2	GCK86	969596.9	
309.230	HCOOH	9R04	WZN73		120
309.500	CD2CL2	10P06	SDFY86		
309.500	CH2CHF	N2O	GRBF84		
309.700	CD2CL2	N2O	BF86		
309.714	HCN	ELEC	HJ67	967965.8	89
309.800	CD3OD	13CO2	DVPA81		94
310.000	CD3OH	10R28	DSF75A		
310.000	H13COOH	10R28	DG82		
310.000	CH3OD	13CO2	DVPA81		
310.000	CHFCHF	10R34	AD80		
310.000	DCOOD	10R24	DSF76		
310.000	CH3OH	9R16	DFSY81B		

λ [μm]	Molecule	Pump	Reference	Freq. [MHz]	Footnotes
310.100	CD3OH	9P36	CIMPS87		
310.350	CD3OH	10R26	CIMPS87		
310.700	CD3OH	10P20	PFS86		
310.800	CD3OH	10P48	CIMPS87		
310.800	CHFCHF	9P34	DRH80		
310.887	HCN	ELEC	HJRFS67	964312.3	43, 89
311.000	CH3NO2	10P16	DFSY81A		
311.000	CD3OD	10P10	PVSEP85		
311.000	ND2ND2	10P30	SDEF85		
311.070	CH3BR	10R12	CM76		99
311.075	NH2NH2	9P20	RPJM77	963731.4	
311.100	13CH3OH	10P28	PS87		
311.100	CH3BR	10P20	CM76		99
311.200	CH3OH	9P24	DSF74A		
311.200	CH3BR	*10P40	CM76		
311.210	CH3BR	*10R50	CM76		
311.213	13CH2F2	9R08	STPVE85	963302.2	
311.554	HCOOH	10R22	EWTR79	962250.0	121
311.900	CH3CH2OH	9P26	BS85		
312.000	CH3OH	9P28	DFSY81B		
312.000	DCOOH	10R24	DSF76		
312.100	SO2	9R40	BS81		
312.100	CH3OH	C18O2	PD82A		
312.276	13CH2F2	9P24	STPVE85	960024.5	
312.500	CD3OH	10R08	CIMPS87		
312.700	CD3OD	9R22	PVSEP85		
312.900	CD3OH	9P48	PFS86		
312.910	COF2	10P38	GRF88		
313.500	CH3OH	13C18O2	PD84		
313.600	O3	9P30	DDM83		
313.797	H13COOH	9P06	DG82	955370.3	
313.797	H13COOH	9P06	DG82	955368.1	
313.880	CH3OH	9P30	Henni86		115
313.900	CD2CL2	C18O2	PD82B		
314.300	CD3OD	10R44	PVSEP85		
314.646	CD2F2	9R10	VPE81	952793.3	
314.841	CD3OD	10R10	VSPE81	952203.9	103
314.847	CH3NH2	9R04	RPJM77	952185.0	
316.000	CD3OD	9R26	YKYSF81		
316.000	CH3SH	9P44	Lands80B		
316.329	13CH2F2	9R24	STPVE85	947723.7	

λ [μm]	Molecule	Pump	Reference	Freq. [MHz]	Footnotes
316.600	CDF3	10R20	TF80		
316.700	CH2CF2	10P38	HW84		
317.000	CHD2OH	9R14	FPVSF89		
317.052	CD2F2	9P40	VPE81	945562.5	
317.500	SIH2F2	C18O2	DFS85		
318.000	CH3NO2	10P18	DFSY81A		
318.000	CD3CL	10R28	DFBS75		26
318.080	13CH2F2	9R12	STPVE85	942507.4	
318.600	CD2F2	10R18	VPE81		
319.000	CH3SH	9P38	Lands80B		
319.000	CH3OH	9R14	DFSY81B		
319.000	CH3CHF2	CO2H	FBRG85		
319.268	D2CO	9P32	DDG80	939000.3	
319.400	CD3OD	9R16	PVSEP85		88
319.480	HCOOH	*10R22	WZN73		
319.700	13CH3OH	10R14	PS87		
320.000	NH2NH2	N2O	JTT82		
320.000	HCOOH	10R24	DSF76		
320.000	CH3OD	9P30	DSF74A		
320.000	D2CO	9P34	Lands80A		
320.400	13CH3OH	10R20	PS87		
320.597	CD2F2	10R44	VPE81	935107.5	
320.700	CH3OD	13CO2	PD85		
321.000	CH2CHF	CO2H	RGF84		
321.000	CD3OH	9R16	DSF75A		
321.410	13CD3OD	10R12	VE85	932741.3	
322.100	CD3OH	10P12	CIMPS87		
322.350	CD3OH	10P12	CIMPS87		
322.452	CH2DOH	9P12	SPEJ80	929726.8	
322.500	SIHF3	10R32	DFS85		
322.800	CH2CHF	10R26	RGF84		
323.000	DCOOD	10R30	DSF76		
323.179	CD2F2	9P40	VPE81	927636.8	
323.300	CD3F	10P08	TSW79B		
323.500	CD3OH	9P24	CIMPS87		
323.500	CD2CL2	10P14	SDFY86		
324.000	D2CO	10P22	Lands80A		
324.000	HCOOD	10R10	DSF76		
324.000	CH3SH	10R16	Lands80B		
324.000	CH2CLF	9R04	ADF84		
324.000	CHCLF2	9R08	DFP78		

λ [μm]	Molecule	Pump	Reference	Freq. [MHz]	Footnotes
324.140	13CD3OD	10R14	VE85	924885.8	
324.400	CH2CF2	13CO2	MPD83		
324.423	D2CO	10P24	DF78	924078.4	
325.000	DCOOD	10R28	DSF76		
325.170	13CH3OH	9P36	HP78		
325.300	CH2CF2	10P08	HW84		
325.900	13CD3F	9R22	TF81		
326.000	CHCLF2	9R12	DFP78		
326.000	HCOOD	10R30	DSF76		
326.000	CH2CF2	13CO2	MPD83		
326.423	CH2F2	9R14	PSE80	918417.0	
326.500	CD2CL2	N2O	BF86		
326.600	CHFCHF	10P06	BH82		
327.000	NH2NH2	9R22	DSF74C		
327.500	CH318OH	9P32	IMPSG89		
327.600	CH3CH2BR	10R10	BS83		
327.770	CH3OH	*9R12	Henni86		
327.800	CD3OD	9R28	VSPE81		
328.000	CH3CHO	9R22	LSB81		
328.457	DCOOH	10P22	DSF76	912729.7	
328.900	13CH3OH	9P38	PS87		
328.960	CHCLF2	9R28	DFP78	911332.7	
329.000	CH3OH	10R26	DFSY81B		
329.200	CD3OD	9P18	PVSEP85		
329.500	CD3OH	10P32	PFS86		
329.500	CH2CF2	10R36	HW84		
329.800	CH2CHF	N2O	GRBF84		
329.900	CD3OD	10P16	PVSEP85		
330.000	SIHF3	13CO2	DFS85		
330.000	SIH3F	10P22	DS82		mm4
330.000	SIH2F2	10P22	DFS85		mm4
330.019	CDF3	10R08	TLD83	908408.6	
330.100	CH2CHF	10R22	RGF84		
330.200	CH3CH2F	*9R22	WZN73		
330.991	CD2F2	9P34	VPE81	905742.6	
331.088	HDCO	10P30	DDG80	905477.0	
331.500	NH2NH2	9P30	DSF74C		
331.669	NH2NH2	9P12	RPJM77	903889.4	
331.700	CD3OD	10P22	FK86		
331.720	CH3I	N2O	GRF87		
331.790	CD3I	N2O	GRF87		

λ [μm]	Molecule	Pump	Reference	Freq. [MHz]	Footnotes
332.000	CH3OH	9P36	DFSY81B		
332.600	CH3OD	9R04	PD85		92
332.603	13CH3OH	10R16	HPPJE79	901351.2	
332.860	CH3BR	*10R06	CM76		
333.000	NH2NH2	9P12	DSF74C		
333.150	CH3BR	10P08	CM76		99
333.261	13CD3OH	10P16	IEPSV84	899571.7	
333.600	CH2CHF	N2O	GRBF84		
333.900	CD3OH	9P36	CIMPS87		
333.926	13CH2F2	9P36	STPVE85	897781.9	
333.935	CH3CL	9P42	GCK86	897758.2	107
334.000	SIHF3	10R12	DFS85		
334.600	13CH3OH	10R40	PS87		
334.820	HCOOH	*9P18	WZN73		
334.910	HCOOH	*9R14	WZN73		
335.000	CH2CHF	10P06	TW82		
335.000	CH2CF2	10P14	TW82		
335.183	HCN	ELEC	HJ67	894414.2	
335.467	CHCLF2	9R16	DFP78	893656.9	
335.709	DCOOD	9R08	DSF76	893013.6	
335.850	COF2	N2O	GRF88		
336.000	HCOOH	9R12	DDSB77B		
336.000	NH2NH2	10P24	DSF74C		
336.000	CD3OH	10R30	DSF75A		
336.000	CH2CHF	10P38	TW82		
336.000	HCCCHO	10P26	DJ80		
336.246	CH2DOH	9P36	SPEJ80	891586.3	
336.300	HCOOH	9R14	DSF76		52
336.380	CD3F	10P50	TD86	891231.4	
336.500	13CD3F	CO2H	TF81		
336.500	CD3OH	9R32	PFS86		
336.500	13CD3OH	9R28	IEPSV84		
336.558	HCN	ELEC	HJRFS67	890759.5	43, 89
336.700	CH3CH2F	9R16	WZN73		116
336.800	CD3OH	9P34	PFS86		
337.000	NH3(D)	10R30	GDEFJ83		
337.000	CLO2	9R26	DSF75B		
337.040	CH3OH	*10P16	Henni86		
337.094	CHCLF2	9R32	DFP78	889343.5	
337.300	CD3OH	10R30	PFS86		
337.500	CD3OH	9P24	PFS86		

λ [μm]	Molecule	Pump	Reference	Freq. [MHz]	Footnotes
337.775	CH2F2	9R14	IIMSD86	887551.1	
338.900	CH3OCH3	9R22	BSKK82		
338.964	13CH3OH	9P22	HP78	884438.0	
339.000	COF2	10R08	TW82		
339.000	CHFCHF	10P26	BH82		
339.100	CH2CF2	10P18	HW84		
339.300	CH2CF2	10P10	HW84		
339.900	13CH3OH	10R36	PS87		pp8
340.000	CLO2	9R40	DSF75B		
340.000	HCOOD	9P28	DSF76		
340.000	CH3NO2	9P08	DFSY81A		
340.000	13CH3OH	10R32	IMSD85		pp8
340.000	SIH3F	10R22	DS82		
340.300	CHCL2F	9P20	VWPE83	880965.6	
340.357	CH2DOH	10R32	SPEJ80	880818.6	
340.600	CD2CL2	N2O	BF86		
340.627	13CD3OH	10P16	IEPSV84	880120.4	
340.700	CD3OD	10R26	PVSEP85		
341.000	CH3SH	9P16	Lands80B		
341.000	D2CO	9P34	Lands80A		
341.000	CD3BR	9P36	Lands80A		
342.000	DCOOH	10R16	DSF76		
342.127	CD2F2	9P44	VPE81	876261.3	
342.700	CD3OD	9P16	PVSEP85		
342.740	HCOOH	*9R14	WZN73		
342.800	CD3OD	9P06	PVSEP85		
342.800	CH318OH	10P42	IMPSG89		
343.000	SIH2F2	10R14	DFS85		mm5
343.000	CH3CHO	9R30	LSB81		
343.260	CH2CHCN	10P10	GRBKF85		
343.300	CD2CL2	10P16	VHHR86		71
343.500	SIH3F	10R14	DS82		mm5
344.000	CH3NO2	10P22	DFSY81A		
344.000	C2H4O2H2	*9P22	PCD73		
344.000	CH3NO2	10P20	DFSY81A		
344.000	CH2CLF	9P32	ADF84		
344.521	13CH2F2	9R16	STPVE85	870171.9	
344.778	CD3OD	10R04	VSPE81	869522.7	103
344.800	CH2CHF	N2O	GRBF84		
344.900	CHD2OH	10R02	FPVSF89		
345.000	CHCLF2	10P34	DFP78		

λ [μm]	Molecule	Pump	Reference	Freq. [MHz]	Footnotes
345.000	SIHF3	9P32	DFS85		
345.500	CH2CHF	10P38	RGF84		
345.500	COF2	N2O	GRF88		
345.800	CDF3	9R22	TF80		17
346.000	CH3OH	9R26	DFSY81B		
346.000	CHD2OH	9P20	ZD78		
346.000	D2CO	10P22	Lands80A		
346.000	CD3OH	9R14	GW78		
346.320	CH3CN	*9P16	CM76		
346.488	CH3OH	9P22	PEJS80	865233.1	
346.670	CD3I	CO2H	GRF87		
346.670	13CH3I	CO2H	GRF87		
347.000	CH3NH2	*10R20	PCD73		
347.000	HCOOD	10R08	DSF76		
347.600	CH2CF2	13CO2	MPD83		
347.640	CH3OH	*9R24	Henni86		
348.100	CD3OD	9P28	PVSEP85		
348.300	CH3OH	C18O2	PD82A		
348.600	CD3OD	C18O2	PD85		
348.899	CD3F	9P34	TD86	859252.2	
349.000	CH2CLF	9R10	ADF84		
349.000	CH3NH2	10R18	Lands80D		
349.100	SO2	9R40	SL84		
349.387	CH3CL	10R18	GCK86	858053.3	22, 107
349.500	CH2CF2	10P10	HW84		
349.937	CH3NH2	10R20	IMEJ86	856703.7	
350.000	DCOOD	10R40	DSF76		
350.000	CH2CF2	N2O	FGRD84		
350.500	CD3OH	9P32	PFS86		
351.000	CH3SH	9R28	Lands80B		
351.000	CH2CHF	N2O	GRBF84		
351.000	CH3NH2	9P46	Lands80D		
351.000	HCOOD	9P36	DSF76		
351.000	CH2CF2	10R16	HW84		
351.000	CH3NO2	10P20	DFSY81A		
351.000	CH3OH	13C18O2	PD84		
351.100	CH3OH	13CO2	PD82A		
351.200	CD3OH	10R30	PFS86		
351.300	CH3OD	9R04	FK86		
351.300	CH3OD	10R22	FK86		
351.400	CD3OH	9R22	PFS86		

λ [μm]	Molecule	Pump	Reference	Freq. [MHz]	Footnotes
351.400	CD3OH	9R32	PFS86		
351.500	CH3CHDOH	9P46	BS85		
352.000	DCOOD	10R14	DSF76		
352.000	HCOOD	10R22	DSF76		
352.300	CD3OH	9R06	PFS86		
352.500	SIH2F2	13CO2	DFS85		
352.500	CH3OD	9P30	DSF74A		
352.503	CD3OH	9R14	SEJZS87	850468.0	75
352.750	CH3BR	*9P18	CM76		
352.800	CH2CHF	N2O	GRBF84		
352.902	CD2F2	10R34	VPE81	849506.4	
353.000	CH2CHF	N2O	GRBF84		
353.000	HCOOD	10P28	DSF76		
353.000	HCOOD	10R06	DSF76		
353.100	13CD3OD	9R26	VE85		
353.300	13CD3OD	9R14	VE85		
353.800	CD3OD	9P12	PVSEP85		
354.000	DFCO	C18O2	JDL83		
354.176	CD3OD	10R16	VSPE81	846450.3	102
354.500	ND2ND2	10P32	SDEF85		
354.500	CH2CHF	N2O	GRBF84		
354.630	COF2	N2O	GRF88		
355.000	SIH2F2	C18O2	DFS85		
355.000	CHD2OH	10P18	ZD78		
355.000	HCOOD	9P38	DSF76		
355.000	CH2CHF	10P20	TW82		
355.126	CH2F2	9P08	PSE80	844185.9	
355.500	SIHF3	13CO2	DFS85		
355.500	CD3OD	10R16	DVPA81		
355.550	13CH3I	10P30	GRF87		
355.900	CD2CL2	N2O	BF86		
356.000	CH2CHF	N2O	GRBF84		
356.000	HCOOD	10P30	DSF76		
356.000	CH2CHBR	10R20	DESF76		
356.400	CD3OD	9P14	PVSEP85		88
356.500	CD3OD	9P26	PVSEP85		
356.500	CH3OH	C18O2	PD84		
356.600	CH2CHF	N2O	GRBF84		
357.000	COF2	10P32	TW82		
357.867	13CH2F2	9R20	STPVE85	837719.4	
357.901	CH2F2	C18O2	GCK86	837640.8	

λ [μm]	Molecule	Pump	Reference	Freq. [MHz]	Footnotes
358.000	DFCO	C18O2	JDL83		
358.000	C2H4O2H2	*9P34	PCD73		
358.000	CH2CF2	10R18	HW84		
358.000	HCOOD	10R16	DSF76		
358.100	CD2CL2	10R16	BF86		
358.111	COF2	10P32	DPK87	837149.1	
358.400	13CD3OD	9P32	VE85		
358.920	13CH3OH	9P40	HP78		
359.200	CH318OH	9P10	IMPSG89		
359.362	13CH2F2	9R22	STPVE85	834235.9	
359.810	HCOOH	9R34	WZN73		120
360.000	CD3I	CO2H	GRF87		
360.000	CD2CL2	N2O	BF86		
360.000	CHFCHF	10R38	BH82		
360.053	CH2F2	C18O2	GCK86	832635.0	
360.200	CD2CL2	N2O	BF86		
360.500	CHFCHF	9P30	BH82		
360.504	13CH2F2	9P16	STPVE85	831592.7	
360.606	CHCLF2	9R14	DFP78	831356.9	
360.900	CH2CHF	N2O	GRBF84		
361.000	HCOOD	9P38	DSF76		
361.200	CD2CL2	10R10	BF86		
361.231	CDF3	10P24	TLD83	829918.3	
361.500	SIHF3	9P34	DFS85		
361.800	CH2CHF	N2O	GRBF84		
362.000	DCOOH	10P20	DSF76		
362.000	CH3OH	*10R32	HL82		
362.100	CH3CH2F	*9R18	WZN73		
362.200	CH2CHF	10P24	RGF84		
362.423	CDF3	9R24	TLD83	827188.4	
362.650	CH318OH	10R20	IMPSG89		
362.800	CD3OH	9P28	PFS86		
362.800	CH2CHF	10R28	RGF84		
363.000	CHD2OH	10R16	ZD78		mm6
363.000	CH3COOD	10R28	DFSY81A		
363.000	CH3OH	9P34	DFSY81B		
363.000	CH2DOH	10R16	ZD78		mm6
363.860	CH318OH	10R20	IMPSG89		
363.900	CH2CF2	10R14	HW84		
364.300	CH318OH	10R18	IMPSG89		
364.484	D2CO	10P28	DG81	822512.2	

λ [μm]	Molecule	Pump	Reference	Freq. [MHz]	Footnotes
364.500	CH318OH	9P34	IMPSG89		
364.600	CH2CF2	N2O	FGRD84		
365.000	DCOOH	10R12	DSF76		
365.725	CHCL2F	9P18	VWPE83	819720.5	
365.830	CH3OH	*9P40	TOH88		
365.866	CD2F2	9R34	VPE81	819405.8	
366.000	TRIOX	10R10	DWC81		
366.273	CHCLF2	9R30	DFP78	818494.6	
366.420	CH3OH	*9R20	TOH88		
366.625	CD3BR	9P14	DF78	817708.3	
366.920	13CH3I	CO2H	GRF87		
367.000	DCOOD	9R04	DSF76		
367.399	CD2F2	10R14	VPE81	815985.9	
367.600	CH2CF2	10P38	HW84		
368.000	CH3OH	10P12	DFSY81B		
368.000	HCOOH	*9R18	PCD73		
368.448	CD3F	10R48	TD86	813662.6	
368.862	NH2NH2	9R18	IMEJ86	812750.0	156
368.900	CH3OH	13C18O2	PD84		
369.000	SIH3F	10R16	DS82	.	
369.100	CH3CF3	10R14	FR87		
369.114	CH3OH	9P16	PEJWG75	812195.4	
369.400	CD2CL2	N2O	BF86		
369.550	CD3OH	10P08	CIMPS87		
369.620	COF2	CO2S	GRF88		
369.700	CD3OH	10P38	CIMPS87		
369.968	HCOOD	10R28	DSF76	810320.5	
370.000	CD2CL2	10R10	SDFY86		
370.000	CH2CHBR	10P28	DESF76		
370.000	CHCLF2	9R40	DFP78		
370.000	CH3SH	10R34	Lands80B		
370.000	CD3OH	9P28	DSF75A		
370.400	14NH3	10R02	WHH80		51
370.483	CD3OH	9R28	SEJZS87	809193.2	75
370.800	CH3CHF2	10R06	FBRG85		
371.000	NH2NH2	N2O	JTT82		
371.300	CH3CHF2	N2O	FBRG85		
372.000	CH2CHF	10R20	TW82		
372.000	CD3OD	10R14	PVSEP85		
372.000	CH3OH	9P32	DFSY81B		
372.000	HCOOD	10R36	DSF76		

λ [μm]	Molecule	Pump	Reference	Freq. [MHz]	Footnotes
372.000	HCOOD	10P26	DSF76		
372.360	CD3OH	10P12	CIMPS87		
372.400	CH3OD	9R02	FK86		
372.500	NH2NH2	10P18	DSF74C		
372.528	HCN	ELEC	HJ67	804750.9	89
372.680	CH3F	*9P50	CM71		
372.800	13CH3I	CO2H	GRF87		
372.814	CH3CN	10P20	DF78	804134.8	110
372.870	CHCLF2	9R22	DFP78	804012.9	
373.000	CD2CL2	10P10	SDFY86		
373.000	NH2NH2	10R12	DSF74C		
373.400	15NH3	10R42	FWSGW82		56
373.400	CH2CF2	C18O2	MPD83		
373.400	CD3OD	9P20	PVSEP85		
374.000	HCOOD	9P34	DSF76		
374.086	CH2DOH	10P46	SPEJ80	801399.6	
374.600	CD3OD	13CO2	DVPA81		
375.000	CH3OCH3	*10P20	PCD73		
375.000	NH3	9R42	GDEFJ83		
375.100	CH2CF2	10P18	HW84		
375.300	CD3OD	10P32	PVSEP85		
375.400	CH2CF2	10R28	HW84		
375.407	CHD2F	9R06	Tobin84	798579.5	
375.500	SIH2F2	C18O2	DFS85		
375.545	CH2CF2	10P12	RPJM77	798286.6	
375.980	CHCL2F	9P16	VWPE83	797362.5	
376.000	CH3CH2F	*9R14	WZN73		
376.000	CH3NO2	9P06	DFSY81A		
376.600	CH2CF2	10P18	HW84		
376.700	CHFCHF	10R16	BH82		
376.800	13CD3F	10P38	TF81		
376.900	CD2CL2	10R16	BF86		
377.000	CH3NH2?	9P18	DFSY81B		
377.400	CH2CHF	10P14	CD81B		
377.450	CH3I	9R16	CM76		101
377.500	CH2CF2	10R22	HW84		
377.718	13CH2F2	9R10	STPVE85	793693.1	
378.000	CH3NO2	10P22	DFSY81A		
378.000	CH3CH2F	*9R32	WZN73		
378.200	CH3OCH3	10P34	BSKK82		
378.400	OCS	9R08	Lands80C		

λ [μm]	Molecule	Pump	Reference	Freq. [MHz]	Footnotes
378.500	CH2CF2	10P48	HW84		
378.600	CD2CL2	C18O2	PD82B		
378.880	CD2F2	10P22	VPE81	791260.4	
379.000	CH3SH	9P22	Lands80B		
379.242	COF2	10R40	TD86	790504.6	
379.500	CH3CHDOH	9P32	BS85		
379.590	COF2	N2O	GRF88		
380.000	CLO2	9R22	DSF75B		
380.000	CHCLF2	9R30	DFP78		
380.020	CH3BR	10R18	CM76		99
380.565	DCOOD	10R12	DSF76	787755.5	
380.710	CH3CN	*10P16	CM71		
380.800	CD3OH	10R32	PFS86		
381.000	TRIOX	10P12	DWC81		
381.600	CH3I	N2O	GRF87		
381.615	H13COOH	9P24	DG82	785587.2	
381.820	CH3OH	*10R44	Henni86		114
381.996	CH2F2	9R36	PSE80	784806.0	
382.000	CD2CL2	10R22	SDFY86		
382.357	H13COOH	9P26	DG82	784063.1	
382.639	CH2F2	9P10	PSE80	783486.0	
382.766	CHCLF2	10R40	DFP78	783226.7	
382.880	CH318OH	10R18	IMPSG89		
383.000	CH2CF2	CO2H	FGRD84		
383.200	CH3CF3	10P14	FR87		
383.285	CD3CL	9R34	DFBS75	782166.1	26
384.000	DFCO	10P20	JDL83		
384.000	CH2CF2	10R28	TW82		
384.000	DFCO	9R12	JDL83		
384.000	CH3SH	9P16	Lands80B		
384.100	CD2CL2	C18O2	PD82B		
384.319	CHD2F	10P28	Tobin84	780061.5	36
384.869	TRIOX	9R30	DWC81	778946.7	
384.916	COF2	10R16	DPK87	778852.3	
385.000	CH3CHO	9R30	LSB81		
385.400	CHD2OH	9P30	FPVSF89		
385.400	CD2CL2	N2O	BF86		
385.400	CH2CF2	CO2H	FGRD84		
385.687	CHCLF2	9R18	DFP78	777294.8	
385.700	CD3OH	9P30	PFS86		
385.800	CH2CHCN	10R22	GRBKF85		

λ [μm]	Molecule	Pump	Reference	Freq. [MHz]	Footnotes
385.909	CH2CHCL	10P22	RPJM77	776847.1	
386.037	CD3OH	10R34	SEJZS87	776589.1	75
386.339	CH3OH	9P14	PEJS80	775982.4	
386.410	CH3CN	*9P46	CM76		
386.500	CHFCHF	10R40	BH82		
386.500	ND2ND2	10R18	SDEF85		
386.600	CD3OH	9P16	PFS86		
386.900	CD3OH	9R22	PFS86		
386.966	CHCLF2	9R18	DFP78	774726.1	
387.000	CH3NH2?	9P34	DFSY81B		
387.000	CHCLF2	9R16	DFP78		
387.200	13CD3OH	9R38	IEPSV84		
387.310	CH3CN	9R12	CM76		108
387.500	CD3OD	9R08	FK86		
387.559	CH2DOH	9P40	SPEJ80	773539.9	
387.800	CH3CHF2	10R28	FBRG85		
388.000	CHCLF2	9R36	DFP78		
388.000	CD3OH	10R14	PFS86		62
388.000	HCOOH	*9R16	PCD73		
388.000	HCOOD	10R04	DSF76		
388.000	C2H4O2H2	*9P36	PCD73		
388.060	CH3CH2OH	9P32	VJE86	772542.0	
388.273	CDF3	10R32	TLD83	772117.0	
388.390	CH3CN	*9P22	CM76		
388.500	15NH3	13CO2	DVPA81		35, 53
388.652	CDF3	10R42	TLD83	771365.4	
388.900	CH3CF3	N2O	FR87		
389.000	ND2ND2	10P22	SDEF85		
389.000	TRIOX	9P18	DWC81		
389.000	NH3	9R30	GDEFJ83		
389.300	CHFCHF	10R20	BH82		
389.600	13CD3OH	9R08	IEPSV84		
389.907	DCOOD	10R12	DSF76	768882.0	
390.000	CH3OH	10R32	DFSY81B		
390.000	CD3I	9P26	DFBS75		
390.100	CH3OH	CO2S	WGS77		
390.200	CH3OH	13C18O2	PD84		
390.400	CH2CHCL	10P24	FBM84		
390.530	CH3I	10P42	CM76		115
390.780	COF2	10R38	TD86	767165.1	
391.300	CD3OD	10R46	PVSEP85		

λ [μm]	Molecule	Pump	Reference	Freq. [MHz]	Footnotes
391.461	13CH2F2	9R30	STPVE85	765829.0	
391.689	HCOOD	10R38	DSF76	765384.6	
392.000	HCOOH	*9R18	PCD73		
392.000	CD2CL2	N2O	BF86		
392.000	CH3OD	9P02	FK86		
392.069	CH3OH	9P36	PEJWG75	764642.6	
392.480	CH3I	*9R14	CM76		
393.000	CD2F2	10R08	VPE81		
393.000	HCOOD	9R16	DSF76		
393.000	TRIOX	10R26	DWC81		
393.300	CH3CF3	10P20	FR87		
393.300	CH3CF3	10R14	FR87		
393.330	COF2	10R32	GRF88		
393.485	H13COOH	9P32	DG82	761888.8	
393.500	HCOOH	9R14	DSF74D		pp3, 48
393.631	HCOOH	9R18	KW76	761608.3	pp3, 4, 121
394.701	CH2F2	9P06	PSE80	759543.3	
395.000	HCOOD	9P10	DSF76		
395.000	HCOOD	9P16	DSF76		
395.000	13CH3I	CO2S	GRF87		
395.149	DCOOD	10R10	DSF76	758682.5	
395.712	HCOOD	10R12	DSF76	757601.9	
396.000	CH3CH2OH	9P32	JEJ75		
396.000	HCOOH	9R42	DDSB77B		
396.000	CH2CHBR	9P34	DESF76		
396.000	CH2DOH	9P18	ZD78		
396.000	DCOOD	10R20	DSF76		
396.000	CHCLF2	9R10	DFP78		
396.400	CD3OH	10R28	PFS86		
397.000	DCOOD	10R22	DSF76		
397.510	CH3F	*9P50	CM71		
397.700	CH3CHF2	10R22	FBRG85		
398.000	CD3OH	10R28	DSF75A		
398.000	HCOOD	10R22	DSF76		
398.000	CH3NO2	9R36	DFSY81A		
398.900	CD2CL2	N2O	BF86		
398.960	CH2CHCN	N2O	GRBKF85		
399.288	13CH2F2	9R34	STPVE85	750817.6	
399.300	CH2CF2	C18O2	MPD83		
399.420	CH2CHCN	10R16	GRBKF85		
399.800	CH2CHCN	10P42	GRBKF85		

λ [μm]	Molecule	Pump	Reference	Freq. [MHz]	Footnotes
399.800	13CD3OH	9P34	IEPSV84		
400.100	13CH3OH	9R20	PS87		
400.200	CD2CL2	N2O	BF86		
401.000	HCOOH	*9R16	PCD73		
401.250	CH2CHCN	9P42	GRBKF85		
401.300	CH2CF2	10P20	HW84		
401.444	CH2F2	C18O2	GCK86	746784.7	
402.000	CH3NC	10R12	LSB81		
402.300	CH2CF2	N2O	FGRD84		
402.915	COF2	10P06	TD86	744058.0	27
403.000	CH3SH	9P12	Lands80B		
403.000	HCOOH	*9R18	PCD73		
403.600	CH2CF2	C18O2	MPD83		
403.710	CH2F2	C18O2	GCK86	742593.9	
403.777	13CH2F2	9P32	STPVE85	742470.4	
404.000	CH3CH2F	9P34	WZN73		116
404.000	TRIOX	10R12	DWC81		
404.000	CH3CH2F	9R30	Radfo75		
404.000	CH3NC	10R04	DJ80		
404.100	HCOOH	10R42	WZN73		120
404.300	CD3OD	9P40	PVSEP85		
404.600	CHD2OH	9P28	FPVSF89		
405.486	HDCO	9P16	DDG80	739340.3	
405.504	CH3CH2F	9R30	RPJM77	739307.5	121
405.585	HCOOH	9R18	RPJM77	739161.0	3, 121
405.750	HCOOH	9P26	WZN73		116
405.950	CH2CHCN	9R06	GRBKF85		
406.360	CH2CHCN	10P04	GRBKF85		
406.878	CHD2F	10P28	Tobin84	736812.0	
407.000	CD3OH	9R44	DSF75A		
407.100	13CD3OD	9P28	VE85		
407.294	CH2CF2	10P14	RPJM77	736059.6	
407.500	CH318OH	10P26	IMPSG89		
407.600	CH2CHF	N2O	GRBF84		
407.720	CH3BR	*9P28	CM76		
407.900	CD3OH	9R34	PFS86		
408.800	CD3OD	10R18	PVSEP85		
409.000	CLO2	10P20	DSF75B		
409.100	CD3OH	10R40	CIMPS87		
409.300	CH2CF2	10P18	HW84		
409.600	CD2CL2	N2O	BF86		

λ [μm]	Molecule	Pump	Reference	Freq. [MHz]	Footnotes
409.800	CD2CL2	10R02	BF86		
410.000	CD3OH	9P32	DSF75A		
410.100	CH3CF3	N2O	FR87		
410.200	CD3OD	9P14	FK86		
410.712	CD3OD	10R12	VSPE81	729932.8	103
411.000	CH2CHBR	10R26	DESF76		
411.000	HCOOD	9P40	DSF76		
411.100	CHFCHF	9P16	BH82		
411.200	CH3CF3	N2O	FR87		
411.600	CD3OD	9R34	PVSEP85		
412.000	CD3OH	10R12	DSF75A		
412.000	SIHF3	13CO2	DFS85		
412.200	CH2CHF	CO2S	RGF84		
413.000	HCOOH	*9R16	PCD73		
414.000	DCOOD	10P44	DSF76		
414.000	HCOOH	*9R20	PCD73		
414.000	CH3NO2	9R04	DFSY81A		
414.000	CD3OD	10R12	HP79		
414.000	HCOOH	*9R22	PCD73		
414.351	CHCLF2	9R36	DFP78	723522.9	
414.800	CD2F2	10R18	TSD82		
414.980	CH3BR	*10R02	CM76		
415.000	DCOOD	10R14	DSF76		
415.000	C2H4O2H2	*9P14	PCD73		
415.000	CH3CHO	9R40	LSB81		
415.000	CH2CF2	10P14	HRB73		
415.075	CHCLF2	9R30	DFP78	722260.5	
415.363	13CH2F2	9P04	STPVE85	721759.8	
416.000	CH2CHBR	10R22	DESF76		
416.522	CH3OH	9P14	PEJS80	719751.1	
416.700	CH3OD	9P10	FK86		
416.710	CH3OH	9P28	Henni86		115
417.000	CD3OH	10R32	CIMPS87		
417.000	CH2F2	9R12	DW78C		
417.000	HCOOD	9R22	DSF76		
417.100	CH3OD	9P06	BP77		
417.244	CD2F2	9P20	VPE81	718505.6	
417.300	13CD3OD	9R20	VE85		
417.400	CD3OD	9R08	FK86		
418.000	CLO2	10R08	DSF75B		
418.000	CH3OH	9P32	DFSY81B		pp9

λ [μm]	Molecule	Pump	Reference	Freq. [MHz]	Footnotes
418.083	CH3OH	9P36	PEJS80	717065.0	pp9
418.100	CD3OH	10R38	PFS86		
418.200	CD3OD	10R36	FK86		mm12
418.270	CH2F2	9R12	PSE80	716743.3	
418.300	HCOOH	9R18	SJFWM87		60
418.310	CH3BR	10P26	CM76		99
418.510	HCOOH	9R24	WZN73		pp10, 120
418.613	HCOOH	9R22	RPJM77	716156.8	pp10, 60, 6
418.712	CD3OH	10R36	SEJZS87	715987.6	mm12, 77
418.790	CH3OH	9P06	Henni86		115
419.000	CH2CHBR	10R32	DESF76		
419.000	CH3OH	9R16	DFSY81B		
419.300	CH3OH	13CO2	PD82A		
419.550	HCOOH	*9R22	WZN73		117
419.839	TRIOX	10P22	DWC81	714065.8	
420.000	CH2CHF	10P40	TW82		
420.000	CD3OH	10R36	DSF75A		59
420.200	CH3OD	9P30	FK86		
420.300	CD3OH	10R32	PFS86		59, 78
420.311	CDF3	10R26	TLD83	713263.1	
420.391	HCOOH	9R08	DSF76	713127.6	121
420.980	CDF3	10R46	TLD83	712130.6	
421.000	CH2CHCL	10P04	LSB81		
421.000	HCOOH	*9R18	WZN73		
421.053	13CH2F2	9P14	STPVE85	712005.8	
421.300	CH3CHF2	N2O	FBRG85		
421.800	CH2CHF	N2O	GRBF84		
422.000	CD3OH	9P20	DSF75A		
422.000	CH3CF3	10R14	FR87		
422.117	CH3CN	10P24	IMEJ86	710212.3	110
422.151	CH2DOH	9R08	SPEJ80	710154.3	
422.500	CHFCHF	9P28	BH82		
422.780	CH3BR	*10R26	CM76		
423.000	CH2CHF	10R46	RGF84		
423.354	CH2CHCL	10R30	DSFE74	708137.1	
424.000	CH2CHBR	10P20	DESF76		
424.000	CHFCHF	10R14	BH82		
424.000	CH2CHCL	10R28	Radfo75		
424.000	CH3NO2	10P24	DFSY81A		
424.130	COF2	N2O	GRF88		
424.550	CD3I	10P02	GRF87		

λ [μm]	Molecule	Pump	Reference	Freq. [MHz]	Footnotes
425.000	DCOOD	10P18	DSF76		
425.650	CH2CHCN	10P18	GRBKF85		
425.800	13CH3OH	9P26	PS87		
425.870	CH2CHCN	N2O	GRBKF85		
426.000	CHD2OH	10R38	ZD78		
426.000	CH3NO2	9R22	DFSY81A		
426.800	CH2CF2	10P18	HW84		
427.000	CH2CHBR	10R24	DESF76		
427.040	CH3CN	*9P26	CM76		
427.100	CHD2OH	10P20	FPVSF89		
427.200	CH2DOH	10P36	SPEJ80		
427.700	CH2CF2	13CO2	MPD83		
427.807	CHCLF2	9R26	DFP78	700766.2	
427.890	CH3CCH	*10P10	CM71		
428.000	HCOOH	*9R20	PCD73		
428.000	HCOOH	*9R18	PCD73		
428.000	CD3BR	10R02	Lands80A		
428.870	CH3CCH	*9R38	CM76		
429.690	HCOOD	10P24	DSF76	697695.1	
429.900	CH2CHF	N2O	GRBF84		
430.000	CD3BR	9P18	Lands80A		
430.000	CH3OH	9R12	DFSY81B		
430.000	CH2CHF	10P18	TW82		
430.100	CH2CF2	10P36	HW84		
430.438	HCOOD	10P06	DSF76	696482.3	
430.482	CH3CN	10P18	DF78	696410.9	110
430.910	COF2	N2O	GRF88		
430.927	CD3OH	9R34	SEJZS87	695691.5	
431.140	CH2CHCN	N2O	GRBKF85		
431.400	CD3OH	10R34	PFS86		
431.736	CD3BR	9P18	DF78	694388.4	
432.000	CHCLF2	10P06	DFP78		
432.000	HFCO	9R32	JD81		
432.109	HCOOH	9R22	DDSB77B	693788.5	34
432.244	CHCLF2	9R40	DFP78	693572.9	
432.300	CH3CHDOH	9P24	BS85		
432.400	CH2F2	9R06	DW78C		
432.631	HCOOH	9R20	KW76	692951.4	7, 121
432.667	HCOOH	9R20	DSF76	692895.0	8
432.987	CDF3	10P20	TLD83	692381.5	
433.000	TRIOX	10P44	DDSB77A		

λ [μm]	Molecule	Pump	Reference	Freq. [MHz]	Footnotes
433.000	CH2CHF	C18O2	TJD86		
433.000	HCOOD	10R16	DSF76		
433.000	DCOOH	10P16	DSF76		
433.000	CH3COOD	10P20	DFSY81A		
433.100	HCOOH	*9R22	WZN73		117
433.104	CD3I	9P28	DFBS75	692195.5	
433.235	DCOOH	10R14	DSF76	691985.3	
433.438	CHCLF2	10R34	DFP78	691662.4	
433.500	CH2CHF	N2O	GRBF84		
433.600	CD3OH	10P20	PFS86		
433.800	CHFCHF	9P16	BH82		
433.900	CH3CHF2	N2O	FBRG85		
434.000	ND2ND2	9P14	SDEF85		
434.400	CD2CL2	C18O2	PD82B		
434.950	CH318OH	10R16	IMPSG89		
434.951	CH2F2	9R06	PSE80	689255.1	
435.000	CD3OH	9P28	GW78		
435.000	HCOOH	*9P16	PCD73		
435.100	CD3OH	10R36	CIMPS87		
435.300	CD3OH	10R34	CIMPS87		
435.427	CHD2F	10R38	Tobin84	688503.0	
435.772	NH2NH2	10P24	RPJM77	687957.4	
435.900	CH2CHCL	N2O	GRBF84		
435.900	CH3CHF2	10R02	FBRG85		
436.500	SIHF3	10R30	DFS85		
437.000	COF2	10P22	TW82		
437.400	CHD2OH	9P26	FPVSF89		
437.451	HCOOH	9P16	DSF76	685316.6	121
437.600	CHFCHF	10R06	BH82		
437.600	CH2CF2	C18O2	MPD83		
438.000	CH2CHCL	10P42	FBM84		
438.022	13CH2F2	10R18	STPVE85	684423.3	
438.100	CH318OH	9P38	IMPSG89		
438.400	CD2CL2	10R06	BF86		
438.507	CH2CHBR	10P28	DESF76	683666.5	
438.800	CD3OD	10R14	PVSEP85		
438.870	CD3OH	10R44	CIMPS87		
439.000	SIHF3	13CO2	DFS85		
439.063	CH2F2	C18O2	GCK86	682800.4	
440.000	CD3BR	9R18	Lands80A		
440.000	COF2	10P36	TW82		

λ [μm]	Molecule	Pump	Reference	Freq. [MHz]	Footnotes
440.010	CH2CHCN	10R46	GRBKF85		
440.200	CD2CL2	N2O	BF86		
440.884	CD2F2	10P32	VPE81	679979.8	
441.000	HCOOH	*9R18	PCD73		
441.150	CH3CN	*9R16	CM76		
441.300	CH3OCH3	10P08	BSKK82		
441.300	CH2CF2	N2O	FGRD84		
441.674	CD3BR	9R18	DF78	678764.4	24
441.700	CH2CHF	CO2S	RGF84		
441.810	HCOOH	9R20	DH78A		
442.100	CHFCHF	10R30	BH82		
442.168	CH2CHCL	10P16	RPJM77	678006.1	
443.000	SIH2F2	13CO2	DFS85		
443.000	DCOOD	10P14	DSF76		
443.265	CD3CL	9P10	DFBS75	676328.5	26
443.500	CH2CHBR	10P24	DESF76		
443.600	CH3CHF2	CO2S	FBRG85		
443.800	CH3OH	13C18O2	PD84		
444.000	CHCLF2	9R40	DFP78		
444.386	CD3I	9R32	DFBS75	674621.3	
444.400	CH2CHF	10R20	CD81B		
444.745	COF2	10P36	TD86	674077.8	
445.000	CH2CF2	N2O	FGRD84		
445.000	CH2CHCL	10R18	Radfo75		
445.000	CH2CHBR	10P22	DESF76		
445.200	CH2CHF	N2O	GRBF84		
445.210	HCOOH	10P14	WZN73		120
445.400	CH3OD	9P52	FK86		
445.663	CDF3	10P20	TLD83	672689.5	
445.900	HCOOH	9R20	DSF76	672331.8	9, 121
446.100	CD2F2	10R16	TSD82		
446.505	HCOOH	9R22	DDSB77B	671419.5	121
446.700	CH2CHF	N2O	GRBF84		
446.700	CH2CHF	10R50	RGF84		
446.873	HCOOH	9R16	DSF76	670867.2	121
447.000	CH3CH2CL	10R28	DK82		
447.000	HCOOD	9P28	DSF76		
447.080	CH3OH	*9P28	Henni86		
447.142	CH3I	10P18	KW76	670463.0	10, 65, 107
447.600	CD2CL2	N2O	BF86		
448.080	CH3OH	9R12	Henni86		115

λ [μm]	Molecule	Pump	Reference	Freq. [MHz]	Footnotes
448.300	CH3OH	13CO2	PD82A		
448.455	CH3OH	9P12	VWE87	668500.1	
448.534	H13COOH	9R26	DG82	668383.6	45
448.700	CD3I	13CO2	MPD83		
449.000	CH3CH2OH	9P22	BS85		
449.200	CH2CF2	N2O	FGRD84		
449.300	CD3OD	10R02	FK86		
449.300	CH3CHF2	10R18	FBRG85		
449.500	CH3CHF2	N2O	FBRG85		
449.800	CD3CL	10R20	DFBS75	666502.0	26
450.000	HCOOD	10R26	DSF76		
450.000	DFCO	9P12	JDL83		
450.000	CH3NO2	9R34	DFSY81A		
450.200	CD2CL2	C18O2	PD82B		
450.400	CH3OH	9P12	IIMMS81		
450.700	CD3OD	9P04	FK86		
450.980	HCOOD	10P12	DSF76	664757.9	
451.000	CH3COOD	9P32	DFSY81A		
451.475	CH2DOH	9P32	SPEJ80	664028.4	
452.000	CH3CH2F	9R22	Knigh81		
452.000	DCOOD	10R10	DSF76		
452.380	CD3I	N2O	GRF87		
452.400	CH2DOH	9P46	SPEJ80		
452.400	13CH3OH	9R36	PS87		
452.425	13CH2F2	9P16	STPVE85	662635.0	
452.500	CHD2OH	10R08	FPVSF89		
452.900	CD3OH	10R20	PFS86		
453.100	CD3OD	9P30	PVSEP85		
453.397	CH3CN	9R16	RPJM77	661213.4	31, 107
453.570	CH2CHCN	10R34	GRBKF85		
453.600	CD3OD	9P18	PVSEP85		
453.600	CH3CH2BR	10R34	BS83		
453.800	CH2CHBR	9P36	BGRF84		
454.000	CH3NC	10R04	LSB81		
454.000	CH3NO2	10P26	DFSY81A		
454.000	ND2ND2	10R40	SDEF85		
454.300	CH2CHF	CO2H	RGF84		
454.500	CH2CF2	10R16	HW84		
454.600	CH3CF3	N2O	FR87		
454.800	CH3CF3	10R52	FR87		
455.073	CD3CN	9P08	DF78	658778.6	

λ [μm]	Molecule	Pump	Reference	Freq. [MHz]	Footnotes
455.500	SIHF3	10R28	DFS85		
455.600	CD3OH	9P18	PFS86		
456.000	CH3SH	9P44	Lands80B		
456.000	CH2CHF	13CO2	TJD86		
456.000	CH3CN	9R16	Radfo75		
456.100	CD3OD	9R32	PVSEP85		
456.200	CD2F2	10R42	VPE81		
457.250	CH3I	*10P18	CM76		
457.300	CH2CF2	10P30	HW84		95
457.341	DCOOD	10P30	DSF76	655511.9	
457.500	CD3OD	9R16	PVSEP85		87
457.900	CH2CHBR	N2O	GRBF84		mm15
458.000	CH3CHF2	10P20	HRB73		
458.000	CH2CHF	CO2H	RGF84		
458.000	CH2CHF	N2O	TJD86		mm15
458.430	HCOOH	*9R36	WZN73		pp16
458.523	HCOOH	9R38	DDSB77B	653822.2	pp16, 11
458.662	CH3BR	10R20	DF78	653624.5	
459.000	CH2CHCN	10R06	GRBKF85		
459.180	CH3I	*10P08	CM76		
459.310	CH2CHCN	N2O	GRBKF85		
459.400	CH2CHCL	10P52	FBM84		
459.400	CDF3	10R18	TSD82		
459.428	TRIOX	9R22	DWC81	652533.9	
459.600	CDF3	10R26	TSD82		
459.800	CH2CHF	10R20	RGF84		
459.886	CLO2	10R24	DSF75B	651884.5	31
460.000	HCOOH	9R20	SJFWM87		
460.440	CH3OH	*9R24	Henni86		81
460.510	HCOOH	*9R10	WZN73		
460.562	CD3I	9R12	DFBS75	650927.5	
461.000	CH2CHF	CO2H	RGF84		
461.000	CH3OCH3	*10P34	PCD73		
461.000	CH2CHF	10R50	RGF84		
461.000	CH3OH	9R08	DFSY81B		
461.000	CH3OH	9R26	DFSY81B		
461.072	NH2NH2	10P16	RPJM77	650207.7	
461.200	CH3CL	*9R42	CM76		
461.261	HCOOD	10P16	DSF76	649941.0	
461.385	13CH3OH	9P12	HP78	649767.0	
461.700	CH3OH	9R24	PS87		81

λ [μm]	Molecule	Pump	Reference	Freq. [MHz]	Footnotes
462.800	13CD3OH	10P08	IEPSV84		
462.920	CH3CH2F	*9P32	WZN73		
463.000	CH3CF3	10R12	FR87		
463.624	CH2CF2	10R20	DSFE74	646628.1	25
464.400	CH2CHCN	10P36	GRBKF85		
464.412	CH2F2	9P06	PSE80	645530.9	
464.627	H13COOH	9R26	DG82	645231.8	
464.700	13CD3OD	10R08	VE85		
464.757	CD3CL	10R20	DFBS75	645052.4	26
464.800	CH3CHF2	N2O	FBRG85		
465.000	CH3COOD	10R20	DFSY81A		
465.000	SIHF3	10R16	DFS85		
465.000	NH3(D)	10R34	GDEFJ83		
465.500	CD3I	N2O	GRF87		
465.500	CD2F2	9P28	VPE81		
465.500	CH318OH	9P18	IMPSG89		
465.700	CH318OH	9R34	IMPSG89		
466.000	CD3I	N2O	GRF87		
466.250	CH3CN	*9R16	CM76		
466.546	DCOOH	10P14	DSF76	642578.4	
466.643	CD3BR	9R26	DF78	642445.1	
467.000	TRIOX	10P22	DWC81		
467.200	CH2CHF	N2O	GRBF84		
467.515	CHCL2F	9R06	VWPE83	641246.9	
467.700	CHCLF2	9R44	TD80		
467.850	CH3OH	10P10	Henni86		115
468.236	CH2DOH	9P26	SPEJ80	640259.5	
468.400	CH2CF2	13CO2	MPD83		
468.965	13CD3OH	10R26	IEPSV84	639264.6	
469.000	CD2CL2	10R04	ZD78		
469.000	DCOOD	10P12	DSF76		
469.023	CH3OH	10R38	PEJWG75	639184.6	54
469.500	CH2CF2	C18O2	MPD83		
470.000	CH2CHCN	10R30	GRBKF85		
470.000	CH3NO2	9P14	DFSY81A		
470.000	CH3OH	9P24	DFSY81B		
470.065	13CD3F	10P34	TD86	637768.5	
470.386	CHCL2F	9R34	VWPE83	637332.6	
470.900	CH3OH	10R40	PS87		
471.000	SIH2F2	10R20	DFS85		
471.200	CH3CF3	N2O	FR87		

λ [μm]	Molecule	Pump	Reference	Freq. [MHz]	Footnotes
471.500	CD2CL2	N2O	BF86		
471.800	SO2(I)	9R18	CD81B		165
472.000	HCOOD	10P16	DSF76		
472.000	CH3NO2	9R08	DFSY81A		
472.400	CD3OH	9R18	PFS86		
472.400	CH2CHF	10P20	CD81B		
473.000	HCOOD	9R38	DSF76		
473.200	CD2CL2	C18O2	PD82B		
473.680	CD3I	CO2S	GRF87		
474.600	CH2CHCL	10P48	FBM84		
475.100	CH2CF2	10P32	HW84		
475.100	CH2CF2	N2O	FGRD84		
475.200	CD2CL2	C18O2	PD82B		
475.300	CH2CHBR	N2O	GRBF84		
476.000	CD3I	N2O	GRF87		
476.000	CH2CHF	N2O	GRBF84		
476.000	CHCLF2	10R30	DFP78		
476.250	CD3OH	10R34	CIMPS87		81
476.300	CH2CF2	10R22	FGRD84		
476.800	CH2CHCL	CO2H	FBM84		
477.000	HCOOD	9P14	DSF76		
477.000	CH2CHF	10P36	TW82		
477.100	CH3CF3	10R50	FR87		
477.300	CD3OH	10R34	PFS86		81
477.300	CH2CF2	10R20	HW84		
477.870	CH3I	*9P26	CM76		
477.963	H13COOH	9P30	DG82	627229.1	
478.072	COF2	10P24	TD86	627086.5	27
479.000	DCOOD	10R04	DSF76		
479.123	13CH2F2	9R40	STPVE85	625711.5	
479.150	CH3OH	10P16	Henni86		115
479.904	DCOOH	10P14	DSF76	624692.6	
480.000	CD3OH	9P16	DSF75A		
480.000	H13COOH	10R46	DG82		
480.000	CH3OCH3	*10P34	PCD73		
480.010	CH3CN	*9R16	CM76		
480.310	CD3CL	9P36	DFBS75	624164.3	26
481.000	CH3NC	10P42	LSB81		
481.452	CHCLF2	9R30	DFP78	622683.6	
482.120	CH318OH	9P14	IMPSG89		
482.200	CH3CHF2	10R20	FBRG85		

λ [μm]	Molecule	Pump	Reference	Freq. [MHz]	Footnotes
482.500	CD3OD	9R06	PVSEP85		
482.700	CD3OH	10P26	PFS86		
482.700	CD3OH	9R26	PFS86		
482.900	CHD2OH	9P06	FPVSF89		71
482.961	CH2CHBR	10P26	DESF76	620737.8	
483.000	CD3OD	10R12	FK86		
483.160	CD3I	N2O	GRF87		
483.500	NH2NH2	9P20	DSF74C		
483.500	CD3OD	9P28	FK86		
483.500	CD2CL2	10R12	BF86		
483.800	CH2CHF	10P10	RGF84		
484.300	CD2CL2	N2O	BF86		
484.400	CHD2OH	9P22	FPVSF89		
484.774	CH2CF2	10R28	DSFE74	618417.5	25
485.270	COF2	10P22	GRF88		
485.400	CH3CF3	10P08	FR87		
485.600	CH3CF3	10R38	FR87		
485.800	CH3CF3	10P12	FR87		
486.000	CH3CH2F	9R24	Radfo75		
486.100	CH3OH	CO2S	WGS77		
486.100	CH2CF2	10R30	HW84		
486.100	CH3CF3	10R32	FR87		
486.500	CD3OD	10R24	VSPE81		102
486.768	CH2CF2	10P22	DSFE74	615883.3	
487.000	CH2CHCL	9P10	Radfo75		
487.000	CH3NO2	9R06	DFSY81A		
487.000	NH2NH2	N2O	JTT82		
487.000	SIHF3	C18O2	DFS85		
487.000	CH3NO2	9R12	DFSY81A		
487.144	CHCLF2	9R32	DFP78	615408.5	
487.226	CD3I	9P10	DFBS75	615304.6	
487.500	CH2CHF	N2O	GRBF84		
487.600	CH3OD	9P06	FK86		
487.700	CH2CHF	CO2H	RGF84		
487.800	CH2CHF	10P20	RGF84		
487.800	CH2CHCL	9P52	FBM84		
488.000	SIHF3	10R24	DFS85		
488.000	SIHF3	13CO2	DFS85		
488.110	COF2	10P24	GRF88		
488.276	CD2F2	9P42	VPE81	613981.5	
488.528	CDF3	10R38	TLD83	613665.3	

λ [μm]	Molecule	Pump	Reference	Freq. [MHz]	Footnotes
488.880	CH3CCH	*10P12	CM71		
489.000	CH2CHCN	10P08	Radfo75		
489.000	CH3NO2	10P28	DFSY81A		
489.038	O3	9R32	DDM83	613025.0	
489.238	CD2F2	10R26	VPE81	612774.8	
489.300	CH2CHF	N2O	GRBF84		
490.000	CH2CHF	CO2H	RGF84		
490.000	CH2CHF	10P22	TW82		
490.083	CH2CHBR	10P16	DESF76	611717.8	
490.391	CD3I	9R22	DFBS75	611333.6	
490.700	CH2CF2	N2O	FGRD84		
491.200	CD3OD	9R08	PVSEP85		
491.376	TRIOX	10P30	DWC81	610108.3	
491.800	CH3CD2OH	9P40	BS85		
491.800	CD2F2	10R18	Tobin80		
491.891	DCOOD	10P08	DSF76	609469.8	
492.000	CH3OCH3	*10P34	PCD73		
492.000	HCOOH	*9P42	WZN73		
492.040	CHCL2F	9R36	VWPE83	609284.6	
493.000	CD2CL2	10P28	SDFY86		
493.000	CH3CH2I	10P34	BS83		
493.156	HCOOD	10R40	DSF76	607905.7	
493.280	HCOOH	*9P14	WZN73		
493.500	CD3OD	10P02	FK86		
493.500	CH2CHF	N2O	GRBF84		
493.541	CH3OH	10R04	VWE87	607431.2	
494.000	SIH2F2	10P12	DFS85		
494.000	SIH2F2	10R18	DFS85		
494.100	CD3OD	9R22	PVSEP85		
494.646	CH3CN	9P06	RPJM77	606074.7	107
494.700	CD3OD	10R18	PVSEP85		mm8
495.000	CH3OCH3	*10P12	PCD73		
495.000	CD3OH	10R18	DSF75A		mm8
495.963	CHCL2F	9P08	VWPE83	604465.0	
496.000	HCOOH	*9R18	PCD73		
496.070	CH3F	9P20	IIMSD86	604334.7	
496.101	CH3F	9P20	KW76	604297.3	110
496.151	CH3F	9P20	IIMSD86	604236.9	
496.300	13CH3OH	10R16	PS87		
496.400	13CH3OH	9P30	PS87		
496.500	CH3OCH3	10P20	BSKK82		

λ [μm]	Molecule	Pump	Reference	Freq. [MHz]	Footnotes
496.660	13CH2F2	9P14	STPVE85	603617.2	
496.890	CH3F	C18O2	DFHAL83		72
497.000	CH2CF2	CO2H	SWTRD80		
497.000	TRIOX	9R26	DWC81		
497.000	CD2CL2	10R28	SDFY86		
497.200	CD3OD	9R06	PVSEP85		
497.300	CD2CL2	10P14	BF86		
497.400	CH3OCH3	9R26	BSKK82		
497.500	CH2CHBR	CO2S	BGRF84		
497.677	CD2F2	9P28	VPE81	602383.9	
498.000	HCOOD	9P44	DSF76		
498.000	CH3OD	9P32	DSF74A		
498.000	CD3OH	10R34	DW78B		
498.500	SIHF3	10R20	DFS85		
498.700	CH2CF2	N2O	FGRD84		
498.700	CD3OH	9R26	SEJZS87		
500.000	TRIOX	9P16	DWC81		
500.000	CD2CL2	10P14	SDFY86		
500.577	CD2F2	10R24	VPE81	598893.7	
501.164	TRIOX	10P18	DWC81	598192.4	
501.600	CH3CF3	10R48	FR87		
501.900	CHD2OH	9P20	FPVSF89		
502.262	CH3CH2F	9R24	RPJM77	596884.2	121
503.000	CH2CHCN	9R12	Radfo75		
503.057	CH2F2	9R06	PSE80	595941.7	
504.000	CH3CH2F	9R24	Radfo75		
504.000	CH3CH2I	10P32	BS83		
504.500	CH2CHBR	N2O	GRBF84		
504.752	CDF3	10R38	TLD83	593940.1	
505.000	SO2(l)	9R24	BS84		
505.000	CH2CHF	10P22	RGF84		
505.000	CH3CHF2	N2O	FBRG85		
505.800	CH318OH	9P42	IMPSG89		
505.829	COF2	10P22	TD86	592676.0	27
506.000	CH2CHBR	10R38	DESF76		
506.250	CH318OH	9P36	IMPSG89		
506.300	CH2CHF	CO2H	RGF84		
507.480	CH3OH	*9P26	Henni86		
507.584	CH2CHCL	10P22	DSFE74	590626.3	30
507.591	CH2CHCL	10P22	DSFE74	590618.4	
508.000	DCOOD	10R28	DSF76		

λ [μm]	Molecule	Pump	Reference	Freq. [MHz]	Footnotes
508.000	CH2CHF	10P38	TW82		
508.330	CH2CHCN	10P40	GRBKF85		
508.370	CH3I	9P34	CM76		119
508.480	CH3BR	*10R42	CM76		
508.791	DCOOD	10P08	DSF76	589225.0	
509.000	CH3CHO	9R36	LSB81		
509.160	CH2CHCN	10R36	GRBKF85		
509.372	CH2DOH	10P46	IMEJ86	588553.4	
509.440	COF2	10R10	GRF88		
509.500	CD3OH	9P06	PFS86		
509.700	CH2CHBR	CO2H	BGRF84		
509.859	CLO2	9R36	DSF75B	587991.1	
509.890	TRIOX	10P40	DWC81	587955.6	
510.000	CH3OD	9P06	KFK82		
510.160	CH3CN	*9P06	CM76		
510.400	CH3CF3	N2O	FR87		
510.500	CD2CL2	10R06	SDFY86		
510.700	CH3CF3	10P18	FR87		
511.445	CH2F2	9R28	PSE80	586167.4	
511.900	CH3CL	*10R52	CM76		
511.900	CH3OCH3	10P52	BSKK82		
512.000	HCOOH	*9R26	PCD73		
512.000	TRIOX	10P40	DDSB77A		
512.800	CHD2OH	9P04	FPVSF89		
513.002	HCOOH	9R28	DDSB77B	584388.2	12, 121
513.016	HCOOH	9R28	DSF76	584372.9	
513.400	CH3CHF2	N2O	FBRG85		
514.000	DFCO	10P16	JDL83		
514.000	CH3NO2	9P28	DFSY81A		
514.951	DCOOD	10P34	DSF76	582177.0	
515.000	CD3OD	13CO2	PD85		
515.170	HCOOH	9P16	DSF76	581929.7	13
515.800	CH3OD	9P06	FK86		
515.800	CH2CHBR	10P04	BGRF84		
516.000	SIH3F	10R20	DFS85		46
516.000	HCCCHO	10P14	DJ80		
516.000	CH3OH	9P26	DFSY81B		
516.253	CD3CN	9P30	DF78	580708.2	
516.382	COF2	10R08	TD86	580562.9	27
516.500	CD3OH	10P42	PFS86		
516.770	CH3CCH	*9R12	CM76		

λ [μm]	Molecule	Pump	Reference	Freq. [MHz]	Footnotes
517.330	CH3I	*10P14	CM76		
517.500	CH2CHBR	10R48	BGRF84		
517.800	CHD2OH	10P10	FPVSF89		
518.000	CHD2OH	9P30	ZD78		
518.400	CH2CHF	CO2S	RGF84		
518.600	CH2CHF	10P52	RGF84		
518.800	CH3CF3	10R04	FR87		
518.830	HCOOH	*9P16	WZN73		
518.900	CH3CHF2	N2O	FBRG85		
519.000	CH2CHCL	10P34	Radfo75		
519.075	CH3CH2F	9R04	RPJM77	577551.1	
519.200	CH3CF3	N2O	FR87		
519.303	CD3CL	9P36	DFBS75	577297.5	26
519.600	CH2CHF	N2O	GRBF84		
519.700	CD2CL2	N2O	BF86		
520.000	CH3OCH3	*10P12	PCD73		
520.000	CD2CL2	10R12	ZD78		
520.300	CD3OH	10R18	PFS86		
521.110	13CH3I	CO2H	GRF87		
521.237	CDF3	10R24	TLD83	575156.1	
521.400	CH3CHF2	N2O	FBRG85		
523.091	CH2DOH	9P40	SPEJ80	573116.8	
523.120	CH3OH	10P18	Henni86		115
523.406	CD3I	10P38	DFBS75	572772.1	
523.500	SIHF3	C18O2	DFS85		
523.800	CH2CF2	10P16	HW84		97
523.900	CD3I	13CO2	MPD83		
524.000	CH3NO2	9R28	DFSY81A		
524.600	CD3OH	10P24	PFS86		
524.800	CH2CHCL	N2O	GRBF84		
524.900	15NH3	CO2S	FWSGW82		51
525.000	CLO2	9R36	DSF75B		
525.000	CH3COOD	10R12	DFSY81A		
525.300	SO2(I)	9R30	BS84		
525.320	CH3I	*9P04	CM76		
525.560	CH2CHCN	10P30	GRBKF85		
526.300	CH3OCH3	10P12	BSKK82		
526.486	DCOOD	10P34	DSF76	569421.9	
527.000	COF2	10R22	TW82		
527.215	DCOOD	10P34	DSF76	568634.6	
527.700	CH2CF2	10P40	HW84		

λ [μm]	Molecule	Pump	Reference	Freq. [MHz]	Footnotes
527.873	NH2NH2	9P12	RPJM77	567925.4	
527.900	CH3CH2BR	10R30	BS83		
528.497	CH2CHBR	10R40	DESF76	567255.3	
528.880	CH2F2	C18O2	GCK86	566843.6	
529.280	CH3I	*10P36	CM76		
529.300	CH3CH2OH	9R04	BS85		
529.880	CD3CN	9R04	DF78	565774.2	
530.000	CH3NO2	9R16	DFSY81A		
530.000	HCOOH	*9R26	PCD73		
530.000	HCOOH	*9R28	PCD73		
530.132	CD3BR	10R10	DF78	565505.1	
530.400	CD3OH	13CO2	DVPA81		mm7
530.533	CH2CHCL	9P16	DSFE74	565077.8	
530.700	CH3OCH3	9R28	BSKK82		
530.854	CHCL2F	9R06	VWPE83	564736.1	
531.000	HCOOD	9P40	DSF76		
531.038	CH3BR	10P24	DF78	564540.7	107
531.080	CH3CCH	*9P06	CM76		
531.300	CH2CF2	10R14	HW84		
531.363	13CH2F2	9P08	STPVE85	564195.3	148
531.900	CD3OD	13CO2	PD85		mm7, 50
533.000	CH3CHF2	10P20	HRB73		
533.000	ND2ND2	10P38	SDEF85		
533.137	CHCLF2	10R16	DFP78	562317.4	
533.330	CD3I	N2O	GRF87		
533.573	CH2F2	C18O2	GCK86	561858.6	
533.655	NH2NH2	10R08	IMEJ86	561772.0	70
533.678	HCOOH	9P16	DSF76	561747.5	13
533.701	HCOOH	9R28	DSF76	561724.0	
534.200	CH3CHF2	N2O	FBRG85		
534.430	CHCLF2	9R34	DFP78	560957.0	
535.000	CH2CF2	CO2H	SWTRD80		
536.096	H13COOH	9P24	DG82	559214.1	
537.060	CH2CHCN	N2O	GRBKF85		
537.410	13CD3F	CO2H	TD86	557847.0	
537.650	CH2CHCN	10P22	GRBKF85		
538.000	CH2CHCL	10R04	Radfo75		
538.415	COF2	10P16	TD86	556805.5	27
538.600	CH2CHF	10P56	RGF84		
539.100	COF2	N2O	GRF88		
539.100	CD2CL2	C18O2	PD82B		

λ [μm]	Molecule	Pump	Reference	Freq. [MHz]	Footnotes
540.000	CH2CHF	10P32	TW82		
540.000	CD3I	9R06	DFBS75		
540.736	CDF3	10R36	TLD83	554415.6	
540.900	CH3CH2F	*9P38	WZN73		
540.986	CH2F2	9R42	PSE80	554159.0	
542.000	CH3CH2I	10P30	BS83		
542.990	CH3I	*10P26	CM76		
543.200	CHFCHF	9P34	BH82		
544.100	CH2CHBR	N2O	GRBF84		
545.000	CH3I	10P26	DFBS75		
545.000	NH2OH	C18O2	TJD86		
545.000	CH2CHCN	10R20	GRBKF85		
545.279	CH3BR	10P38	DF78	549796.0	107
545.412	CH3BR	10R32	DF78	549662.8	107
545.500	CH2CHCL	10P18	FBM84		
545.560	CD3I	N2O	GRF87		
545.880	CD3I	CO2S	GRF87		
546.000	CH2CF2	13CO2	MPD83		
546.800	CHFCHF	9P20	BH82		
546.800	CH318OH	10R36	IMPSG89		
547.529	CHCL2F	9R10	VWPE83	547537.6	
548.700	CD2F2	10R28	TSD82		
548.843	H13COOH	9P20	DG82	546225.3	
549.258	CHCL2F	9R08	VWPE83	545813.2	
549.500	CHFCHF	10P06	BH82		
549.686	CH2CHCN	10P14	DSFE74	545388.2	
550.000	CH3OD	9R04	FK86		
550.000	CH3NO2	10P30	DFSY81A		
550.100	CD3OD	9R28	PVSEP85		
550.200	CD3OH	9P22	PFS86		
551.100	13CH2F2	9R06	STPVE85		
551.200	CH2CF2	N2O	FGRD84		
551.500	CH2CHF	10P56	RGF84		
551.900	CD3OH	9R16	PFS86		
552.000	CH3OH	C18O2	PD82A		
552.000	CD3BR	9P32	Lands80A		
552.000	CH3NO2	9P06	DFSY81A		
552.000	ND2ND2	10R12	SDEF85		
552.000	CH3CH2OH	9P26	BS85		
552.400	CD3OD	9R10	PVSEP85		
552.600	CH3OH	13C18O2	PD84		

λ [μm]	Molecule	Pump	Reference	Freq. [MHz]	Footnotes
552.940	COF2	N2O	GRF88		
553.000	CD3OH	9R34	PFS86		
553.300	CH2CHBR	9P40	BGRF84		
553.696	CH2CHBR	10P40	DESF76	541438.5	
553.883	CD3BR	9P32	DF78	541256.2	
554.000	CD3OH	10R08	PFS86		62
554.365	CH2CF2	10P14	RPJM77	540785.1	28
554.560	CH2CHCN	10P04	GRBKF85		
554.700	13CD3I	13CO2	MPD83		
555.100	CH2CHBR	CO2H	BGRF84		
555.200	CD3I	13CO2	MPD83		
555.750	CH318OH	10R20	IMPSG89		
555.900	CH2CF2	10R12	FGRD84		29
556.097	CHCLF2	9R34	DFP78	539101.3	
556.200	CD3I	13CO2	MPD83		
556.470	CH2CHCN	10P44	GRBKF85		
556.800	CH2CHCL	9P52	FBM84		
556.800	CH2CHCL	CO2H	FBM84		
556.803	CD3BR	9R30	DF78	538417.8	
556.876	CD3I	10P36	DFBS75	538347.3	
557.000	CHFCHF	9P40	BH82		
557.000	CH2CHF	10P36	TW82		
557.100	CD3I	13CO2	MPD83		
557.700	CH2CF2	10P22	TSD82		
558.000	CD2CL2	10P24	SDFY86		
558.500	CD3OD	9P06	PVSEP85		
558.577	TRIOX	9R10	DWC81	536707.3	
558.800	CH3CF3	N2O	FR87		
558.800	CHD2OH	10P28	FPVSF89		
558.820	13CH3I	10P26	GRF87		
560.000	CH3OD	9P06	FK86		
560.703	CDF3	10R36	TLD83	534672.7	
560.803	CDF3	10R40	TLD83	534577.4	
561.028	CHCL2F	9R40	VWPE83	534362.8	
561.294	DCOOD	10P20	DSF76	534109.6	
561.410	CH3CN	*9R08	CM76		
562.000	CD2CL2	10R22	SDFY86		
562.400	CD3OH	10R36	CIMPS87		
562.450	CHCLF2	9R34	DFP78	533011.8	
563.000	CH2CHF	10R36	TW82		
563.130	CH3CCH	*10P24	CM71		

λ [μm]	Molecule	Pump	Reference	Freq. [MHz]	Footnotes
563.440	CH2CHCN	10R14	GRBKF85		
563.700	CH2CF2	10P24	HW84		pp4, 96
564.000	CH3NO2	10P30	DFSY81A		
564.000	CH3OH	10R16	DFSY81B		
564.680	CH3BR	*10P28	CM76		
564.700	CH3OCH3	10P16	BSKK82		
564.700	CH2CHCN	N2O	GRBKF85		
565.000	CH2CHF	13CO2	TJD86		
566.100	CH3CH2OH	9R28	BS85		
566.440	CH3CCH	*9P18	CM76		
566.750	CH3OH	10P16	Henni86		115
567.107	HCOOD	10P14	DSF76	528635.2	
567.532	CH2F2	9R28	PSE80	528239.2	
567.700	CD3I	13CO2	MPD83		
567.800	CD3OD	9R40	PVSEP85		
567.868	DCOOD	10R26	DSF76	527926.0	
567.945	CH2CHCL	10P16	DSFE74	527854.1	30
568.500	CH2CF2	10P22	TSD82		pp4, 85
568.810	CH3CL	*10R26	CM76		
569.000	DFCO	C18O2	JDL83		
569.400	CH3CHF2	N2O	FBRG85		
569.477	CD3I	10P36	DFBS75	526434.4	
569.700	CH3OH	13C18O2	PD84		
570.300	SO2(I)	9R16	BS84		167
570.332	13CH2F2	9R20	STPVE85	525645.3	
570.569	CH3OH	9P16	PEJWG75	525427.5	15
572.330	H13COOH	9R32	DG82	523810.4	
572.510	COF2	10R12	GRF88		
572.692	CH2CHCN	10R20	DSFE74	523479.7	
573.000	CH2CHF	13C18O2	TJD86		
573.750	13CH3I	10P26	GRF87		
574.000	CH2CHCL	10P16	Radfo75		
574.027	CH2CHCN	10R16	DSFE74	522262.2	pp5
574.380	CH2CHCN	10R18	GRBKF85		pp5
574.600	13CD3I	C18O2	MPD83		
575.000	NH2NH2	N2O	JTT82		
575.300	CH3CH2OH	9P34	BS85		
576.170	CH3I	10P16	CM76		115
577.000	HCOOH	*9P38	PCD73		
577.001	CH2CF2	10P30	HW84		
577.800	CD3OD	10R04	PVSEP85		

λ [μm]	Molecule	Pump	Reference	Freq. [MHz]	Footnotes
578.000	CH2CHCN	10R14	Radfo75		
578.900	CH3I	*10R34	CM76		
579.000	CH2CHF	13CO2	TJD86		
579.761	CH2CHCL	10P16	DSFE74	517096.5	
580.300	CD2CL2	10R14	BF86		
580.387	HCOOH	9R22	DDSB77B	516538.7	
580.600	CH3CF3	10R46	FR87		
580.800	CH2CHCL	10R50	FBM84		
580.800	CH3CF3	10R30	FR87		
580.801	HCOOH	9P38	DSF76	516170.7	121
580.869	CHCL2F	9R12	VWPE83	516110.2	
581.300	CH2CHCL	CO2H	FBM84		
581.500	CD2CL2	10R20	SDFY86		
581.600	CH3CD2OH	10R24	BS85		
581.984	CDF3	10R28	TLD83	515121.1	
582.100	CDF3	10R12	TSD82		
582.500	CH2CHF	10R32	RGF84		
582.500	CH3CHF2	10P28	FBRG85		
582.554	HCOOD	9P18	DSF76	514617.8	
582.800	CD2F2	10R22	TSD82		
583.100	CH2CHF	N2O	GRBF84		
583.300	CH3CHF2	N2O	FBRG85		
583.300	CD3OH	9R22	PFS86		
583.700	CHFCHF	9P42	BH82		
583.770	CH3CCH	*9P20	CM76		
583.870	CH3I	*9P04	CM76		
583.872	CH2CHCN	10P12	DSFE74	513455.5	
584.000	CH2CHCL	N2O	GRBF84		
584.800	CH2CF2	13CO2	MPD83		
585.100	CD2CL2	N2O	BF86		
585.500	CD3OD	9R22	PVSEP85		
585.777	CH3BR	9P40	DF78	511785.8	107
585.800	CH2CHF	CO2S	RGF84		
586.000	CH2CHF	13CO2	TJD86		
586.382	CH2CHCN	10P20	DSFE74	511258.1	
586.720	CH2CHCN	10P08	GRBKF85		
586.800	CH2CHCN	10R14	GRBKF85		
587.000	CH2CHCL	13CO2	TJD86		
587.500	ND2ND2	10P34	SDEF85		
587.884	CH2F2	C18O2	GCK86	509951.3	
588.028	CH2F2	9R46	PSE80	509827.2	

λ [μm]	Molecule	Pump	Reference	Freq. [MHz]	Footnotes
588.440	CH2CHCN	10P02	GRBKF85		
588.600	CD2CL2	N2O	BF86		
588.700	CH3OH	10R10	PS87		
589.321	CH3CN	9P30	IMEJ86	508708.2	
590.000	HCOOD	10P14	DSF76		
590.000	HCCF	C18O2	DJ80		
590.000	CH2CHBR	CO2H	BGRF84		
590.000	CH2CF2	CO2H	FGRD84		
590.000	CHCLF2	10P12	DFP78		
590.369	CH2CHCL	9P18	DSFE74	507804.8	
591.130	CHCLF2	10R24	DFP78	507151.1	
591.165	CH2F2	C18O2	GCK86	507121.4	
591.441	CH2CF2	10P26	DSFE74	506885.1	
591.616	DCOOD	10R26	DSF76	506735.1	
591.700	CH2CF2	C18O2	MPD83		
592.000	DCOOD	10P18	DSF76		
592.441	CHCLF2	9R40	DFP78	506029.4	
592.759	CH2F2	C18O2	GCK86	505758.1	
593.000	TRIOX	10P38	DWC81		
593.100	CD3OH	10R06	PFS86		
593.279	CD2F2	9P30	VPE81	505314.1	
593.506	CH3CH2F	9P36	RPJM77	505121.4	121
593.900	CH2CHCL	N2O	GRBF84		
594.000	CH3NO2	9R28	DFSY81A		
594.000	HCOOD	9R30	DSF76		
594.729	CH2CHBR	10P32	DESF76	504082.8	
597.000	CH2CHCN	10P08	GRBKF85		
597.000	CH2CHCN	10P12	GRBKF85		
597.330	CH2CHCN	N2O	GRBKF85		
598.000	CH3NO2	9R16	DFSY81A		
598.300	CHD2OH	9R14	FPVSF89		
598.300	CH2CHCL	N2O	GRBF84		
598.400	CD3OD	10R24	PVSEP85		
598.600	CD3OH	10P28	PFS86		
599.000	CD3OH	10R16	DSF75A		
599.000	CH2CHCN	10P20	GRBKF85		
599.550	CD3I	10R22	DFBS75	500029.2	
600.000	CH3NH2?	9P40	DFSY81B		
600.700	CH3BR	10R20	PRP83		
601.670	COF2	N2O	GRF88		
601.897	CH2CHCL	10P38	RPJM77	498079.1	

λ [μm]	Molecule	Pump	Reference	Freq. [MHz]	Footnotes
602.000	CH2CF2	10R10	HW84		
602.487	CH3OH	9P24	PEJS80	497591.6	
602.500	CD3I	N2O	GRF87		
605.000	CH2CHF	10R02	RGF84		
605.000	CH2CF2	10P48	SWTRD80		123
605.400	CH3CHF2	10P40	FBRG85		
605.600	CDF3	10R20	TSD82		
605.700	CH2CHBR	N2O	GRBF84		
606.000	CH2CHCL	10P14	FBM84		
606.600	CH2CHF	N2O	GRBF84		
606.700	CHD2OH	9P34	FPVSF89		
606.700	CH2CHBR	10R46	BGRF84		
606.700	CH2CHCL	10P44	FBM84		
606.800	CH3CF3	10R26	FR87		
606.800	CH2CHF	10P42	RGF84		
607.300	CHD2OH	9P34	FPVSF89		
607.714	TRIOX	10P36	DWC81	493311.8	
608.000	DFCO	C18O2	JDL83		
610.300	CD3OH	10R08	PFS86		
613.000	SIH2F2	10R22	DFS85		
613.500	CH3OD	9R04	FK86		
614.110	CD3I	10R22	DFBS75	488174.0	
614.285	CH3OH	9P24	PEJS80	488034.7	
614.300	CF2CL2	10P32	LPMD81		
615.329	CHCLF2	10R14	DFP78	487206.6	
615.900	CH2CF2	13CO2	MPD83		
616.335	CH2DOH	9P26	SPEJ80	486411.5	
617.000	CH2CHF	CO2H	RGF84		
617.656	CHCLF2	9R40	DFP78	485371.3	
617.700	CH2CHBR	CO2H	BGRF84		
617.700	CH2CF2	CO2H	SWTRD80		104
617.700	CH2CF2	N2O	FGRD84		
617.900	CH2CF2	CO2H	FGRD84		104
618.000	CH2CHF	10P32	RGF84		
618.446	CH2CHBR	10R30	DESF76	484751.1	
618.896	13CH2F2	9R08	STPVE85	484398.7	
619.000	TRIOX	10R22	DDSB77A		
619.300	CH2CHBR	10R28	BGRF84		
620.000	CH2CHCL	9P46	FBM84		
620.000	CH3NO2	9R18	DFSY81A		
620.300	CH3CH2OH	9R12	BS85		

λ [μm]	Molecule	Pump	Reference	Freq. [MHz]	Footnotes
620.340	CH2CHCN	10P22	GRBKF85		
620.400	CH3CH2F	9P22	WZN73		116
621.000	CH2CHCL	9P48	FBM84		
621.700	CH318OH	9R08	IMPSG89		
622.000	CH3OH	10R22	DFSY81B		
622.000	SIH3F	13CO2	DFS85		
622.000	CH3OH	9P34	DFSY81B		
622.300	CH2CHCL	10R34	FBM84		
623.000	CH2CHCN	10R12	Radfo75		
623.000	HCCCH2F	9P24	TJD86		
624.096	CH2CHBR	10R18	DESF76	480362.9	
624.430	CH3OH	9P38	PEJS80	480105.7	
624.700	CH2CHBR	10R26	BGRF84		
625.700	CH2CHBR	CO2H	BGRF84		
626.800	CH3CH2I	10R06	BS83		
627.340	CH3OH	*9P16	WZN73		
628.000	CD3I	N2O	GRF87		
628.000	CH3COOD	9R06	DFSY81A		
629.000	CD3I	N2O	GRF87		
629.300	CH3CF3	10R28	FR87		
629.844	13CH3OH	9P12	HP78	475979.0	
630.166	HCOOD	10R10	DSF76	475735.6	
630.400	CH2CF2	10R14	FGRD84		
630.700	CH2CHBR	N2O	GRBF84		
631.000	CD2CL2	10R18	ZD78		
631.000	CH2CHCN	10R06	Radfo75		
631.000	CH3NO2	10P32	DFSY81A		
631.500	CH2CF2	CO2H	FGRD84		
631.800	CD3OD	9P04	FK86		
631.930	CH3BR	10P16	CM76		99
632.000	CD3I	N2O	GRF87		
632.050	CH3BR	10P22	DF78	474318.0	107
632.900	CH3CHF2	10P10	FBRG85		mm9
633.000	CH3CHF2	10P40	BT77		
633.400	CH3CF3	N2O	FR87		
634.000	CH3NO2	9R30	DFSY81A		
634.000	CH3CF3	10R32	FR87		
634.471	CH2CHCL	10P20	RPJM77	472507.8	
634.471	CH2CHCL	9P20	DSFE74	472507.5	
634.700	CH3CF3	10P10	FR87		mm9
635.000	CD3OD	10R12	FK86		

[μm]	Molecule	Pump	Reference	Freq. [MHz]	Footnotes
.355	CH2CHBR	10R26	DESF76	471850.5	
.300	CD3OD	10P38	FK86		
036.300	CH2CHCL	CO2H	FBM84		
637.100	CH2CHCL	N2O	GRBF84		
637.500	CH3CHF2	10P26	FBRG85		
638.000	CH2CHCL	10P06	Radfo75		
638.394	13CH2F2	9P26	STPVE85	469604.1	
638.400	CF2CL2	10P36	LPMD81		
639.128	DCOOH	10P08	DSF76	469064.7	
639.730	CH3I	*9P06	CM76		
640.000	CD3I	10R18	DFBS75		
640.350	COF2	10R18	GRF88		
640.700	CH2CHBR	10P56	BGRF84		
640.700	CH2CHBR	CO2H	BGRF84		
641.000	ND2ND2	9P22	SDEF85		
641.000	ND2ND2	10P38	SDEF85		
641.430	CH2CHCN	9R22	GRBKF85		
642.280	CH2CHCN	N2O	GRBKF85		
642.600	CH2F2	9R44	PSE80	466530.5	
642.860	CH2CHCN	10R06	GRBKF85		
643.516	CD2F2	10P22	VPE81	465866.4	
644.000	CD3I	10P16	DFBS75		
644.500	CD2F2	10R20	TSD82		
644.500	CH3CHF2	N2O	FBRG85		
644.640	CH2CHCN	CO2S	GRBKF85		
644.640	CH2CHCN	10R44	GRBKF85		
645.000	CH2CHCN	10P44	GRBKF85		
645.000	DCOOD	10R24	DSF76		
645.200	CD3OH	10R36	KK83		42
645.289	CH2CHCL	9P20	DSFE74	461586.2	
645.500	CH3OCH3	9R34	BSKK82		
646.000	CH2CHBR	10R26	DESF76		
646.000	CH3NO2	9R18	DFSY81A		
646.477	CD3OH	10R08	SEJZS87	463732.4	42
646.500	CH3CHF2	10P40	FBRG85		
647.348	DCOOH	10R30	DSF76	463108.3	
647.890	CH3CCH	*10P14	CM71		
648.600	CH2CHBR	N2O	GRBF84		
649.425	CH2CHBR	10P18	DESF76	461627.2	
649.590	CH3CCH	*10P34	CM71		
649.600	CD2CL2	10P04	BF86		

λ [μm]	Molecule	Pump	Reference	Freq. [MHz]	Footnotes
650.700	COF2	N2O	GRF88		
651.790	CH2CHCN	10R30	GRBKF85		
651.900	CD3OD	9P10	PVSEP85		
652.680	CH3CN	9P30	CM76		109
653.220	CH318OH	9P14	IMPSG89		
653.700	CH2NOH	13CO2	DP84		
654.000	HFCO	9P18	JD81		
654.920	CH3OH	*9P22	Henni86		
655.000	CH2CHF	N2O	GRBF84		
655.400	CH2CHCL	10R40	FBM84		
655.900	CH2CHCN	10R02	GRBKF85		
656.000	TRIOX	10P36	DWC81		
656.000	CH3NO2	9R20	DFSY81A		
657.000	HCOOD	9P22	DSF76		
657.239	CH2F2	9P10	PSE80	456139.1	
657.590	CH2CHCN	9P44	GRBKF85		
657.900	CH2CF2	10P10	HW84		
657.938	CDF3	10P12	TLD83	455654.7	
657.989	CDF3	10R10	TLD83	455619.1	
658.152	CDF3	10P06	TLD83	455506.2	
658.260	CH2CHCN	10R44	GRBKF85		
658.500	ND2ND2	10P10	SDEF85		
658.530	CH3BR	*9P56	CM76		
658.570	CD3I	N2O	GRF87		
658.570	CD3I	10R02	GRF87		
658.900	CH3OH	9P22	PS87		
659.000	NH2OH	C18O2	TJD86		
659.690	CH2CHCN	CO2S	GRBKF85		
660.000	HCOOD	10R12	DSF76		
660.000	CH3CH2F	9R16	Knigh81		
660.000	CH2CHF	10P24	CD81B		
660.200	CH2CHF	N2O	GRBF84		
660.200	CD3I	13CO2	MPD83		
660.328	CH3CH2I	10P26	TD86	454005.5	
660.340	CH2CHCN	10R38	GRBKF85		
660.582	CD3I	10P46	DFBS75	453830.6	
660.700	CH3BR	10R20	CM76		99, mm19
660.882	13CH3BR	10R20	IMEJ86	453624.6	149, mm19
661.000	TRIOX	9R12	DWC81		
661.153	CHCL2F	9R30	VWPE83	453438.9	
662.000	CH2CF2	N2O	FGRD84		

λ [μm]	Molecule	Pump	Reference	Freq. [MHz]	Footnotes
662.200	CH2CHBR	10P18	BGRF84		
662.816	CH2CF2	10P24	RPJM77	452301.5	28
663.000	CD3OD	9P22	PVSEP85		
663.080	13CH3I	CO2S	GRF87		
663.670	CH3OH	10P18	Henni86		115
664.000	DFCO	C18O2	JDL83		
665.700	COF2	N2O	GRF88		
665.885	CHCLF2	10P18	DFP78	450216.4	
666.000	DCOOD	10P30	DSF76		
666.604	CH2CHCL	10R20	DSFE74	449731.0	
667.232	CD3I	10P10	DFBS75	449307.5	
668.000	HCOOD	9R40	DSF76		
668.100	CH2CHCL	10R18	FBM84		
669.531	HCOOH	9R30	DDSB77B	447765.0	16
670.094	CD3I	10R08	DFBS75	447388.6	
670.114	CD3I	10R08	DFBS75	447375.1	
670.400	CD3OD	10R26	PVSEP85		
670.790	CH2CHCN	10R08	GRBKF85		
670.990	CH3I	*10P28	CM76		
671.000	CH2CHF	CO2H	RGF84		
671.150	CH2CHCN	10R10	GRBKF85		
671.430	CH2CHCN	CO2S	GRBKF85		
671.500	CH3CHF2	N2O	FBRG85		
672.100	CH2CHF	10P36	CD81B		
673.000	CH3NO2	9R30	DFSY81A		
674.061	13CH2F2	9P04	STPVE85	444755.4	
674.800	CH3OH	13CO2	PD82A		
675.000	CH3COOD	10R22	DFSY81A		
675.000	CH3NO2	9R14	DFSY81A		
675.200	CH2CF2	10P44	HW84		
675.290	CH3CCH	*9P40	CM76		
676.000	CH3OH	9P18	DFSY81B		
676.000	CH3COOD	10P18	DFSY81A		
676.700	CH3CF3	10P02	FR87		
677.962	CH2F2	C18O2	GCK86	442196.7	
678.570	CD3I	10R02	GRF87		
679.100	CH2CF2	10P24	HW84		
679.766	TRIOX	10P34	DWC81	441023.2	
680.000	CD3I	N2O	GRF87		
680.000	CH3CHF2	N2O	FBRG85		
680.000	CD3OH	9P06	DSF75A		

λ [μm]	Molecule	Pump	Reference	Freq. [MHz]	Footnotes
680.541	CH2CHBR	10R16	DESF76	440520.5	
681.000	CH2CHCL	13CO2	TJD86		
681.400	CH2CHBR	10R48	BGRF84		
681.500	CH2CHCL	CO2H	FBM84		
682.175	CHCLF2	10P14	DFP78	439465.6	
682.600	CH2DOH	9R24	SPEJ80		
683.600	CH2CHBR	CO2S	BGRF84		
683.738	CH2CHCL	10R20	DSFE74	438460.7	
684.300	CD3OD	9P40	PVSEP85		
684.500	CD3OH	10R34	PFS86		
684.600	CH2CHBR	N2O	GRBF84		
684.700	CD3OD	9P34	PVSEP85		
684.700	CF2CL2	10P42	LPMD81		
684.740	CF2CL2	10P34	LPMD81		
684.800	SO2	9R28	BS81		
685.190	CH2CHCN	10R46	GRBKF85		
687.200	CHFCHF	9P38	BH82		
687.300	CD3OD	10P02	FK86		
687.837	CDF3	10R10	TLD83	435848.1	
688.300	CH2CF2	N2O	FGRD84		
689.000	SIH3F	13CO2	DFS85		
689.178	13CH2F2	9R22	STPVE85	435000.0	
689.998	HCOOD	10R26	DSF76	434483.0	
690.000	CD2CL2	10R24	SDFY86		
690.000	CD3OH	9P12	PFS86		
690.000	13CD3I	10P10	DH78A		
690.400	CH3CHF2	10P30	FBRG85		
691.119	CD3I	9R20	DFBS75	433778.2	
691.250	CHD2F	10R26	Tobin84	433695.8	
692.000	HCOOD	10R12	DSF76		
692.025	CD3BR	9P26	DF78	433210.6	
693.000	CH2CHBR	CO2H	BGRF84		
693.140	CH2CHBR	10R16	DESF76	432513.8	
693.800	CH2CHBR	10P50	BGRF84		
694.000	NH3(D)	10R14	GDEFJ83		
694.100	CD2CL2	10R06	BF86		
694.189	CH3OH	9P24	PEJS80	431859.8	
694.428	TRIOX	9R16	DWC81	431711.4	
695.000	TRIOX	10R22	DWC81		
695.000	NH3(D)	10R26	GDEFJ83		
695.000	CD3OH	9P12	DSF75A		

λ [μm]	Molecule	Pump	Reference	Freq. [MHz]	Footnotes
695.202	CH2CHCL	9P22	DSFE74	431230.7	
695.350	CH3OH	10R16	PEJS80	431139.0	
695.672	HCOOD	10R36	DSF76	430939.4	
696.000	C2H4O2H2	*9P34	PCD73		
697.000	CH3NO2	9R30	DFSY81A		
697.000	CH3OH	9P40	DFSY81B		
697.455	DCOOH	10R36	DSF76	429837.6	
697.800	CH2CF2	10P24	DH78B		
698.000	CH3CH2CL	10R38	DK82		
698.555	CD3CL	9P06	DFBS75	429160.6	26
698.600	CH3CHF2	10P34	FBRG85		
699.000	CH2CHCL	9P22	Radfo75		
699.000	ND2ND2	10R12	SDEF85		
699.000	CD3OH	10R18	PFS86		
699.423	CH3OH	9P34	PEJWG75	428628.5	
699.636	COF2	10R22	TD86	428497.6	27
700.300	CH2CHF	10P36	RGF84		
700.400	CH2CHCL	9P48	FBM84		
701.000	CH3COOD	10R18	DFSY81A		
701.100	CH2CHBR	CO2H	BGRF84		
701.500	CD3OH	9P24	PFS86		
704.200	CD3OH	10R36	KK83		42
704.530	CH3CN	*9R34	CM76		
704.600	CD2CL2	10P12	BF86		
704.925	CH2CHCL	9P18	DSFE74	425282.7	
705.000	CHFCHF	10R16	BH82		
705.000	HCOOH	9R06	DSF76		
705.000	NH2NH2	C18O2	JTT82		
705.300	CH2CHBR	10R52	BGRF84		
706.600	CD3OH	9P10	PFS86		
707.100	CH2CHBR	N2O	GRBF84		
707.221	CH2CHBR	10R24	DESF76	423902.1	
707.800	CH3CH2BR	10P14	BS83		
709.200	CH3CF3	N2O	FR87		
709.500	CH3CF3	10R36	FR87		
709.800	CH3CF3	10R40	FR87		
710.000	TRIOX	9P10	DWC81		
710.000	DCOOH	10P06	DSF76		
710.000	CD3OD	9R08	FK86		
711.000	CD3OH	9P08	DSF75A		
711.752	TRIOX	9R32	DWC81	421203.7	

λ [μm]	Molecule	Pump	Reference	Freq. [MHz]	Footnotes
712.000	CH2CHBR	10R10	DESF76		
712.760	CH2CHCN	9R24	GRBKF85		
713.106	DCOOH	10R34	DSF76	420404.0	
713.200	CD2CL2	10P20	BF86		
713.720	CH3CN	*10P32	CM71		
715.300	CH3OD	9P02	FK86		
715.390	CH3BR	10R14	DF78	419061.7	107
715.400	CH2CF2	N2O	FGRD84		
715.800	CD2CL2	10P06	BF86		
717.000	CH3NO2	9R14	DFSY81A		
718.000	CH3CHF2	10P30	FBRG85		
718.000	CD2CL2	10P20	SDFY86		
718.700	CH2F2	C18O2	PD82B		
718.900	CD2F2	10R18	TSD82		
719.300	CH3I	10P22	CM76		115
719.500	CD2CL2	10P06	SDFY86		
720.200	CD2CL2	10P14	BF86		
720.800	CH2CHF	10P04	RGF84		
721.000	NH2NH2	10P12	DSF74C		
722.000	CD3OH	10P20	DSF75A		
722.000	CH2CHCN	10P42	Radfo75		
722.359	18O3	9P10	DDM83	415018.4	
723.000	CD2CL2	10R26	SDFY86		
723.080	CH3I	CO2H	GRF87		
723.100	CD2CL2	N2O	BF86		
723.100	CD2CL2	10P20	BF86		
724.000	ND2ND2	10R30	SDEF85		
724.140	CH2CHBR	10P14	DESF76	413998.0	
724.920	CH2F2	9P04	PSE80	413552.3	
726.000	CD2CL2	10P20	SDFY86		
726.920	DCOOD	10P10	DSF76	412414.5	
727.000	TRIOX	10R22	DWC81		
727.570	CH2CHCN	10P16	DSFE74	412046.3	
727.949	HCOOD	10R42	DSF76	411831.6	
728.100	CD3I	C18O2	MPD83		
728.900	CH2CF2	10R06	HW84		
730.323	CD3I	9R28	DFBS75	410492.7	
731.000	CH2CF2	10P26	DW84		
733.000	TRIOX	10P20	DWC81		
733.000	HCOOD	10R36	DSF76		
733.574	D2CO	9P32	DDSB77A	408673.8	

λ [μm]	Molecule	Pump	Reference	Freq. [MHz]	Footnotes
733.597	D2CO	9P32	DDG80	408661.1	
734.162	NH2NH2	10R38	Knigh81	408346.7	
734.262	CD3I	9P22	DFBS75	408290.6	
734.600	CH3OD	10R44	FK86		
734.800	CH2CHBR	CO2H	BGRF84		
734.959	13CH2F2	9P36	STPVE85	407903.8	
735.000	CH3NO2	10P34	DFSY81A		
735.130	CD3CL	9P06	DFBS75	407808.9	26
735.820	CH3BR	10R12	IMEJ86	407426.4	
737.000	DCOOD	10P04	DSF76		
737.113	D2CO	10R32	DDG80	406711.7	31
738.000	CH2CHCN	10P16	Radfo75		
738.414	13CH2F2	9R12	STPVE85	405995.1	
740.000	13CH2F2	9P28	STPVE85		
741.115	CH2CHBR	10P20	DESF76	404515.5	
741.200	CD3OD	9P52	FK86		
741.620	CH3CN	*9R08	CM76		
741.700	CD3OD	10R10	FK86		
742.573	HCOOH	9R40	DSF76	403721.5	17
743.700	CD2CL2	10R12	BF86		
744.050	HCOOH	9R24	DDSB77B	402919.6	18
745.000	CD3I	10P08	DFBS75		
745.000	CD3OH	9R26	DSF75A		
745.000	CD3OD	10R40	FK86		
745.500	13CD3I	13CO2	MPD83		
746.400	CH3CHF2	N2O	FBRG85		
746.500	CD2CL2	10R12	SDFY86		
747.050	CHCLF2	9R10	DFP78	401301.7	
747.700	CD3OD	9P34	FK86		
749.368	CH3BR	10R14	DF78	400060.3	107
749.371	CH3BR	10P14	MD78	400058.7	
749.372	TRIOX	9P30	DWC81	400058.4	
749.390	CH3BR	10R14	DF78	400048.7	
750.000	CH3CN	9P16	Knigh81		
750.000	TRIOX	10P18	DDSB77A		61
750.000	DFCO	9P20	JDL83		
750.380	CH2CHCN	9P26	GRBKF85		
750.606	TRIOX	10P20	DWC81	399400.6	61
751.400	CF2CL2	10P30	LPMD81		
751.831	CH2CHCN	10P28	DSFE74	398749.7	
752.681	D2CO	9R32	DDSB77A	398299.6	44

λ [μm]	Molecule	Pump	Reference	Freq. [MHz]	Footnotes
752.748	DCOOH	10R34	DSF76	398263.8	
753.800	CH2CHBR	10P40	BGRF84		
754.000	CH2CHCN	CO2H	GRBKF85		
756.000	CH3COOD	10P18	DFSY81A		
757.410	CH3CCH	*10P10	CM71		
758.200	CH2CHCL	CO2H	FBM84		
758.460	CH2CHCN	N2O	GRBKF85		
758.700	CH2CHF	10P50	RGF84		
759.600	CH3OD	9P06	FK86		
760.000	CD3OH	10P18	DSF75A		
761.000	HCOOH	9R24	Radfo75		
761.200	CH3CHF2	N2O	FBRG85		
761.670	CH2CHCN	N2O	GRBKF85		
761.762	DCOOD	10P10	DSF76	393551.5	
762.500	CH2DOH	9P18	SPEJ80		
764.100	CH2CF2	10P10	BP77		
764.200	CH2CHF	CO2H	RGF84		
765.200	CF2CL2	10P42	LPMD81		
765.420	COF2	N2O	GRF88		
766.600	CH3CF3	10P06	FR87		
767.800	CH3CF3	10R20	FR87		
768.012	CHD2F	10R26	Tobin84	390348.8	
768.800	CH2CHCL	N2O	GRBF84		
769.053	CH2CHCN	10R44	DSFE74	389820.1	
769.100	CH2CHBR	N2O	GRBF84		
769.500	CD2CL2	N2O	BF86		
769.800	CH3CH2BR	10R22	BS83		
770.000	CH2CHCN	N2O	GRBKF85		
770.000	CH3CHF2	10P14	BT77		
770.000	CH2CF2	10P10	Knigh81		
770.400	CH2CHCL	10R48	FBM84		
771.038	TRIOX	10P20	DWC81	388816.9	
771.200	CH2CHCL	CO2H	FBM84		
771.300	CH2CHCL	10R20	FBM84		
774.000	CD3OH	9P30	DSF75A		
774.100	CD2CL2	C18O2	PD82B		
774.600	CD2CL2	N2O	BF86		
774.900	CH2CHF	10P22	CD81B		
775.000	CH2CHCN	10R42	Radfo75		
775.000	CLO2	9R14	DSF75B		
777.920	CH2CHCN	N2O	GRBKF85		

λ [μm]	Molecule	Pump	Reference	Freq. [MHz]	Footnotes
778.000	CH3NO2	9R40	DFSY81A		
778.200	CD3OD	9R08	FK86		
779.874	DCOOD	10P26	DSF76	384411.2	
780.000	CH3NO2	9R32	DFSY81A		
780.133	CH2CHBR	10R14	DESF76	384283.8	
780.830	CH2CHCN	10P02	GRBKF85		
781.000	CH3OH	*9R24	Henni86		pp17
781.000	CH3OH	9P12	DFSY81B		
781.000	CH3OH	9R26	DFSY81B		pp17
782.000	TRIOX	10R22	DWC81		
782.700	CH3CF3	CO2S	FR87		
783.000	CH2CHF	10P32	TW82		
783.200	CH3CF3	N2O	FR87		
783.700	CH2CHBR	N2O	GRBF84		
784.268	CH2CHBR	10P24	DESF76	382257.6	
784.400	CH3OD	10P46	FK86		
784.400	13CH3OH	9P36	PS87		
784.500	CH3CF3	10R30	FR87		
784.600	CH2CHBR	N2O	GRBF84		
785.000	CH2CHCL	13CO2	TJD86		
786.162	HCOOH	9R40	DDSB77B	381336.9	
786.942	HCOOH	9R32	DSF76	380958.8	
787.500	CH2CHCN	10P06	GRBKF85		
788.000	DFCO	C18O2	JDL83		
788.330	CH2CHCN	10P08	GRBKF85		
788.482	CD3I	10P12	DFBS75	380214.9	
788.919	H13COOH	9P12	DDSB77B	380004.0	
789.000	TRIOX	9R14	DWC81		
789.420	DCOOD	10R20	DSF76	379762.8	
789.840	HCOOH	9R36	DSF76	379561.2	
790.800	CH3CHF2	10P02	FBRG85		
791.060	H2O	ELEC	MF66		
792.000	CD3CL	9P28	DH78A		26
793.000	CH2CHCN	10R40	Radfo75		
793.200	CH2CF2	N2O	FGRD84		
795.000	NH2NH2	10P32	DSF74C		
795.000	DCOOD	10P04	DSF76		
796.300	CH2CHF	N2O	GRBF84		
796.500	CH2CHF	CO2H	RGF84		
796.700	CH2CHBR	CO2H	BGRF84		
797.500	CH3OD	9R04	FK86		

λ [μm]	Molecule	Pump	Reference	Freq. [MHz]	Footnotes
798.550	CH3CCH	*10P20	CM71		
799.170	COF2	N2O	GRF88		
802.400	NH2NH2	10R24	DSF74C		
802.500	CH2CHBR	N2O	GRBF84		
805.800	CH3CF3	10P06	FR87		
806.000	13CD3I	10P12	DH78A		
807.500	CH2CHBR	10P46	BGRF84		
809.000	CH3NO2	10P40	DFSY81A		
812.000	DCOOD	10P04	DSF76		
812.600	CD3OH	9P06	PFS86		
813.000	TRIOX	10P32	DWC81		
813.654	TRIOX	9P32	DWC81	368452.2	
813.757	HCOOD	9P12	DSF76	368405.3	
815.123	CH2CHCL	9P24	DSFE74	367788.0	
816.100	CD2F2	10R16	TSD82		
816.195	CH2CHCN	10P16	DSFE74	367305.1	
817.300	CH2CF2	CO2H	FGRD84		
817.500	COF2	N2O	GRF88		
819.000	HCOOD	10P36	DSF76		
820.000	CH3OD	9R04	FK86		
820.000	13CH3I	10P48	GRF87		
821.400	CD3OD	9P30	FK86		
822.300	CH3CHF2	10P56	FBRG85		
822.300	CH3CHF2	CO2H	FBRG85		
823.000	CH3NC	10P30	GB81		
823.500	CF3BR	9R08	PLM79		
825.000	CH3OD	9R08	FK86		
826.000	HCOOD	9P12	DSF76		
826.944	CH2CHBR	10P22	DESF76	362530.4	
828.000	CH2CHCN	10R18	Radfo75		
828.000	CH2CHCL	9P24	Radfo75		
829.000	CD2CL2	10P06	ZD78		
829.540	CH2CHCN	N2O	GRBKF85		
830.450	CH2CHCN	10R26	GRBKF85		
831.267	CH3BR	10P28	DF78	360645.1	107
832.700	CH2CHBR	10R42	BGRF84		
832.757	CHCL2F	9R04	VWPE83	359999.9	
832.770	CH2CHCN	N2O	GRBKF85		
833.200	CH2CHCL	N2O	GRBF84		
833.300	CH3CF3	10R18	FR87		
835.000	DCOOD	10R20	DSF76		

λ [μm]	Molecule	Pump	Reference	Freq. [MHz]	Footnotes
836.800	CH2CHBR	N2O	GRBF84		
837.270	COF2	10R14	GRF88		
837.730	CH2CHCN	N2O	GRBKF85		
838.200	CH2CHF	10P10	RGF84		
838.300	CH3CH2BR	10R20	BS83		
838.369	H13COOH	9P12	DG82	357589.7	
839.400	COF2	10R16	GRF88		
841.000	CH3NO2	9R10	DFSY81A		
842.125	CHCLF2	9R16	DFP78	355995.2	
842.623	CH2CF2	10P30	DSFE74	355784.7	86
842.900	CD3I	C18O2	MPD83		
843.237	DCOOD	9P12	DSF76	355525.8	
845.000	CH3NO2	9R42	DFSY81A		
847.700	CH2CHF	10P18	RGF84		
848.000	13CH3I	CO2H	GRF87		
848.600	CH2CHBR	10P02	BGRF84		
851.000	CH3CF3	10R16	FR87		
851.324	CD3BR	10R18	DF78	352148.4	
851.900	CH3CH2F	9P30	WZN73		116
852.500	CH3CF3	10P10	FR87		
853.300	CH3CF3	10R20	FR87		
853.438	CH2CHBR	10P10	DESF76	351276.2	
854.585	CH3CN	9P16	DF78	350804.8	107
854.700	CD3OH	10R18	KK83		mm14
855.900	CD3OD	10R18	FK86		mm14
858.254	CD3OH	10R18	SEJZS87	349305.1	
858.730	CF2CL2	10P34	LPMD81		
859.500	CH2CHBR	CO2H	BGRF84		
861.100	CD3OH	10R12	CIMPS87		
862.000	CD3OH	10R18	PFS86		
862.544	CHD2F	10R20	Tobin84	347568.0	
863.000	NH2NH2	13C18O2	JTT82		
863.100	CH2CHCL	10R16	FBM84		
863.600	CH2CHBR	10P46	BGRF84		
865.500	CH3CHF2	N2O	FBRG85		
866.400	CH2CF2	N2O	FGRD84		
866.400	CD2CL2	10P08	BF86		
867.200	CH2CF2	10P50	FGRD84		
867.270	COF2	N2O	GRF88		
869.000	CH3NO2	10P36	DFSY81A		
870.800	CH3CL	*9P52	CM76		

λ [μm]	Molecule	Pump	Reference	Freq. [MHz]	Footnotes
871.360	CH2CHCN	N2O	GRBKF85		
871.585	CD3OH	9R14	IMEJ86	343962.4	
872.270	CH2CHCN	10P40	GRBKF85		
875.000	CH2CHF	13CO2	TJD86		
875.000	CH2CHCN	10P12	GRBKF85		
876.800	CH2CHCL	N2O	GRBF84		
877.200	CH3CHF2	CO2H	FBRG85		
877.300	CH2CHBR	N2O	GRBF84		
877.548	DCOOD	10P26	DSF76	341625.1	
878.100	CH3CF3	10R08	FR87		
878.500	CH3CF3	N2O	FR87		
880.410	CH2CHCN	10P10	GRBKF85		
881.300	CH3OH	10R10	PS87		
883.000	CF3BR	9R10	PLM79		
883.598	CD3CL	9P34	DFBS75	339286.0	26
884.000	CH2CF2	10P12	HRB73		
886.300	CH2CHBR	10P42	BGRF84		
888.000	CD2CL2	N2O	BF86		
888.862	CH2CF2	10P22	DSFE74	337276.7	
889.000	CH3CHDOH	9R12	BS85		
889.086	CH2CF2	10P22	DSFE74	337191.9	
889.466	TRIOX	9R20	DWC81	337047.8	
889.716	TRIOX	9R26	DWC81	336952.9	
890.200	CH2CF2	CO2H	SWTRD80		
891.087	H13COOH	10R32	DG82	336434.4	
892.000	CH3CHF2	CO2S	FBRG85		
894.000	CH2CHF	CO2H	RGF84		
894.021	13CH2F2	9R16	STPVE85	335330.3	
895.000	CD3I	10P30	DFBS75		
896.500	CH3CH2BR	10P10	BS83		
897.521	13CH2F2	9P14	STPVE85	334022.9	
899.000	CH2CHCN	CO2H	GRBKF85		
899.384	CHCLF2	9R38	DFP78	333330.9	
900.134	CH2CHBR	10R18	DESF76	333053.2	
901.191	CH2CHCN	10R14	DSFE74	332662.6	
901.300	13CD3I	13CO2	MPD83		
902.500	CH2CHCL	10P36	FBM84		
902.789	13CH2F2	9R30	STPVE85	332073.8	
903.000	CH2CHCN	10P10	GRBKF85		
905.428	CHCL2F	9R04	VWPE83	331105.8	
906.000	DFCO	13C18O2	JDL83		

λ [μm]	Molecule	Pump	Reference	Freq. [MHz]	Footnotes
906.500	CH2CHBR	N2O	GRBF84		
910.000	CH2CHCN	10R12	Radfo75		
910.500	CD3OD	9P10	FK86		
912.500	CH2CHBR	10P38	BGRF84		
912.500	CH2CHBR	N2O	GRBF84		
914.721	CLO2	9R14	DSF75B	327742.0	
914.735	CLO2	9R14	DSF75B	327736.8	
914.755	CLO2	9R14	DSF75B	327729.7	
914.780	CLO2	9R14	DSF75B	327720.8	
917.000	CH3OD	13CO2	DVPA81		
917.881	CH2CF2	10P22	DSFE74	326613.8	
918.000	CD2CL2	10P16	BF86		
918.148	CH2CF2	10P22	DSFE74	326518.7	
918.610	CD3I	9R28	DFBS75	326354.4	
919.936	HCOOD	10R32	DSF76	325884.2	
921.500	CH2CHCL	N2O	GRBF84		
923.000	CD2CL2	10P16	SDFY86		
925.520	CH3BR	*10R46	CM76		
926.209	HCOOD	10R14	DSF76	323677.0	
926.660	CD3I	N2O	GRF87		
927.000	CH2CHCN	10P42	GRBKF85		
927.981	DCOOD	10P20	DSF76	323058.7	
929.000	CH2CHCN	CO2H	GRBKF85		
929.800	13CD3I	13CO2	MPD83		
930.000	HCOOH	10R14	Knigh81		
930.000	HCOOH	10R32	Knigh81		146
934.000	CH2CHCL	N2O	GRBF84		
934.200	CH3OCH3	10P20	BSKK82		
934.223	CH2CHBR	9P28	DESF76	320900.3	
935.000	CH2CHCL	10P46	Radfo75		
935.009	DCOOD	9P16	DSF76	320630.4	
935.604	13CH2F2	9R10	STPVE85	320426.8	
936.000	13CH2F2	9R04	STPVE85		
936.159	CH2CHBR	10R32	DESF76	320236.7	
936.602	DCOOD	10P26	DSF76	320085.1	
937.900	CH3CHF2	N2O	FBRG85		
938.000	CH3NC	10P32	GB81		
939.000	CH2CHF	N2O	GRBF84		
939.500	CH2CHCN	N2O	GRBKF85		159
939.500	CH2CHCN	N2O	GRBKF85		159
939.500	CH3CHF2	10R08	FBRG85		

λ [μm]	Molecule	Pump	Reference	Freq. [MHz]	Footnotes
940.000	CH2CHCN	10P28	Radfo75		
941.486	CD3BR	10R18	DF78	318424.6	
942.000	CH2CHCL	13CO2	TJD86		
944.019	CH3CL	9R12	GCK86	317570.3	22, 107
944.028	CH3CL	9R12	GCK86	317567.5	22
945.000	CH2CHCL	N2O	GRBF84		
945.000	NH2NH2	13CO2	JTT82		
948.250	TRIOX	10R06	DWC81	316153.4	
948.925	TRIOX	9R24	DDSB77A	315928.6	
949.000	CH2CHF	N2O	GRBF84		
949.685	CLO2	10P16	DSF75B	315675.7	
950.600	CH2CF2	13CO2	MPD83		
952.000	TRIOX	10P08	DWC81		
952.200	CH2CHF	10P54	RGF84		
953.880	CD3I	9R28	DFBS75	314287.4	
954.467	CH2CF2	10P30	DSFE74	314094.2	
957.000	CH2CHCN	10P42	GRBKF85		
958.250	CH3CL	*9P38	CM76		
959.000	CH2CHCN	10P28	GRBKF85		
960.200	CD3I	C18O2	MPD83		
961.000	CH2CHF	N2O	TJD86		
963.487	CH2CHBR	10P10	DESF76	311153.5	
964.000	CH3I	10P22	DFBS75		
967.500	CH2CHBR	CO2S	BGRF84		
967.800	CH2CHBR	N2O	GRBF84		
967.900	CH3CF3	10R32	FR87		
968.000	CD3OH	9R20	DSF75A		
968.900	CH3CF3	CO2S	FR87		
969.000	CH3CF3	N2O	FR87		
971.000	CH2CHF	13CO2	TJD86		
971.806	DCOOH	10R28	DSF76	308489.9	
972.000	NH3(D)	10P40	GDEFJ83		
972.000	CD3OD	10R24	PVSEP85		
973.000	CH3NO2	9P08	DFSY81A		
973.000	CH3OH	9P26	DFSY81B		
973.000	CH3OH	13C18O2	PD84		
973.000	CH3CF3	N2O	FR87		
976.800	CH2CHCL	CO2H	FBM84		
980.000	CF2CL2	10P36	LPMD81		
980.100	CD2F2	10R14	TSD82		
981.100	CH2CHF	N2O	GRBF84		

λ [μm]	Molecule	Pump	Reference	Freq. [MHz]	Footnotes
981.709	CD3I	10P22	DFBS75	305378.0	
984.795	CHD2F	10P46	Tobin84	304421.1	
985.859	CH2CHBR	10R02	DESF76	304092.7	
986.070	CHD2F	10R20	Tobin84	304027.5	
986.313	HCOOD	10R32	DSF76	303952.8	
986.349	CH2CHCN	10P28	DSFE74	303941.5	
988.000	FCN	9R28	TJD86		158
988.100	CD3OH	10R18	PFS86		
988.259	CH2CHCL	10P24	DSFE74	303354.0	
988.695	CH2CHCL	10R28	DSFE74	303220.3	
988.900	CD2CL2	10R14	BF86		
988.900	CD3OD	9P40	FK86		
989.190	CH2CHBR	10P16	DESF76	303068.5	
990.000	CH2CF2	10P22	DSF72		
990.500	CH2CHCN	10P28	GRBKF85		
990.569	CH3BR	10P10	DF78	302646.6	107
990.630	CH2CHBR	10R04	DESF76	302628.0	
992.000	CH3F	CO2S	DW78D		
994.900	CH2CHBR	10R06	BGRF84		
995.000	CH2CHCL	10R26	Radfo75		
998.514	DCOOD	9P12	DSF76	300238.6	
1001.000	CH3NO2	9R26	DFSY81A		
1005.000	CH2CHCL	9P44	FBM84		
1005.000	DFCO	C18O2	JDL83		
1005.348	CD3I	10P34	DFBS75	298197.8	
1005.800	CH3CHF2	N2O	FBRG85		
1006.000	HCCCH2F	9P32	TJD86		
1007.000	NH2NH2	10P22	DSF74C		
1008.558	CDF3	10R12	TLD83	297248.6	
1009.409	DCOOD	10R18	DSF76	296997.9	
1010.000	CD3OD	9P30	FK86		
1010.000	CH3CD2OH	10R20	BS85		
1011.000	CH2CHF	10P38	RGF84		
1013.000	CH3CH2F	9P28	Radfo75		
1014.000	SIH3F	10R30	DS82		
1014.000	CH2CLF	9P08	ADF84		
1014.890	CH3CN	*9R14	CM76		
1016.009	CH2CHCN	10P32	DSFE74	295068.8	
1016.330	CH3CN	*9P08	CM76		
1016.700	CH2CHBR	CO2S	BGRF84		
1017.800	CH2CHCN	10R24	GRBKF85		

λ [μm]	Molecule	Pump	Reference	Freq. [MHz]	Footnotes
1018.258	CH2CF2	10P14	DSFE74	294416.9	
1018.300	CH2CHBR	10P36	BGRF84		
1018.400	CH2CHBR	N2O	GRBF84		
1022.000	CD3OD	10P46	FK86		
1025.000	CF2CL2	10P34	LPMD81		
1026.680	CH2CHCL	10R38	DSFE74	292001.8	
1026.709	CH2CHCL	10R28	DSFE74	291993.6	
1027.400	CD2CL2	10P08	BF86		
1028.000	HCCF	9R18	DJ80		
1028.300	CH2CHBR	N2O	GRBF84		
1030.378	H13COOH	9R30	DDSB77B	290953.8	
1032.000	CD2CL2	10P08	SDFY86		
1035.600	CH3CF3	N2O	FR87		
1039.855	CH2CHCL	9P24	DSFE74	288302.1	
1041.000	CH2CHCL	10R36	Radfo75		
1042.200	CH3CHF2	N2O	FBRG85		
1043.000	CF3BR	9R12	LPM81		
1044.400	CH3CH2I	10P20	BS83		
1044.800	CH3CHF2	10R06	FBRG85		
1047.000	CD3I	10P34	GRF87		
1047.200	CH2CHCL	N2O	GRBF84		
1047.579	DCOOH	10R12	DSF76	286176.6	
1047.600	CH2CHBR	10P36	BGRF84		
1049.810	CH3CH2I	10R04	TD86	285568.4	
1053.000	SIH2F2	13CO2	DFS85		mm10
1053.477	TRIOX	10R22	DWC81	284574.4	
1055.000	TRIOX	10P18	DWC81		
1056.000	SIH3F	13CO2	DFS85		mm10
1056.852	CH3BR	10P18	DF78	283665.4	
1058.000	SIH3F	13CO2	DFS85		
1059.000	CH3CH2BR	10P08	BS83		
1062.000	CH2CF2	CO2H	SWTRD80		
1063.000	CH2CHF	10P32	RGF84		
1063.290	CH3I	*10P38	CM76		
1065.000	CH3CHF2	CO2S	FBRG85		
1067.000	CH3I	10P38	DFBS75		
1069.000	CH3CH2F	9R10	Radfo75		
1070.000	CH3NO2	9R34	DFSY81A		
1070.231	DCOOD	9P12	DSF76	280119.5	
1071.300	CH2CHCL	N2O	GRBF84		
1075.200	CH2CF2	10P22	HW84		

λ [μm]	Molecule	Pump	Reference	Freq. [MHz]	Footnotes
1079.300	CH2CHBR	10P36	BGRF84		
1079.380	COF2	10P10	GRF88		
1080.537	CDF3	10R12	TLD83	277447.6	
1080.600	CH2CHBR	10P48	BGRF84		
1082.000	13CH2F2	9R10	STPVE85		
1082.600	CH2CHF	10P48	RGF84		
1083.000	CF3BR	9R28	LPM81		
1083.100	CH2CHF	10P32	RGF84		
1086.890	CH3CN	*9P40	CM76		
1091.637	CH2F2	C18O2	GCK86	274626.6	
1092.800	CD3OH	9P12	PFS86		
1093.100	CH2CHCL	N2O	GRBF84		
1094.000	CH2CF2	N2O	FGRD84		
1094.000	CH2CHF	10P24	RGF84		
1097.110	CH3CCH	*9P08	CM76		
1099.544	CD3I	10P22	DFBS75	272651.6	
1100.000	CH2CF2	N2O	FGRD84		
1100.000	CD3OH	9P12	DSF75A		
1116.483	H13COOH	9R30	DG82	268514.8	
1127.752	CH2CHCN	10P26	DSFE74	265831.9	
1133.800	CH3CHF2	10P20	FBRG85		
1134.000	CH3OD	9R04	FK86		
1134.113	CLO2	10P16	DSF75B	264340.9	
1135.070	COF2	10R14	TD86	264118.1	27
1137.500	CD3I	10P04	GRF87		
1146.000	CD3OH	9P24	DSF75A		
1151.000	CF3BR	9R34	LPM81		
1155.500	CD3OH	10R38	CIMPS87		
1156.000	CH2CHCN	10P26	Radfo75		
1157.318	HCOOD	10R38	DSF76	259040.7	
1160.000	DCOOD	10P08	DSF76		
1161.676	HCOOD	10R20	DSF76	258068.8	
1162.200	CD3OD	10R14	PVSEP85		
1164.000	CF2CL2	10P36	LPMD81		
1164.800	CH2CHCL	10P36	FBM84		
1164.830	CH3CN	*9P10	CM76		
1165.700	CH2CHF	10P46	RGF84		
1167.600	CH2CHF	10P28	RGF84		
1169.300	CD2CL2	10R12	BF86		
1170.700	CH2CHF	N2O	GRBF84		
1170.700	CH2CHF	10P32	RGF84		

λ [μm]	Molecule	Pump	Reference	Freq. [MHz]	Footnotes
1173.700	CH2CHCN	10P26	GRBKF85		
1173.900	CH2CHBR	CO2S	BGRF84		
1174.870	CH3CCH	*10P44	CM71		
1182.200	13CD3I	13CO2	MPD83		
1184.000	CH2CHCN	10R38	Radfo75		
1184.380	COF2	10P10	GRF88		
1185.079	TRIOX	10R22	DWC81	252972.6	
1188.000	CH3I	N2O	GRF87		
1191.563	COF2	10P32	TD86	251596.0	
1194.300	CH2CHCL	N2O	GRBF84		
1197.100	CH2CHCN	10R04	GRBKF85		
1198.600	CH2CHBR	N2O	GRBF84		
1201.400	CH3CHF2	10P18	FBRG85		
1202.200	CH2CHCN	9R34	GRBKF85		
1204.300	CH2CHBR	N2O	GRBF84		
1205.000	CF2CL2	10P36	LPMD81		
1210.700	CH3CHF2	10P32	FBRG85		
1213.362	HCOOH	9P28	DSF76	247075.8	
1217.000	CH3OH	*9P16	TTMYY74		
1218.600	CH2CHCN	CO2H	GRBKF85		
1218.600	CH2CHCN	10R38	GRBKF85		
1221.893	13CH3F	9P32	TLD83	245350.7	110, 23
1223.660	CH3OH	9P16	PEJS80	244996.6	98
1234.300	CH2CHF	N2O	GRBF84		
1237.100	CH3CHF2	10P32	FBRG85		
1237.966	DCOOH	10R24	DSF76	242165.4	
1239.480	CD3CL	9P12	DFBS75	241869.6	26
1245.710	13CH3I	CO2H	GRF87		
1247.594	CH2CHBR	10R12	DESF76	240296.5	
1250.000	CH2CHBR	10P06	BGRF84		
1253.738	CH3I	10P32	DFBS75	239118.9	65, 107
1253.815	CH3I	10P32	DBMF78	239104.2	
1253.822	CH3I	10P32	DBMF78	239102.8	
1253.859	CH3I	10P32	DBMF78	239095.8	
1255.700	CH2CHF	10P34	RGF84		
1257.100	CH2CHCN	10R02	GRBKF85		
1260.561	CDF3	10R16	TLD83	237824.7	
1264.300	CH2CHF	N2O	GRBF84		
1264.300	CH3CF3	N2O	FR87		
1267.100	CH2CHCN	10P12	GRBKF85		
1278.600	CH2CHBR	N2O	GRBF84		

λ [μm]	Molecule	Pump	Reference	Freq. [MHz]	Footnotes
1281.649	DCOOD	9P38	DSF76	233911.6	
1286.000	SIH3F	10R32	DFS85		
1290.000	CD3OH	10R20	DSF75A		
1292.100	CH3CHF2	N2O	FBRG85		
1292.200	CH2CHCL	N2O	GRBF84		
1292.400	CH2CHBR	10P02	BGRF84		
1292.743	TRIOX	10R10	DWC81	231904.1	
1293.000	CH2CHBR	10P38	BGRF84		
1296.400	CH2CHBR	10P12	BGRF84		
1306.000	CH3CH2CL	10R28	DK82		
1310.569	CH3BR	10R04	IMEJ86	228749.8	149, 107
1310.748	CLO2	10P14	DSF75B	228718.6	
1315.000	CH2CHCN	10P34	GRBKF85		
1320.800	CD2CL2	10P08	BF86		
1322.100	CH2CHBR	10P36	BGRF84		
1324.300	CH3CF3	N2O	FR87		
1325.000	CH3OD	9P26	FK86		
1325.000	CH2CHBR	N2O	GRBF84		
1340.000	CH3I	N2O	GRF87		
1345.000	CD2CL2	10P06	SDFY86		
1351.780	CH3CN	*9R20	CM76		
1372.000	CH2CF2	N2O	FGRD84		
1374.200	CH2CHCN	10P34	GRBKF85		
1377.000	CDF3	10R22	TSD82		
1383.882	CH2CHBR	10P24	DESF76	216631.5	
1387.500	CD2CL2	10R14	BF86		
1388.300	CH2CHCN	10P06	GRBKF85		
1394.063	CH2CHBR	10R20	DESF76	215049.5	
1405.000	CH3CHF2	N2O	FBRG85		
1406.000	CH2CHCL	10R12	FBM84		
1427.500	CH2CHCN	10P10	GRBKF85		
1428.600	CH2CF2	10P16	HW84		
1432.500	CH2CHCN	10P50	GRBKF85		
1434.000	CH2CF2	N2O	FGRD84		
1440.000	CH3CH2F	9P08	Knigh81		
1448.096	CH2F2	9R44	PSE80	207025.3	
1450.000	CD3F	9R36	TSW79A		
1480.000	CD3I	N2O	GRF87		
1485.000	CD3F	10P46	TSW79A		
1491.846	H13COOH	9P14	DG82	200953.9	
1492.500	CH2CHBR	CO2H	BGRF84		

λ [μm]	Molecule	Pump	Reference	Freq. [MHz]	Footnotes
1503.400	CH3CHF2	N2O	FBRG85		
1504.000	CH2CHBR	N2O	GRBF84		
1521.376	CH3CH2F	9P10	IMEJ86	197053.5	
1523.000	CH2CHF	CO2H	RGF84		
1526.000	CF3BR	9R16	LPM81		
1541.750	HCOOD	9P30	DSF76	194449.4	
1543.000	CD3I	10R18	GRF87		
1546.000	CH3CH2F	9P10	Radfo75		
1547.000	HCCCH2F	9P18	TJD86		
1549.505	CD3I	9R10	DFBS75	193476.3	
1555.000	CH2CHCN	10R28	GRBKF85		
1556.000	CF3BR	9R38	LPM81		
1570.200	SO2(I)	9R16	BS84		167
1572.640	CH3BR	*10P04	CM76		
1577.000	CH2CHCN	10R30	GRBKF85		
1579.000	CH2CHCN	CO2H	GRBKF85		
1579.903	TRIOX	10R12	DWC81	189753.7	
1581.705	TRIOX	9P26	DWC81	189537.5	
1600.000	CH2CF2	N2O	FGRD84		
1612.000	CH2CHF	13CO2	TJD86		
1613.000	CH3CF3	10R14	FR87		
1614.888	CH2CHBR	10P26	DESF76	185642.9	
1624.000	13CH3I	N2O	GRF87		
1650.312	COF2	10P06	TD86	181658.1	
1669.000	CH3CH2CL	10R26	DK82		
1671.000	CH3OD	9P26	FK86		
1676.000	CD3OH	10R36	PFS86		
1687.000	CF3BR	9R40	LPM81		
1714.130	CD2F2	9P30	VPE81	174894.8	
1730.833	HCOOD	10R24	DSF76	173207.0	
1733.000	CH2CHF	13CO2	TJD86		
1814.370	CH3CN	*10P46	CM71		
1827.424	CLO2	10P20	DSF75B	164051.9	
1880.000	CH2CHCL	CO2H	FBM84		
1886.695	CH3CL	9P26	GCK86	158898.2	107
1891.062	COF2	10R08	TD86	158531.3	
1895.000	CF3BR	9R20	LPM81		
1899.889	CH2CHBR	10P20	DESF76	157794.7	
1900.000	COF2	10R08	Tobin82		
1930.000	CD3OH	10R36	PFS86		
1941.700	CH2CF2	10P14	HW84		

λ [μm]	Molecule	Pump	Reference	Freq. [MHz]	Footnotes
1965.652	CH3BR	10P28	DF78	152515.5	107
1990.757	CD3CL	9P14	DFBS75	150592.2	26
2031.281	TRIOX	10R30	DWC81	147587.9	
2042.500	CH3CHF2	10P22	FBRG85		
2070.188	CH2CF2	10P14	DSFE74	144814.1	
2085.000	CD3I	N2O	GRF87		
2140.000	CF3BR	9R34	LPM81		
2140.000	CH3NC	10P14	GB81		
2206.000	CH2CHCL	9P32	FBM84		
2216.000	DFCO	C18O2	JDL83		
2347.500	CH2CHBR	CO2H	BGRF84		
2356.400	CH2CHBR	CO2S	BGRF84		
2388.800	CH3CHF2	10P22	FBRG85		
2453.000	CH2CHF	CO2S	RGF84		
2525.000	CH2CHF	10P42	RGF84		
2650.000	CH3BR	10P10	MD78		
2923.000	CD3OD	10R24	PVSEP85		
3030.000	CD3OH	10R36	PFS86		

Footnotes

In the following, only when it is stated that the value of a wavelength or frequency *includes* other determinations does the value differ from that given by the reference in the table

1 earlier frequency determinations were made by PEJWG75 and RPJM77; BCKEP80 obtained the same value as that listed
2 an earlier frequency determination was made by RPJM77; BCKEP80 obtained a slightly higher value than that listed
3 other frequency determinations were made by DSF76 and DDSB77B
4 this value includes determinations made by DSF76, KW76, DDSB77B, RPJM77 and DWDB79
5 line was reported by CBB70A; isotopic identification is from FWSGW82
6 this value includes determinations made by DSF76, RPJM77, DDSB77B and DWDB79
7 other frequency determinations were made by DSF76, RPJM77, DDSB77B and DWDB79
8 also measured by BDMF77
9 remeasured by DDSB77B and DWDB79
10 other frequency determinations were made by DFBS75 and RPJM77

11 other frequency determinations were made by DSF76, RPJM77 and DWDB79

12 an earlier determination was made in DSF76

13 also measured by DDSB77B

14 an earlier wavelength measurement by DJ80 was in error

15 also measured by KW76

16 also measured by DSF76, RPJM77

17 this value includes determinations made by DDSB77B and DWDB79

18 an earlier determination was made by DSF76, a measurement by DWDB79 is in disagreement

19 see also PH84

20 the wavelength value includes a determination made by BS81

21 previously noted by SE79 and attributed to a conference paper

22 an earlier wavelength determination was made by JEJ75

23 this was attributed to $^{13}CH_2F_2$ by STPVE85, whose sample evidently contained a small amount of $^{13}CH_3F$

24 assigned to $9R16$ by DF78, corrected by GDB84

25 these lines were assigned to next highest pump transition by DSFE74, corrected by DH78B and Duxbu84B

26 the Cl isotopic number is given by DH78A

27 this value icludes a determination made by DPK86

28 an earlier, slightly lower, frequency determination was made by DSFE74

29 the wavelength value includes a determination made by HW84

30 this value includes a determination made by RPJM77

31 also reported in DF78 (a conference paper)

32 lines at 141 and 193 μm were reported by DSV68, Hard69 reported lines at 140.85 and 192.67 μm

33 the frequency was remeasured by HJ68A and DHJRS69

34 DSF76 reported this line in HCOOD, presumably as a result of contamination by HCOOH

35 also lases in natural isotopic NH_3 (see DVPA81)

36 reported earlier as a weak line in CD_3F at 384.7 μm (TSW79B, Tobin p.c. 1987)

37 earlier wavelength determinations were made by MC64, PBT69 and/or DHJRS69

38 frequency had been measured earlier; see EWME70 and references in BBEK73

39 earlier wavelength determinations were made by Edwar78, DW78B, KK83 and CIMPS87

40 previously reported in a D, C, N mixture by MF67

41 The 249 μm line belongs to $9P22$ not $9P20$ as stated in reference (authors' private communication to D.J.E. Knight)

42 an earlier wavelength determination was made by DSF75A

43 reported earlier by FM66 or MF66; frequency measured by HJ67

44 assigned to $9R30$ by DF78

45 frequency measured earlier by DDSB77B but pump wrongly assigned (9R22)

46 reported earlier by DS82 but pump wrongly assigned (10R18)

47 the authors gave 291.27 μm as the wavelength, which is inconsistent with the frequency; the review DH78A repeats this error. note that c / 291.27 μm = 1 029 247.3 MHz, which is typographically similar to the given frequency

48 this is the only line not repeated by the authors in DSF76, it may therefore be in error

49 a somewhat different wavelength measurement is given without further reference by MLL88

50 an assignment is suggested by MLL88

51 improved wavelength or isotopic identification taken from WFGR84

52 wavelength taken from DSF74D (see note under DSF76); both DDSB77B and DSF74D give 9R12 as the pump for this l ine

53 the $^{13}CO_2$ pump line is here reported as 10R10, in DP84 it is reported as 10P10

54 an earlier wavelength determination was made by DTS74, see also for polarization

55 an earlier wavelength determination was made by DW78A

56 an earlier wavelength determination was made by WDVP80

57 also reported in NH3(D) by GDEFJ83

58 reported at 230.5 μm by DV88

59 these seem to be distinct transitions despite disagreements in pump assignments between DSF75A and the references thereto in PFS86, see table 1 in PFS86

60 these are distinct; see table 1 in SJFWM87

61 these are distinct; see table 1 in DWC81

62 a recent, possibly more accurate wavelength determination is given in CIMPS87

63 apparently unpublished; reported in CIMPS87

64 typographical error in pump; presumed to be correct as shown here

65 previously reported by DSF74B with incorrect pump assignment

66 a precise wavelength determination was also reported by VHHR86

67 an earlier wavelength determination was made by Lands80C

68 an earlier wavelength determination was made by KYHH75

69 an earlier wavelength determination was made by NH80

70 an earlier wavelength determination was made by DSF74A or DSF74C

71 an earlier wavelength determination was made by ZD78

72 reported earlier by DJ80. This wavelength determination was from a laser diode spectroscopic measurement

73 previously reported by DW78B

74 previously reported by KK83

75 previously reported by PFS86

76 these are distinct on the basis of polarization. A line at 265.1 μm (PSF86), with the same polarization as the line at 264.759 μm, has been eliminated

77 a line at 418.7 μm (PFS86) has been eliminated since SEJZS87 regard it as identical to this one, despite disagreement in the reported polarization

78 presumed identical with a line reported by CIMPS87 at 421.0 $\pm$ 0.1μm; see comment in note 79 below

79 presumed identical with a line reported by CIMPS87 at 254.25 $\pm$ 0.2μm; if this is correct then comparison with the frequency measurement shows the actual error to be 0.53 μm

80 a more recent measurement appears in SEJZS87

81 these pairs are distinct on the basis of polarization

82 previous wavelength measurement by CIMPS87

83 no polarization given; presumed identical with lines reported by IMSD85 (q.v. for polarizations)

84 a line at 71.0 μm (DW78B) has been eliminated since SEJZS87 regard it as identical to this one, despite disagreement in the reported polarization

85 according to DH78B 10P24 is correct here. However DW84 report 10P22 also

86 DH78B suggests that pump may actually be 10P28

87 previously reported by YKYSF81

88 previously reported by DW84

89 also reported in long-pulsed mode in dimethylamine ($(CH_3)_2NH$) by MCW68

90 previously reported in Lands80C (221.1 μm) and LD79 (225 μm)

91 previously reported in LD79

92 previously reported in BP77 and assumed by PD85 to be identical even though polarizations differ

93 previously reported in Lands80C

94 see also PD85

95 previously reported in DSF72

96 previously reported in HRB73

97 previously reported in Radfo75

98 appears to be identical with the quasi-cw line reported by TTMYY74 at 1217 μm

99 here reported long-pulsed, but reported cw by IMMS81A and/or JV87

100 previously observed by DW78C

101 here reported long-pulsed, but reported cw by DFBD75

102 previously reported by HP79

103 previously reported by KHYH75

104 these lines are ascribed to different transitions and are therefore distinct

105 previously observed by ZRGB80 in long-pulsed mode and cw by SRD86; both using sequence-band pumping

106 also reported in long-pulsed mode in a mixture of deuterated methane and deuterated ammonia by MCW68

107 previously observed in long-pulsed mode by CM76 (q.v. for polarizations)

108 here reported long-pulsed, but previously reported cw by Radfo75 at 392 μm

109 here reported long-pulsed, but previously reported cw by Radfo75

110 previously reported in long-pulsed mode by CM71

111 previously reported in long-pulsed mode by HHCS71 and/or HC69

112 also reported by YYSF82 but with conflicting polarizations in some cases. Contrary to my normal practice the entries of YYSF82 were deleted in such cases because a systematic effect seemed to be the cause of the disagreements

113 see also CIMPS87 for offset information, etc

114 previously reported in long-pulsed mode by HL82

115 here reported long-pulsed, but reported cw by SMNM83, DFSY81B and/or DFBS75

116 here reported long-pulsed, but reported cw by DDM83, DSF74D and/or Radfo75

117 these 9R22 lines are too far from the 418.613 and 432.109 μm lines respectively to be considered identical to them; the pump identification may be incorrect, as supported by the lines labelled pp10. See also the comments in the references under WZN73

118 here reported long-pulsed, but reported cw by KK82 (q.v. for discussion on assignment)

119 here reported long-pulsed, but reported cw by PS87, SBW84 and/or VHHR86

120 here reported long-pulsed, but reported cw by Lands80A and/or DSF76

121 previously reported in long-pulsed mode by WZN73

122 here reported long-pulsed, but reported cw by DG84A

123 see DG84A for polarization

124 assumed identical with a line reported in cw mode by RGF79, isotopic assignment from WFGR84

125 here reported pulsed, but reported cw by LJ69, also measured in long-pulsed mode by MCW67

126 reported as a personal communication; no experimental details given; assumed cw

127 previously reported by RRG84, however for consistency with oth er entries the wavelengths reported by SRD86 have been given

128 observed in a He-Ne mixture and attributed to Ne

129 this wavelength is from unpublished work by the same authors (see MCW68)

130 reported in long-pulsed mode by MC64

131 here reported long-pulsed, but reported cw by MF67

132 reported in various mixtures by MF67 and MCW68 and attributed to HCN by KS80

133 reported in dimethylamine ($(CH_3)_2NH$); probably due to HCN since the 336.558 μm HCN line was produced in the same discharge (cf note 89)

134 reported in a methane-ammonia mixture; probably due to HCN since four lines seen in the corresponding deuterated mixture have since been attributed to DCN (cf note 106)

135 these are calculated wavelengths since the measured values were not given in the reference

136 identification and calculated wavelength given in the reference as well as in Patel68

137 identified as a doublet; see reference or Patel68

138 see Patel68 for calculated wavelength and likely identification

139 note the misprint in reference; c.f. Patel68

140 these are calculated wavelengths since there is a plethora of measured values in the original references (FMPG64A and others mentioned in Patel68)

141 here reported long-pulsed but reported cw by PBT67

142 previously reported by FM66 or MF66

143 assigned to $HC^{14}N$ by Maki78

144 assigned to $HC^{15}N$ by Maki78

145 the reference given by Wille71, and by BEG78 in a later review, appears to be incorrect or ambiguous; reported cw

146 in this reference the line is misplaced in the wavelength-sorted data, suggesting a possible typographical error

147 the review article IMEJ86 gives a slightly different value for the frequency of this line

148 the review article IMEJ86 gives the value 584295.3 MHz for th e frequency of this line

149 assigned to ^{79}Br in reference

150 here reported long-pulsed, but previously reported cw by MH84B

151 previously reported by Henni86

153 here reported long-pulsed, but previously reported cw using a waveguide laser by HIMS81 or HIMS82

154 here reported long-pulsed, but evidently observed cw by BPT69, who measured a wavelength of 71.944 μm

155 presumed identical with lines measured by DFSY81B or PVSEP85, although polarizations were not given by IMEJ86

156 presumed identical with a line reported by DSF74C with the pump given as $9R16$? ; polarization perpendicular

157 presumed identical with a line reported by DSF74A with the pump given as $9R$? ; polarization parallel

158 attributed to $FC^{15}N$

159 these are distinct lines, being pumped by different transitions

160 the species responsible for the emission is uncertain, the system being N_2-CS_2; for further information see Polla71

161 these lines belong to miscellaneous vibrational transitions; several are accompanied by weak satellite lines under some experimental conditions. For information on the more familiar lines of the CO_2 laser, which are not listed here, see Chapter 5

162 assigned by the authors to $^{34}SO_2$

163 assigned by the authors to $^{34}SO^{18}O$

164 assigned by the authors to $^{36}SO_2$
165 assigned by the authors to $S^{18}O_2$
166 assigned by the authors to $^{34}S^{18}O_2$
167 assigned by the authors to $^{36}S^{18}O_2$

mm# these pairs of lines have wavelengths equal to within 0.3% and are ascribed to molecules which differ only by a single isotopic substitution, such as ^{13}C to C or D to H; they also have the same pump transition and polarization, when this was given. They may therefore be the same line

pp# these pairs of lines have wavelengths equal to within 0.1% and are ascribed to near-lying pump transitions; they also have the same polarizations when this was given. They may therefore be the same line

8. Table B. **Data Arranged by Molecule**

In this table the lines have been grouped into molecules and then listed first in order of pump type or transition and then in order of wavelength. The order in which the molecules are listed is "numero-alphabetic" (i.e. 13CD3F comes before CDF3) and is the same order as that in which the molecule codes are listed in Chapter 10. That section also contains an alphabetical listing of chemical names.

For convenience certain "families" of lines have been grouped together whilst preparing this table. This facilitates reviewing of the data in cases where the isotopic identification is uncertain or incomplete. These groups are:

(i) Ammonia: 14NH3, 15NH3, NH3 and deuterated forms 14ND3, 14NH2D, NH3(D);

(ii) Methylamine: CH3NH2, CH3NH2?.

Within the listing for each molecule the columns are:

1. Pump code – note however that in these tables no distinction is drawn between cw and long-pulsed lines, for which Table A should be consulted;
2. Wavelength in μm;
3. Reference;
4. Frequency in MHz, if available;
5. A footnote to the isotopic identification in the case of the "families" mentioned above.

Only columns 2–4 are therefore identical with the corresponding columns in Table A. In particular, Table A may contain additional information in the form of footnotes.

13CD3F

9P24	209.100	TF81	
9P36	183.400	TF81	
9P48	299.500	TF81	
9R22	325.900	TF81	
10P20	280.800	TF81	
10P34	470.065	TD86	637768.5
10P38	376.800	TF81	
10R34	249.900	TF81	
CO2H	336.500	TF81	
	537.410	TD86	557847.0

13CD3I

10P10	690.000	DH78A
10P12	806.000	DH78A
13CO2	554.700	MPD83
	745.500	MPD83
	901.300	MPD83
	929.800	MPD83
	1182.200	MPD83
C18O2	574.600	MPD83

13CD3OD

9P12	126.200	VE85	
	247.000	VE85	
9P24	151.000	VE85	
9P28	151.800	VE85	
	407.100	VE85	
9P32	243.000	VE85	
	358.400	VE85	
9P38	129.200	VE85	
9R10	75.500	VE85	
9R14	82.400	VE85	
	118.553	VE85	2528772.8
	353.300	VE85	
9R20	241.600	VE85	
	417.300	VE85	
9R22	175.100	VE85	
9R24	75.275	VE85	3982631.1
9R26	353.100	VE85	
9R28	227.000	VE85	
9R32	150.300	VE85	
9R34	82.100	VE85	
10P12	84.400	VE85	
10P24	124.253	VE85	2412757.9
	216.356	VE85	1385646.1
10R08	464.700	VE85	
10R12	209.233	VE85	1432817.8
	321.410	VE85	932741.3
10R14	324.140	VE85	924885.8
10R16	93.600	VE85	
	109.926	VE85	2727211.7
	109.938	VE85	2726923.5
	128.100	VE85	2340291.8
10R20	173.637	VE85	1726548.5
10R26	272.958	VE85	1098307.9
10R30	148.617	VE85	2017218.5

13CD3OH			
9P10	72.900	IEPSV84	
	126.100	IEPSV84	
9P34	399.800	IEPSV84	
9P38	156.000	IMSD85	
9R08	196.200	IEPSV84	
	221.000	IEPSV84	
	389.600	IEPSV84	
9R10	65.400	IEPSV84	
9R14	119.400	IEPSV84	
9R18	52.200	IEPSV84	
9R28	153.694	IEPSV84	1950581.6
	336.500	IEPSV84	
9R30	150.200	IEPSV84	
9R32	52.100	IEPSV84	
	55.800	IEPSV84	
9R34	98.500	IEPSV84	
9R38	151.000	IEPSV84	
	387.200	IEPSV84	
9R40	177.600	IEPSV84	
10P08	127.021	IEPSV84	2360174.8
	175.260	IMS89	
	462.800	IEPSV84	
10P16	333.261	IEPSV84	899571.7
	340.627	IEPSV84	880120.4
10P22	124.300	IEPSV84	
	291.000	IEPSV84	
10P42	119.100	IEPSV84	
	148.300	IEPSV84	
10R12	67.800	IEPSV84	
10R20	73.467	IEPSV84	4080637.2
10R22	84.406	IEPSV84	3551805.8
	110.000	IEPSV84	
	127.656	IEPSV84	2348438.4
10R24	145.563	IEPSV84	2059531.6
10R26	146.326	IEPSV84	2048803.6
	197.046	IEPSV84	1521430.9
	209.000	IEPSV84	
	468.965	IEPSV84	639264.6

13CH2F2			
9P04	254.802	STPVE85	1176570.0
	415.363	STPVE85	721759.8
	674.061	STPVE85	444755.4
9P08	531.363	STPVE85	564195.3
9P12	164.656	STPVE85	1820715.0
9P14	421.053	STPVE85	712005.8
	496.660	STPVE85	603617.2
	897.521	STPVE85	334022.9
9P16	200.295	STPVE85	1496757.0
	360.504	STPVE85	831592.7
	452.425	STPVE85	662635.0
9P20	197.388	STPVE85	1518795.0
9P22	193.497	STPVE85	1549340.0
	279.014	STPVE85	1074471.0
9P24	312.276	STPVE85	960024.5
9P26	273.764	STPVE85	1095077.0
	638.394	STPVE85	469604.1
9P28	183.289	STPVE85	1635628.0
	245.652	STPVE85	1220395.0
	740.000	STPVE85	
9P32	403.777	STPVE85	742470.4
9P34	180.600	STPVE85	
9P36	333.926	STPVE85	897781.9
	734.959	STPVE85	407903.8
9P38	186.043	STPVE85	1611414.0
9P44	112.000	STPVE85	
9R04	164.815	STPVE85	1818964.0
	936.000	STPVE85	
9R06	551.100	STPVE85	
9R08	311.213	STPVE85	963302.2
	618.896	STPVE85	484398.7
9R10	377.718	STPVE85	793693.1
	935.604	STPVE85	320426.8
	1082.000	STPVE85	
9R12	318.080	STPVE85	942507.4
	738.414	STPVE85	405995.1
9R14	106.400	STPVE85	
9R16	344.521	STPVE85	870171.9
	894.021	STPVE85	335330.3
9R18	234.800	STPVE85	
9R20	195.158	STPVE85	1536153.0
	357.867	STPVE85	837719.4
	570.332	STPVE85	525645.3
9R22	138.281	STPVE85	2168000.0
	359.362	STPVE85	834235.9
	689.178	STPVE85	435000.0
9R24	266.866	STPVE85	1123382.0
	316.329	STPVE85	947723.7
9R26	248.606	STPVE85	1205896.0
9R30	391.461	STPVE85	765829.0
	902.789	STPVE85	332073.8

13CH2F2

9R34	214.597	IMEJ86	1397005.0
	399.288	STPVE85	750817.6
9R36	300.233	STPVE85	998532.1
9R38	140.405	STPVE85	2135193.0
9R40	306.993	STPVE85	976543.7
	479.123	STPVE85	625711.5
9R44	135.523	STPVE85	2212111.0
	213.351	STPVE85	1405163.0
10R04	301.654	STPVE85	993829.9
10R18	206.043	STPVE85	1455000.0
	438.022	STPVE85	684423.3
10R20	182.381	STPVE85	1643769.0
	260.042	STPVE85	1152860.0
10R38	300.246	STPVE85	998487.9

13CH3BR

10R20	660.882	IMEJ86	453624.6

13CH3F

9P32	1221.893	TLD83	245350.7

13CH3I

10P26	558.820	GRF87	
	573.750	GRF87	
10P30	355.550	GRF87	
10P48	820.000	GRF87	
CO2H	346.670	GRF87	
	366.920	GRF87	
	372.800	GRF87	
	521.110	GRF87	
	848.000	GRF87	
	1245.710	GRF87	
CO2S	293.130	GRF87	
	395.000	GRF87	
	663.080	GRF87	
N2O	1624.000	GRF87	

13CH3OH

9P04	190.300	PS87	
9P06	113.400	PS87	
	205.000	IMSD85	
9P08	87.900	HP78	
9P10	86.112	HP78	3481433.0
	105.070	IMSD85	
	146.097	HP78	2052004.0
	208.412	HP78	1438460.0
	236.530	Knigh81	1267459.0
9P12	63.096	HP78	4751340.9
	157.929	HP78	1898280.0
	166.280	IMSD85	
	237.523	HP78	1262162.0
	238.523	HP78	1256872.0
	461.385	HP78	649767.0
	629.844	HP78	475979.0
9P20	81.010	IMSD85	
	122.402	CIMPS88	2449245.9
	203.960	IMSD85	
	306.500	PS87	
9P22	85.317	HP78	3513853.0
	103.481	HP78	2897083.0
	115.000	IMSD85	
	118.013	HP78	2540332.0
	149.272	HP78	2008360.0
	307.780	HP78	
	338.964	HP78	884438.0
9P24	103.000	IMSD85	
9P26	117.920	IMSD85	
	188.960	IMSD85	
	249.100	PS87	
	425.800	PS87	
9P28	98.000	PS87	
9P30	113.600	IMSD85	
	147.970	HP78	
	496.400	PS87	
9P32	71.700	IMSD85	
	95.240	IMSD85	
	281.000	PS87	
9P36	291.620	HP78	
	325.170	HP78	
	784.400	PS87	
9P38	328.900	PS87	
9P40	168.840	HP78	
	358.920	HP78	
9P42	70.000	IMSD85	
9P44	89.000	IMSD85	
9R20	107.800	PS87	
	400.100	PS87	
9R32	216.500	PS87	
9R36	452.400	PS87	

9R40	101.300	PS87	
10P08	155.000	PS87	
10P10	247.400	PS87	
10P12	133.700	PS87	
10P16	41.900	HP78	
	123.260	HP78	
	181.200	PS87	
10P28	214.300	PS87	
	311.100	PS87	
10P30	140.900	PS87	
10P34	80.300	PS87	
10R02	122.000	IMSD85	
	222.800	PS87	
10R06	240.100	PS87	
10R14	60.000	IMSD85	
	269.900	PS87	
	319.700	PS87	
10R16	115.823	HP78	2588363.0
	148.590	HPPJE79	2017576.1
	152.076	HPPJE79	1971337.2
	203.636	HP78	1472199.0
	268.572	HP78	1116245.0
	280.218	HPPJE79	1069853.4
	280.240	HPPJE79	1069771.4
	332.603	HPPJE79	901351.2
	496.300	PS87	
10R18	105.147	HP78	2851169.0
	110.432	HP78	2714715.0
	171.758	HP78	1745439.0
	268.600	PS87	
10R20	253.500	IMSD85	
	282.960	IMSD85	
	294.040	IMSD85	
	320.400	PS87	
10R22	34.790	HP78	
10R26	77.489	HP78	3868819.0
	103.586	HP78	2894132.0
	275.610	IMSD85	
	307.070	IMSD85	
10R28	85.790	HP78	
	121.200	HP78	
10R30	122.885	CIMPS88	2439623.2
	198.790	IMSD85	
10R32	340.000	IMSD85	
10R36	339.900	PS87	
10R40	72.000	IMSD85	
	334.600	PS87	
10R46	110.900	PS87	

18O3

9P10	722.359	DDM83	415018.4

AR

ELEC	12.140	FMPG64
	12.147	Patel68
	15.037	Patel68
	15.042	Patel68
	26.944	Patel68

BCL3

ELEC	18.300	KKPPS68
	18.800	KKPPS68
	19.100	KKPPS68
	19.400	KKPPS68
	20.200	KKPPS68
	20.600	KKPPS68
	22.400	KKPPS68
	23.000	KKPPS68

C2H4O2H2		
10P30	118.000	PCD73
10R16	62.500	PCD73
	69.100	PCD73
	77.400	PCD73
9P12	288.000	PCD73
9P14	117.100	PCD73
	164.000	PCD73
	415.000	PCD73
9P16	164.000	PCD73
9P22	344.000	PCD73
9P24	132.000	PCD73
9P32	75.200	PCD73
	90.800	PCD73
9P34	70.100	PCD73
	118.900	PCD73
	125.800	PCD73
	185.000	PCD73
	189.000	PCD73
	252.000	PCD73
	262.000	PCD73
	299.000	PCD73
	358.000	PCD73
	696.000	PCD73
9P36	118.000	PCD73
	132.000	PCD73
	135.000	PCD73
	169.000	PCD73
	189.000	PCD73
	200.000	PCD73
	388.000	PCD73
9P38	192.000	PCD73
	197.000	PCD73
	277.000	PCD73
	290.000	PCD73
9R08	171.000	PCD73
9R10	95.800	PCD73
	164.000	PCD73
	231.000	PCD73
	240.000	PCD73
9R16	109.100	PCD73
9R18	185.000	PCD73
	250.000	PCD73

CD2CL2		
10P02	184.400	BF86
10P04	649.600	BF86
10P06	309.500	SDFY86
	715.800	BF86
	719.500	SDFY86
	829.000	ZD78
	1345.000	SDFY86
10P08	866.400	BF86
	1027.400	BF86
	1032.000	SDFY86
	1320.800	BF86
10P10	373.000	SDFY86
10P12	704.600	BF86
10P14	323.500	SDFY86
	497.300	BF86
	500.000	SDFY86
	720.200	BF86
10P16	343.300	VHHR86
	918.000	BF86
	923.000	SDFY86
10P18	191.500	SDFY86
10P20	265.600	BF86
	266.500	SDFY86
	713.200	BF86
	718.000	SDFY86
	723.100	BF86
	726.000	SDFY86
10P24	232.000	SDFY86
	558.000	SDFY86
10P26	237.500	SDFY86
10P28	493.000	SDFY86
10R02	241.100	BF86
	409.800	BF86
10R04	469.000	ZD78
10R06	438.400	BF86
	510.500	SDFY86
	694.100	BF86
10R08	289.400	BF86
10R10	361.200	BF86
	370.000	SDFY86
10R12	483.500	BF86
	520.000	ZD78
	743.700	BF86
	746.500	SDFY86
	1169.300	BF86
10R14	580.300	BF86
	988.900	BF86
	1387.500	BF86
10R16	171.700	VHHR86
	248.900	VHHR86
	358.100	BF86

CD2CL2				CD2F2			
	376.900	BF86		9P08	192.790	VPE81	1555020.1
10R18	631.000	ZD78			286.398	VPE81	1046768.2
10R20	178.000	SDFY86		9P10	233.685	VPE81	1282892.0
	581.500	SDFY86		9P18	267.823	VPE81	1119368.0
10R22	382.000	SDFY86		9P20	417.244	VPE81	718505.6
	562.000	SDFY86		9P28	203.300	VPE81	
10R24	690.000	SDFY86			219.600	VPE81	
10R26	723.000	SDFY86			465.500	VPE81	
10R28	497.000	SDFY86			497.677	VPE81	602383.9
10R36	254.000	ZD78		9P30	280.512	VPE81	1068733.7
C18O2	212.200	PD82B			593.279	VPE81	505314.1
	250.700	PD82B			1714.130	VPE81	174894.8
	281.800	PD82B		9P34	330.991	VPE81	905742.6
	313.900	PD82B		9P40	317.052	VPE81	945562.5
	378.600	PD82B			323.179	VPE81	927636.8
	384.100	PD82B		9P42	249.800	VPE81	
	434.400	PD82B			488.276	VPE81	613981.5
	450.200	PD82B		9P44	342.127	VPE81	876261.3
	473.200	PD82B		9R10	166.879	VPE81	1796461.7
	475.200	PD82B			314.646	VPE81	952793.3
	539.100	PD82B		9R34	236.108	VPE81	1269723.6
	774.100	PD82B			365.866	VPE81	819405.8
N2O	219.400	BF86		10P22	378.880	VPE81	791260.4
	267.900	BF86			643.516	VPE81	465866.4
	277.200	BF86		10P32	440.884	VPE81	679979.8
	279.800	BF86		10R08	393.000	VPE81	
	301.100	BF86		10R14	187.819	VPE81	1596174.9
	309.700	BF86			367.399	VPE81	815985.9
	326.500	BF86			980.100	TSD82	
	340.600	BF86		10R16	202.300	TSD82	
	355.900	BF86			248.108	VPE81	1208313.9
	360.000	BF86			446.100	TSD82	
	360.200	BF86			816.100	TSD82	
	369.400	BF86		10R18	150.438	VPE81	1992795.2
	385.400	BF86			318.600	VPE81	
	392.000	BF86			414.800	TSD82	
	398.900	BF86			491.800	Tobin80	
	400.200	BF86			718.900	TSD82	
	409.600	BF86		10R20	139.266	VPE81	2152662.4
	440.200	BF86			229.067	VPE81	1308755.5
	447.600	BF86			644.500	TSD82	
	471.500	BF86		10R22	249.392	VPE81	1202093.2
	484.300	BF86			582.800	TSD82	
	519.700	BF86		10R24	500.577	VPE81	598893.7
	585.100	BF86		10R26	274.776	VPE81	1091044.7
	588.600	BF86			489.238	VPE81	612774.8
	723.100	BF86		10R28	548.700	TSD82	
	769.500	BF86		10R34	189.832	VPE81	1579250.3
	774.600	BF86			352.902	VPE81	849506.4
	888.000	BF86		10R36	120.469	VPE81	2488553.4
					214.714	VPE81	1396238.8

CD2F2

10R38	207.835	VPE81	1442454.3
	218.267	VPE81	1373513.3
10R42	192.000	VPE81	
	456.200	VPE81	
10R44	320.597	VPE81	935107.5
10R48	303.800	VPE81	

CD3BR

9P14	366.625	DF78	817708.3
9P18	430.000	Lands80A	
	431.736	DF78	694388.4
9P26	692.025	DF78	433210.6
9P32	297.000	Lands80A	
	552.000	Lands80A	
	553.883	DF78	541256.2
9P36	341.000	Lands80A	
9R18	440.000	Lands80A	
	441.674	DF78	678764.4
9R26	466.643	DF78	642445.1
9R30	556.803	DF78	538417.8
10P18	290.000	Lands80A	
10R02	428.000	Lands80A	
10R10	530.132	DF78	565505.1
10R18	851.324	DF78	352148.4
	941.486	DF78	318424.6

CD3CL

9P06	698.555	DFBS75	429160.6
	735.130	DFBS75	407808.9
9P10	443.265	DFBS75	676328.5
9P12	1239.480	DFBS75	241869.6
9P14	1990.757	DFBS75	150592.2
9P16	288.000	DFBS75	
9P24	293.648	DFBS75	1020924.7
9P28	792.000	DH78A	
9P32	245.000	DFBS75	
9P34	883.598	DFBS75	339286.0
9P36	480.310	DFBS75	624164.3
	519.303	DFBS75	577297.5
9P38	249.000	DFBS75	
9R28	224.000	DFBS75	
9R34	383.285	DFBS75	782166.1
10R14	246.000	DFBS75	
10R18	288.000	DFBS75	
10R20	449.800	DFBS75	666502.0
	464.757	DFBS75	645052.4
10R28	318.000	DFBS75	

CD3CN

9P08	455.073	DF78	658778.6
9P30	516.253	DF78	580708.2
9R04	529.880	DF78	565774.2

CD3F

9P16	206.000	TSW79B	
9P28	200.000	TSW79B	
9P34	348.899	TD86	859252.2
9P52	265.000	TSW79A	
9R10	247.500	TSW79B	
9R36	1450.000	TSW79A	
10P08	323.300	TSW79B	
10P10	155.600	TSW79B	
10P12	201.500	TSW79B	
10P46	247.300	TSW79B	
	1485.000	TSW79A	
10P50	336.380	TD86	891231.4
10R06	172.800	TSW79B	
10R48	368.448	TD86	813662.6

CD3I

9P10	487.226	DFBS75	615304.6
9P12	272.000	DFBS75	
9P22	734.262	DFBS75	408290.6
9P26	390.000	DFBS75	
9P28	433.104	DFBS75	692195.5
9R06	540.000	DFBS75	
9R10	1549.505	DFBS75	193476.3
9R12	460.562	DFBS75	650927.5
9R20	691.119	DFBS75	433778.2
9R22	490.391	DFBS75	611333.6
9R26	301.000	DFBS75	
9R28	730.323	DFBS75	410492.7
	918.610	DFBS75	326354.4
	953.880	DFBS75	314287.4
9R32	444.386	DFBS75	674621.3
10P02	424.550	GRF87	
10P04	1137.500	GRF87	
10P08	745.000	DFBS75	
10P10	667.232	DFBS75	449307.5
10P12	788.482	DFBS75	380214.9
10P16	644.000	DFBS75	
10P22	981.709	DFBS75	305378.0
	1099.544	DFBS75	272651.6
10P30	895.000	DFBS75	
10P34	1005.348	DFBS75	298197.8
	1047.000	GRF87	
10P36	556.876	DFBS75	538347.3
	569.477	DFBS75	526434.4
10P38	523.406	DFBS75	572772.1
10P46	660.582	DFBS75	453830.6
10R02	658.570	GRF87	
	678.570	GRF87	
10R08	670.094	DFBS75	447388.6
	670.114	DFBS75	447375.1
10R18	640.000	DFBS75	
	1543.000	GRF87	
10R22	599.550	DFBS75	500029.2
	614.110	DFBS75	488174.0
13CO2	448.700	MPD83	
	523.900	MPD83	
	555.200	MPD83	
	556.200	MPD83	
	557.100	MPD83	
	567.700	MPD83	
	660.200	MPD83	
C18O2	298.000	MPD83	
	728.100	MPD83	
	842.900	MPD83	
	960.200	MPD83	
CO2H	346.670	GRF87	
	360.000	GRF87	

CD3I

CO2S	473.680	GRF87
	545.880	GRF87
N2O	331.790	GRF87
	452.380	GRF87
	465.500	GRF87
	466.000	GRF87
	476.000	GRF87
	483.160	GRF87
	533.330	GRF87
	545.560	GRF87
	602.500	GRF87
	628.000	GRF87
	629.000	GRF87
	632.000	GRF87
	658.570	GRF87
	680.000	GRF87
	926.660	GRF87
	1480.000	GRF87
	2085.000	GRF87

9P04	247.500	PVSEP85	
	450.700	FK86	
	631.800	FK86	
9P06	94.300	FK86	
	342.800	PVSEP85	
	558.500	PVSEP85	
9P08	256.400	PVSEP85	
9P10	184.200	PVSEP85	
	651.900	PVSEP85	
	910.500	FK86	
9P12	353.800	PVSEP85	
9P14	356.400	PVSEP85	
	410.200	FK86	
9P16	129.000	YKYSF81	
	210.500	PVSEP85	
	342.700	PVSEP85	
9P18	329.200	PVSEP85	
	453.600	PVSEP85	
9P20	373.400	PVSEP85	
9P22	663.000	PVSEP85	
9P24	113.000	YKYSF81	
	114.000	YKYSF81	
	294.600	PVSEP85	
9P26	137.000	YKYSF81	
	139.500	PVSEP85	
	356.500	PVSEP85	
9P28	62.700	YKYSF81	
	348.100	PVSEP85	
	483.500	FK86	
9P30	453.100	PVSEP85	
	821.400	FK86	
	1010.000	FK86	
9P32	272.500	PVSEP85	
9P34	684.700	PVSEP85	
	747.700	FK86	
9P38	236.100	PVSEP85	
9P40	404.300	PVSEP85	
	684.300	PVSEP85	
	988.900	FK86	
9P42	129.600	PVSEP85	
9P44	152.300	PVSEP85	
9P46	172.000	PVSEP85	
9P48	255.000	PVSEP85	
9P52	135.900	FK86	
	741.200	FK86	
9R02	178.600	PVSEP85	
9R04	124.500	VSPE81	
	152.500	VSPE81	
9R06	249.700	PVSEP85	
	482.500	PVSEP85	
	497.200	PVSEP85	
9R08	141.300	VSPE81	
	270.733	VSPE81	1107337.9
	387.500	FK86	
	417.400	FK86	
	491.200	PVSEP85	
	710.000	FK86	
	778.200	FK86	
9R10	552.400	PVSEP85	
9R12	141.700	PVSEP85	
9R16	150.500	PVSEP85	
	319.400	PVSEP85	
	457.500	PVSEP85	
9R18	82.600	PVSEP85	
9R22	312.700	PVSEP85	
	494.100	PVSEP85	
	585.500	PVSEP85	
9R24	102.200	PVSEP85	
9R26	316.000	YKYSF81	
9R28	327.800	VSPE81	
	550.100	PVSEP85	
9R32	80.100	PVSEP85	
	456.100	PVSEP85	
9R34	53.600	FK86	
	289.600	PVSEP85	
	411.600	PVSEP85	
9R38	64.400	VSPE81	
	111.300	PVSEP85	
9R40	567.800	PVSEP85	
9R42	263.200	FK86	
10P02	493.500	FK86	
	687.300	FK86	
10P04	60.600	FK86	
10P06	235.700	PVSEP85	
10P08	161.300	PVSEP85	
10P10	78.000	HP79	
	151.300	PVSEP85	
	311.000	PVSEP85	
10P14	165.300	PVSEP85	
10P16	329.900	PVSEP85	
10P20	216.900	PVSEP85	
10P22	331.700	FK86	
10P30	83.600	PVSEP85	
10P32	242.900	PVSEP85	
	375.300	PVSEP85	
10P38	76.300	FK86	
	85.500	PVSEP85	
	636.300	FK86	
10P42	189.200	FK86	
10P46	105.400	PVSEP85	
	1022.000	FK86	
10P48	262.100	PVSEP85	

10R02	202.000	PVSEP85				232.400	VSPE81	
	449.300	FK86		10R32	131.500	PVSEP85	
10R04	344.778	VSPE81	869522.7	10R34	181.500	DVPA81	
	577.800	PVSEP85			297.000	FK86	
10R06	130.500	VSPE81		10R36	418.200	FK86	
10R08	208.300	PVSEP85		10R40	210.500	VSPE81	
10R10	86.500	VSPE81			255.300	VSPE81	
	227.661	VSPE81	1316838.7		745.000	FK86	
	314.841	VSPE81	952203.9	10R42	79.600	PVSEP85	
	741.700	FK86		10R44	87.100	PVSEP85	
10R12	104.300	VSPE81			314.300	PVSEP85	
	107.538	VSPE81	2787789.4	10R46	136.959	IMEJ86	2188929.0
	108.700	VSPE81			391.300	PVSEP85	
	410.712	VSPE81	729932.8	10R52	308.000	FK86	
	414.000	HP79		13CO2	68.930	PD85	
	483.000	FK86			143.800	PD85	
	635.000	FK86			167.500	DVPA81	
10R14	289.500	PVSEP85			182.000	DVPA81	
	290.670	IMEJ86	1031384.4		204.800	PD85	
	372.000	PVSEP85			235.200	PD85	
	438.800	PVSEP85			309.800	DVPA81	
	1162.200	PVSEP85			374.600	DVPA81	
10R16	52.400	VSPE81			515.000	PD85	
	82.200	VSPE81			531.900	PD85	
	87.300	VSPE81		C18O2	66.780	PD85	
	354.176	VSPE81	846450.3		79.050	PD85	
	355.500	DVPA81			143.800	PD85	
10R18	408.800	PVSEP85			148.000	PD85	
	494.700	PVSEP85			154.200	PD85	
	855.900	FK86			174.300	PD85	
10R20	80.500	VSPE81			189.900	PD85	
10R22	165.604	VSPE81	1810294.3		348.600	PD85	
	300.100	FK86					
10R24	184.766	VSPE81	1622555.2				
	298.736	VSPE81	1003536.6				
	486.500	VSPE81					
	598.400	PVSEP85					
	972.000	PVSEP85					
	2923.000	PVSEP85					
10R26	97.500	VSPE81					
	119.057	VSPE81	2518067.7				
	124.798	VSPE81	2402224.0				
	222.000	PVSEP85					
	340.700	PVSEP85					
	670.400	PVSEP85					
10R28	35.000	HP79					
	73.800	VSPE81					
	80.500	VSPE81					
	122.304	VSPE81	2451203.1				
10R30	150.000	HP79					
	192.500	VSPE81					

9P04	235.400	CIMPS87	
	250.100	PFS86	
9P06	222.000	DSF75A	
	225.800	PFS86	
	509.500	PFS86	
	680.000	DSF75A	
	812.600	PFS86	
9P08	44.700	CIMPS87	
	145.700	PFS86	
	221.900	PFS86	
	223.000	GW78	
	711.000	DSF75A	
9P10	706.600	PFS86	
9P12	143.800	PFS86	
	690.000	PFS86	
	695.000	DSF75A	
	1092.800	PFS86	
	1100.000	DSF75A	
9P14	133.700	PFS86	
	265.300	PFS86	
	268.000	DSF75A	
9P16	386.600	PFS86	
	480.000	DSF75A	
9P18	82.700	PFS86	
	455.600	PFS86	
9P20	225.000	PFS86	
	258.000	YYSF82	
	266.000	PFS86	
	422.000	DSF75A	
9P22	90.160	CIMPS87	
	109.100	PFS86	
	198.600	CIMPS87	
	550.200	PFS86	
9P24	37.100	SBW84	
	323.500	CIMPS87	
	337.500	PFS86	
	701.500	PFS86	
	1146.000	DSF75A	
9P26	56.870	CIMPS87	
9P28	47.100	SBW84	
	54.100	SBW84	
	87.900	PFS86	
	362.800	PFS86	
	370.000	DSF75A	
	435.000	GW78	
9P30	30.700	SBW84	
	177.400	PFS86	
	385.700	PFS86	
	774.000	DSF75A	
9P32	350.500	PFS86	
	410.000	DSF75A	

9P34	27.700	SBW84	
	336.800	PFS86	
9P36	189.730	SEJZS87	1580101.8
	199.500	CIMPS87	
	310.100	CIMPS87	
	333.900	CIMPS87	
9P38	42.500	SBW84	
	42.920	CIMPS87	
	118.800	PFS86	
9P40	198.682	SEJZS87	1508908.6
	201.000	GW78	
	221.880	CIMPS87	
	284.300	PFS86	
	286.197	SEJZS87	1047502.3
9P42	283.750	CIMPS87	
9P44	53.100	CIMPS87	
	229.100	CIMPS87	
9P46	258.300	CIMPS87	
9P48	116.500	PFS86	
	312.900	PFS86	
9P50	287.950	CIMPS87	
9R06	35.500	SEJZS87	
	48.700	SBW84	
	56.500	SBW84	
	68.450	CIMPS87	
	136.500	PFS86	
	299.000	DSF75A	
	352.300	PFS86	
9R08	184.000	DSF75A	
9R10	140.000	PFS86	
9R14	119.000	YYSF82	
	120.300	PFS86	
	120.661	SEJZS87	2484584.9
	179.000	GW78	
	182.566	SEJZS87	1642101.9
	236.000	GW78	
	346.000	GW78	
	352.503	SEJZS87	850468.0
	871.585	IMEJ86	343962.4
9R16	321.000	DSF75A	
	551.900	PFS86	
9R18	66.800	CIMPS87	
	158.900	PFS86	
	472.400	PFS86	
9R20	31.100	SBW84	
	297.100	PFS86	
	968.000	DSF75A	
9R22	351.400	PFS86	
	386.900	PFS86	
	583.300	PFS86	
9R24	78.600	CIMPS87	

CD3OH

CD3OH

	226.900	PFS86	
	252.300	PFS86	
9R26	50.000	SEJZS87	
	147.280	CIMPS87	
	276.600	PFS86	
	482.700	PFS86	
	498.700	SEJZS87	
	745.000	DSF75A	
9R28	40.000	SEJZS87	
	42.600	SBW84	
	44.300	SBW84	
	49.800	GW78	
	55.560	Knigh81	
	158.000	GW78	
	159.400	SEJZS87	1880754.6
	181.000	YYSF82	
	181.711	SEJZS87	1649830.3
	370.483	SEJZS87	809193.2
9R32	112.100	CIMPS87	
	148.940	CIMPS87	
	151.300	PFS86	
	176.800	CIMPS87	
	284.400	CIMPS87	
	336.500	PFS86	
	351.400	PFS86	
9R34	48.600	SBW84	
	52.900	GW78	
	53.300	YYSF82	
	53.820	Knigh81	
	60.100	Knigh81	
	60.800	GW78	
	407.900	PFS86	
	430.927	SEJZS87	695691.5
	553.000	PFS86	
9R36	196.950	CIMPS87	
	200.870	CIMPS87	
9R40	120.450	CIMPS87	
	219.700	CIMPS87	
9R42	78.780	CIMPS87	
9R44	232.100	PFS86	
	407.000	DSF75A	
9R46	225.000	PFS86	
10P04	127.300	PFS86	
10P06	84.500	PFS86	
10P08	45.660	CIMPS87	
	61.400	CIMPS87	
	196.600	PFS86	
	369.550	CIMPS87	
10P10	108.668	SEJZS87	2758781.7
	124.930	CIMPS87	
10P12	147.650	CIMPS87	
	172.620	CIMPS87	
	219.000	PFS86	
	253.100	PFS86	
	272.300	CIMPS87	
	322.100	CIMPS87	
	322.350	CIMPS87	
	372.360	CIMPS87	
10P14	103.000	PFS86	
10P16	203.500	PFS86	
10P18	144.118	SEJZS87	2080189.3
	286.000	YYSF82	
	287.308	SEJZS87	1043454.5
	290.000	DW78B	
	760.000	DSF75A	
10P20	87.800	PFS86	
	309.000	DSF75A	
	310.700	PFS86	
	433.600	PFS86	
	722.000	DSF75A	
10P22	34.800	DW78B	
	40.100	DW78B	
	132.100	CIMPS87	
	257.000	YYSF82	
	258.436	SEJZS87	1160027.8
	260.000	DSF75A	
	267.200	CIMPS87	
10P24	162.850	CIMPS87	
	187.050	CIMPS87	
	237.100	PFS86	
	238.300	DW78B	
	251.400	CIMPS87	
	285.000	YYSF82	
	286.724	SEJZS87	1045578.0
	524.600	PFS86	
10P26	35.700	SBW84	
	482.700	PFS86	
10P28	188.900	PFS86	
	190.000	SEJZS87	
	276.716	SEJZS87	1083395.1
	598.600	PFS86	
10P32	76.100	DW78B	
	147.349	SEJZS87	2034573.6
	148.000	PFS86	
	215.081	SEJZS87	1393856.9
	329.500	PFS86	
10P36	205.800	PFS86	
	215.600	PFS86	
10P38	369.700	CIMPS87	
10P40	231.100	PFS86	
	234.800	PFS86	
10P42	76.300	SEJZS87	

	188.424	SEJZS87	1591053.2
	516.500	PFS86	
10P46	153.700	CIMPS87	
10P48	76.930	CIMPS87	
	222.700	CIMPS87	
	310.800	CIMPS87	
10P56	86.300	CIMPS87	
10R04	255.200	PFS86	
	264.700	CIMPS87	
10R06	593.100	PFS86	
10R08	41.250	Knigh81	
	41.500	DW78B	
	44.800	SBW84	
	45.000	CIMPS87	
	70.989	SEJZS87	4223062.0
	71.400	YYSF82	
	117.620	CIMPS87	
	140.950	CIMPS87	
	180.750	CIMPS87	
	203.300	SEJZS87	
	312.500	CIMPS87	
	554.000	PFS86	
	610.300	PFS86	
	646.477	SEJZS87	463732.4
10R10	71.700	CIMPS87	
10R12	123.550	CIMPS87	
	143.800	CIMPS87	
	177.000	CIMPS87	
	412.000	DSF75A	
	861.100	CIMPS87	
10R14	68.100	YYSF82	
	68.700	CIMPS87	
	83.700	CIMPS87	
	107.200	CIMPS87	
	136.627	SEJZS87	2194236.9
	185.000	PFS86	
	267.000	DSF75A	
	268.600	PFS86	
	388.000	PFS86	
10R16	80.900	YYSF82	
	81.557	SEJZS87	3675859.9
	86.400	DW78B	
	150.800	PFS86	
	599.000	DSF75A	
10R18	41.355	SEJZS87	7249266.0
	41.800	DW78B	
	43.697	SEJZS87	6860664.2
	54.700	PFS86	
	219.900	DW78B	
	495.000	DSF75A	
	520.300	PFS86	
	699.000	PFS86	
	854.700	KK83	
	858.254	SEJZS87	349305.1
	862.000	PFS86	
	988.100	PFS86	
10R20	49.780	CIMPS87	
	52.800	SBW84	
	55.400	YYSF82	
	151.800	CIMPS87	
	215.250	CIMPS87	
	452.900	PFS86	
	1290.000	DSF75A	
10R22	76.000	CIMPS87	
	135.400	PFS86	
	286.200	PFS86	
10R24	50.300	CIMPS87	
	61.700	SBW84	
	70.600	YYSF82	
	276.900	PFS86	
	278.000	DSF75A	
10R26	144.400	CIMPS87	
	310.350	CIMPS87	
10R28	80.440	CIMPS87	
	111.400	CIMPS87	
	217.200	CIMPS87	
	308.500	PFS86	
	310.000	DSF75A	
	396.400	PFS86	
	398.000	DSF75A	
10R30	56.700	SBW84	
	67.479	SEJZS87	4442724.8
	235.800	CIMPS87	
	336.000	DSF75A	
	337.300	PFS86	
	351.200	PFS86	
10R32	44.550	CIMPS87	
	49.070	CIMPS87	
	65.870	CIMPS87	
	83.600	YYSF82	
	83.900	PFS86	
	93.880	CIMPS87	
	94.900	CIMPS87	
	114.400	CIMPS87	
	131.563	SEJZS87	2278703.0
	165.000	SEJZS87	
	166.760	CIMPS87	
	174.000	CIMPS87	
	282.800	CIMPS87	
	380.800	PFS86	
	417.000	CIMPS87	
	420.300	PFS86	

CD3OH

10R34	34.100	SBW84	
	37.600	DW78B	
	41.460	CIMPS87	
	42.500	SBW84	
	50.100	SBW84	
	66.400	CIMPS87	
	76.900	CIMPS87	
	86.741	SEJZS87	3456161.2
	102.600	DW78B	
	112.300	DW78B	
	119.900	CIMPS87	
	128.034	SEJZS87	2341508.9
	128.700	DW78B	
	138.400	CIMPS87	
	168.083	SEJZS87	1783601.1
	180.741	SEJZS87	1658689.9
	191.356	SEJZS87	1566672.8
	222.217	SEJZS87	1349100.1
	228.300	CIMPS87	
	239.650	CIMPS87	
	264.759	SEJZS87	1132320.1
	265.000	PFS86	
	297.000	DSF75A	
	386.037	SEJZS87	776589.1
	431.400	PFS86	
	435.300	CIMPS87	
	476.250	CIMPS87	
	477.300	PFS86	
	498.000	DW78B	
	684.500	PFS86	
10R36	69.180	CIMPS87	
	252.000	YYSF82	
	253.720	SEJZS87	1181588.9
	418.712	SEJZS87	715987.6
	420.000	DSF75A	
	435.100	CIMPS87	
	562.400	CIMPS87	
	645.200	KK83	
	704.200	KK83	
	1676.000	PFS86	
	1930.000	PFS86	
	3030.000	PFS86	
10R38	34.200	SBW84	
	50.000	SEJZS87	
	122.154	SEJZS87	2454225.9
	161.100	CIMPS87	
	253.800	PFS86	
	418.100	PFS86	
	1155.500	CIMPS87	
10R40	59.600	CIMPS87	
	71.500	CIMPS87	

CD3OH

	409.100	CIMPS87	
10R44	199.810	CIMPS87	
	438.870	CIMPS87	
10R46	68.800	SEJZS87	
10R48	43.700	CIMPS87	
13CO2	143.400	DVPA81	
	221.000	DVPA81	
	234.700	DVPA81	
	530.400	DVPA81	

CDF3

9P44	266.900	TF80	
9R22	345.800	TF80	
9R24	362.423	TLD83	827188.4
10P06	658.152	TLD83	455506.2
10P12	657.938	TLD83	455654.7
10P14	266.000	TF80	
10P20	432.987	TLD83	692381.5
	445.663	TLD83	672689.5
10P24	361.231	TLD83	829918.3
10R04	286.800	TF80	
10R08	330.019	TLD83	908408.6
10R10	286.800	TF80	
	657.989	TLD83	455619.1
	687.837	TLD83	435848.1
10R12	582.100	TSD82	
	1008.558	TLD83	297248.6
	1080.537	TLD83	277447.6
10R16	1260.561	TLD83	237824.7
10R18	459.400	TSD82	
10R20	316.600	TF80	
	605.600	TSD82	
10R22	1377.000	TSD82	
10R24	521.237	TLD83	575156.1
10R26	420.311	TLD83	713263.1
	459.600	TSD82	
10R28	581.984	TLD83	515121.1
10R32	388.273	TLD83	772117.0
10R36	540.736	TLD83	554415.6
	560.703	TLD83	534672.7
10R38	488.528	TLD83	613665.3
	504.752	TLD83	593940.1
10R40	560.803	TLD83	534577.4
10R42	388.652	TLD83	771365.4
10R46	420.980	TLD83	712130.6
10R48	286.300	TF80	

CF2CL2

10P30	751.400	LPMD81
10P32	614.300	LPMD81
10P34	684.740	LPMD81
	858.730	LPMD81
	1025.000	LPMD81
10P36	638.400	LPMD81
	980.000	LPMD81
	1164.000	LPMD81
	1205.000	LPMD81
10P42	684.700	LPMD81
	765.200	LPMD81

CF3BR

9R08	823.500	PLM79
9R10	883.000	PLM79
9R12	1043.000	LPM81
9R16	1526.000	LPM81
9R20	1895.000	LPM81
9R28	1083.000	LPM81
9R34	1151.000	LPM81
	2140.000	LPM81
9R38	1556.000	LPM81
9R40	1687.000	LPM81

CF4

9R12	16.000	Telle83

CH2CF2

10P08	257.400	HW84	
	325.300	HW84	
10P10	291.300	HW84	
	339.300	HW84	
	349.500	HW84	
	657.900	HW84	
	764.100	BP77	
	770.000	Knigh81	
10P12	288.500	DSF72	
	375.545	RPJM77	798286.6
	884.000	HRB73	
10P14	335.000	TW82	
	407.294	RPJM77	736059.6
	415.000	HRB73	
	554.365	RPJM77	540785.1
	1018.258	DSFE74	294416.9
	1941.700	HW84	
	2070.188	DSFE74	144814.1
10P16	306.700	HW84	
	523.800	HW84	
	1428.600	HW84	
10P18	299.900	HW84	
	339.100	HW84	
	375.100	HW84	
	376.600	HW84	
	409.300	HW84	
	426.800	HW84	
10P20	401.300	HW84	
10P22	486.768	DSFE74	615883.3
	557.700	TSD82	
	568.500	TSD82	
	888.862	DSFE74	337276.7
	889.086	DSFE74	337191.9
	917.881	DSFE74	326613.8
	918.148	DSFE74	326518.7
	990.000	DSF72	
	1075.200	HW84	
10P24	289.800	HW84	
	563.700	HW84	
	662.816	RPJM77	452301.5
	679.100	HW84	
	697.800	DH78B	
10P26	591.441	DSFE74	506885.1
	731.000	DW84	
10P30	457.300	HW84	
	577.001	HW84	
	842.623	DSFE74	355784.7
	954.467	DSFE74	314094.2
10P32	475.100	HW84	
10P36	430.100	HW84	
10P38	281.600	HW84	

	CH2CF2		

					CH2CF2		
	316.700	HW84			890.200	SWTRD80	
	367.600	HW84			1062.000	SWTRD80	
10P40	527.700	HW84		N2O	350.000	FGRD84	
10P44	675.200	HW84			364.600	FGRD84	
10P48	378.500	HW84			402.300	FGRD84	
	605.000	SWTRD80			441.300	FGRD84	
10P50	867.200	FGRD84			445.000	FGRD84	
10R06	728.900	HW84			449.200	FGRD84	
10R08	293.800	HW84			475.100	FGRD84	
10R10	602.000	HW84			490.700	FGRD84	
10R12	555.900	FGRD84			498.700	FGRD84	
10R14	363.900	HW84			551.200	FGRD84	
	531.300	HW84			617.700	FGRD84	
	630.400	FGRD84			662.000	FGRD84	
10R16	351.000	HW84			688.300	FGRD84	
	454.500	HW84			715.400	FGRD84	
10R18	358.000	HW84			793.200	FGRD84	
10R20	463.624	DSFE74	646628.1		866.400	FGRD84	
	477.300	HW84			1094.000	FGRD84	
10R22	377.500	HW84			1100.000	FGRD84	
	476.300	FGRD84			1372.000	FGRD84	
10R28	375.400	HW84			1434.000	FGRD84	
	384.000	TW82			1600.000	FGRD84	
	484.774	DSFE74	618417.5				
10R30	486.100	HW84					
10R36	329.500	HW84					
13CO2	259.500	MPD83					
	324.400	MPD83					
	326.000	MPD83					
	347.600	MPD83					
	427.700	MPD83					
	468.400	MPD83					
	546.000	MPD83					
	584.800	MPD83					
	615.900	MPD83					
	950.600	MPD83					
C18O2	373.400	MPD83					
	399.300	MPD83					
	403.600	MPD83					
	437.600	MPD83					
	469.500	MPD83					
	591.700	MPD83					
CO2H	383.000	FGRD84					
	385.400	FGRD84					
	497.000	SWTRD80					
	535.000	SWTRD80					
	590.000	FGRD84					
	617.700	SWTRD80					
	617.900	FGRD84					
	631.500	FGRD84					
	817.300	FGRD84					

CH2CHBR

9P28	934.223	DESF76	320900.3
9P34	396.000	DESF76	
9P36	453.800	BGRF84	
9P40	553.300	BGRF84	
10P02	848.600	BGRF84	
	1292.400	BGRF84	
10P04	515.800	BGRF84	
10P06	1250.000	BGRF84	
10P10	853.438	DESF76	351276.2
	963.487	DESF76	311153.5
10P12	1296.400	BGRF84	
10P14	724.140	DESF76	413998.0
10P16	490.083	DESF76	611717.8
	989.190	DESF76	303068.5
10P18	649.425	DESF76	461627.2
	662.200	BGRF84	
10P20	424.000	DESF76	
	741.115	DESF76	404515.5
	1899.889	DESF76	157794.7
10P22	445.000	DESF76	
	826.944	DESF76	362530.4
10P24	443.500	DESF76	
	784.268	DESF76	382257.6
	1383.882	DESF76	216631.5
10P26	482.961	DESF76	620737.8
	1614.888	DESF76	185642.9
10P28	370.000	DESF76	
	438.507	DESF76	683666.5
10P32	594.729	DESF76	504082.8
10P36	1018.300	BGRF84	
	1047.600	BGRF84	
	1079.300	BGRF84	
	1322.100	BGRF84	
10P38	912.500	BGRF84	
	1293.000	BGRF84	
10P40	553.696	DESF76	541438.5
	753.800	BGRF84	
10P42	886.300	BGRF84	
10P46	807.500	BGRF84	
	863.600	BGRF84	
10P48	1080.600	BGRF84	
10P50	693.800	BGRF84	
10P56	640.700	BGRF84	
10R02	985.859	DESF76	304092.7
10R04	990.630	DESF76	302628.0
10R06	994.900	BGRF84	
10R10	712.000	DESF76	
10R12	1247.594	DESF76	240296.5
10R14	780.133	DESF76	384283.8
10R16	680.541	DESF76	440520.5
	693.140	DESF76	432513.8

CH2CHBR

10R18	624.096	DESF76	480362.9
	900.134	DESF76	333053.2
10R20	283.000	DESF76	
	356.000	DESF76	
	1394.063	DESF76	215049.5
10R22	416.000	DESF76	
10R24	427.000	DESF76	
	707.221	DESF76	423902.1
10R26	411.000	DESF76	
	624.700	BGRF84	
	635.355	DESF76	471850.5
	646.000	DESF76	
10R28	619.300	BGRF84	
10R30	618.446	DESF76	484751.1
10R32	419.000	DESF76	
	936.159	DESF76	320236.7
10R38	506.000	DESF76	
10R40	528.497	DESF76	567255.3
10R42	832.700	BGRF84	
10R46	606.700	BGRF84	
10R48	517.500	BGRF84	
	681.400	BGRF84	
10R52	705.300	BGRF84	
CO2H	509.700	BGRF84	
	555.100	BGRF84	
	590.000	BGRF84	
	617.700	BGRF84	
	625.700	BGRF84	
	640.700	BGRF84	
	693.000	BGRF84	
	701.100	BGRF84	
	734.800	BGRF84	
	796.700	BGRF84	
	859.500	BGRF84	
	1492.500	BGRF84	
	2347.500	BGRF84	
CO2S	497.500	BGRF84	
	683.600	BGRF84	
	967.500	BGRF84	
	1016.700	BGRF84	
	1173.900	BGRF84	
	2356.400	BGRF84	
N2O	457.900	GRBF84	
	475.300	GRBF84	
	504.500	GRBF84	
	544.100	GRBF84	
	605.700	GRBF84	
	630.700	GRBF84	
	648.600	GRBF84	
	684.600	GRBF84	
	707.100	GRBF84	

CH2CHBR		
769.100	GRBF84	
783.700	GRBF84	
784.600	GRBF84	
802.500	GRBF84	
836.800	GRBF84	
877.300	GRBF84	
906.500	GRBF84	
912.500	GRBF84	
967.800	GRBF84	
1018.400	GRBF84	
1028.300	GRBF84	
1198.600	GRBF84	
1204.300	GRBF84	
1278.600	GRBF84	
1325.000	GRBF84	
1504.000	GRBF84	

CH2CHCL			
9P10	487.000	Radfo75	
9P16	530.533	DSFE74	565077.8
9P18	186.000	LSB81	
	590.369	DSFE74	507804.8
	704.925	DSFE74	425282.7
9P20	634.471	DSFE74	472507.5
	645.289	DSFE74	461586.2
9P22	695.202	DSFE74	431230.7
	699.000	Radfo75	
9P24	815.123	DSFE74	367788.0
	828.000	Radfo75	
	1039.855	DSFE74	288302.1
9P32	2206.000	FBM84	
9P44	1005.000	FBM84	
9P46	620.000	FBM84	
9P48	621.000	FBM84	
	700.400	FBM84	
9P52	487.800	FBM84	
	556.800	FBM84	
10P04	421.000	LSB81	
10P06	638.000	Radfo75	
10P14	606.000	FBM84	
10P16	442.168	RPJM77	678006.1
	567.945	DSFE74	527854.1
	574.000	Radfo75	
	579.761	DSFE74	517096.5
10P18	545.500	FBM84	
10P20	634.471	RPJM77	472507.8
10P22	157.000	LSB81	
	385.909	RPJM77	776847.1
	507.584	DSFE74	590626.3
	507.591	DSFE74	590618.4
10P24	390.400	FBM84	
	988.259	DSFE74	303354.0
10P34	519.000	Radfo75	
10P36	902.500	FBM84	
	1164.800	FBM84	
10P38	601.897	RPJM77	498079.1
10P42	438.000	FBM84	
10P44	606.700	FBM84	
10P46	935.000	Radfo75	
10P48	474.600	FBM84	
10P52	459.400	FBM84	
10R04	538.000	Radfo75	
10R12	1406.000	FBM84	
10R16	863.100	FBM84	
10R18	445.000	Radfo75	
	668.100	FBM84	
10R20	666.604	DSFE74	449731.0
	683.738	DSFE74	438460.7
	771.300	FBM84	

CH2CHCL			
10R26	995.000	Radfo75	
10R28	424.000	Radfo75	
	988.695	DSFE74	303220.3
	1026.709	DSFE74	291993.6
10R30	423.354	DSFE74	708137.1
10R34	622.300	FBM84	
10R36	1041.000	Radfo75	
10R38	1026.680	DSFE74	292001.8
10R40	655.400	FBM84	
10R48	770.400	FBM84	
10R50	580.800	FBM84	
13CO2	587.000	TJD86	
	681.000	TJD86	
	785.000	TJD86	
	942.000	TJD86	
CO2H	293.800	FBM84	
	476.800	FBM84	
	556.800	FBM84	
	581.300	FBM84	
	636.300	FBM84	
	681.500	FBM84	
	758.200	FBM84	
	771.200	FBM84	
	976.800	FBM84	
	1880.000	FBM84	
N2O	172.800	GRBF84	
	435.900	GRBF84	
	524.800	GRBF84	
	584.000	GRBF84	
	593.900	GRBF84	
	598.300	GRBF84	
	637.100	GRBF84	
	768.800	GRBF84	
	833.200	GRBF84	
	876.800	GRBF84	
	921.500	GRBF84	
	934.000	GRBF84	
	945.000	GRBF84	
	1047.200	GRBF84	
	1071.300	GRBF84	
	1093.100	GRBF84	
	1194.300	GRBF84	
	1292.200	GRBF84	

CH2CHCN			
9P26	750.380	GRBKF85	
9P42	401.250	GRBKF85	
9P44	657.590	GRBKF85	
9R06	405.950	GRBKF85	
9R12	503.000	Radfo75	
9R22	641.430	GRBKF85	
9R24	712.760	GRBKF85	
9R34	1202.200	GRBKF85	
10P02	588.440	GRBKF85	
	780.830	GRBKF85	
10P04	406.360	GRBKF85	
	554.560	GRBKF85	
10P06	787.500	GRBKF85	
	1388.300	GRBKF85	
10P08	489.000	Radfo75	
	586.720	GRBKF85	
	597.000	GRBKF85	
	788.330	GRBKF85	
10P10	343.260	GRBKF85	
	880.410	GRBKF85	
	903.000	GRBKF85	
	1427.500	GRBKF85	
10P12	583.872	DSFE74	513455.5
	597.000	GRBKF85	
	875.000	GRBKF85	
	1267.100	GRBKF85	
10P14	549.686	DSFE74	545388.2
10P16	727.570	DSFE74	412046.3
	738.000	Radfo75	
	816.195	DSFE74	367305.1
10P18	425.650	GRBKF85	
10P20	586.382	DSFE74	511258.1
	599.000	GRBKF85	
10P22	537.650	GRBKF85	
	620.340	GRBKF85	
10P26	270.600	DSF72	
	1127.752	DSFE74	265831.9
	1156.000	Radfo75	
	1173.700	GRBKF85	
10P28	751.831	DSFE74	398749.7
	940.000	Radfo75	
	959.000	GRBKF85	
	986.349	DSFE74	303941.5
	990.500	GRBKF85	
10P30	525.560	GRBKF85	
10P32	1016.009	DSFE74	295068.8
10P34	1315.000	GRBKF85	
	1374.200	GRBKF85	
10P36	464.400	GRBKF85	
10P40	508.330	GRBKF85	
	872.270	GRBKF85	

CH2CHCN

10P42	399.800	GRBKF85	
	722.000	Radfo75	
	927.000	GRBKF85	
	957.000	GRBKF85	
10P44	556.470	GRBKF85	
	645.000	GRBKF85	
10P50	1432.500	GRBKF85	
10R02	655.900	GRBKF85	
	1257.100	GRBKF85	
10R04	1197.100	GRBKF85	
10R06	459.000	GRBKF85	
	631.000	Radfo75	
	642.860	GRBKF85	
10R08	670.790	GRBKF85	
10R10	671.150	GRBKF85	
10R12	623.000	Radfo75	
	910.000	Radfo75	
10R14	563.440	GRBKF85	
	578.000	Radfo75	
	586.800	GRBKF85	
	901.191	DSFE74	332662.6
10R16	399.420	GRBKF85	
	574.027	DSFE74	522262.2
10R18	574.380	GRBKF85	
	828.000	Radfo75	
10R20	545.000	GRBKF85	
	572.692	DSFE74	523479.7
10R22	385.800	GRBKF85	
10R24	1017.800	GRBKF85	
10R26	830.450	GRBKF85	
10R28	1555.000	GRBKF85	
10R30	470.000	GRBKF85	
	651.790	GRBKF85	
	1577.000	GRBKF85	
10R34	453.570	GRBKF85	
10R36	509.160	GRBKF85	
10R38	660.340	GRBKF85	
	1184.000	Radfo75	
	1218.600	GRBKF85	
10R40	793.000	Radfo75	
10R42	775.000	Radfo75	
10R44	644.640	GRBKF85	
	658.260	GRBKF85	
	769.053	DSFE74	389820.1
10R46	440.010	GRBKF85	
	685.190	GRBKF85	
CO2H	754.000	GRBKF85	
	899.000	GRBKF85	
	929.000	GRBKF85	
	1218.600	GRBKF85	
	1579.000	GRBKF85	

CH2CHCN

CO2S	644.640	GRBKF85
	659.690	GRBKF85
	671.430	GRBKF85
N2O	398.960	GRBKF85
	425.870	GRBKF85
	431.140	GRBKF85
	459.310	GRBKF85
	537.060	GRBKF85
	564.700	GRBKF85
	597.330	GRBKF85
	642.280	GRBKF85
	758.460	GRBKF85
	761.670	GRBKF85
	770.000	GRBKF85
	777.920	GRBKF85
	829.540	GRBKF85
	832.770	GRBKF85
	837.730	GRBKF85
	871.360	GRBKF85
	939.500	GRBKF85
	939.500	GRBKF85

10P04	720.800	RGF84		10R20	203.000	TW82
10P06	335.000	TW82			372.000	TW82
10P08	298.000	TW82			444.400	CD81B
10P10	483.800	RGF84			459.800	RGF84
	838.200	RGF84		10R22	330.100	RGF84
10P14	377.400	CD81B		10R26	322.800	RGF84
10P18	430.000	TW82		10R28	362.800	RGF84
	847.700	RGF84		10R32	582.500	RGF84
10P20	355.000	TW82		10R36	563.000	TW82
	472.400	CD81B		10R44	281.600	RGF84
	487.800	RGF84		10R46	423.000	RGF84
10P22	171.800	CD81B		10R50	148.200	RGF84
	490.000	TW82			222.300	RGF84
	505.000	RGF84			446.700	RGF84
	774.900	CD81B			461.000	RGF84
10P24	362.200	RGF84		13C18O2	573.000	TJD86
	660.000	CD81B		13CO2	456.000	TJD86
	1094.000	RGF84			565.000	TJD86
10P28	1167.600	RGF84			579.000	TJD86
10P32	540.000	TW82			586.000	TJD86
	618.000	RGF84			875.000	TJD86
	783.000	TW82			971.000	TJD86
	1063.000	RGF84			1612.000	TJD86
	1083.100	RGF84			1733.000	TJD86
	1170.700	RGF84		C18O2	433.000	TJD86
10P34	1255.700	RGF84		CO2H	321.000	RGF84
10P36	290.000	TW82			454.300	RGF84
	477.000	TW82			458.000	RGF84
	557.000	TW82			461.000	RGF84
	672.100	CD81B			487.700	RGF84
	700.300	RGF84			490.000	RGF84
10P38	336.000	TW82			506.300	RGF84
	345.500	RGF84			617.000	RGF84
	508.000	TW82			671.000	RGF84
	1011.000	RGF84			764.200	RGF84
10P40	420.000	TW82			796.500	RGF84
10P42	606.800	RGF84			894.000	RGF84
	2525.000	RGF84			1523.000	RGF84
10P46	1165.700	RGF84		CO2S	201.900	RGF84
10P48	275.500	RGF84			412.200	RGF84
	1082.600	RGF84			441.700	RGF84
10P50	244.100	RGF84			518.400	RGF84
.	758.700	RGF84			585.800	RGF84
10P52	518.600	RGF84			2453.000	RGF84
10P54	142.600	RGF84		N2O	121.000	GRBF84
	194.200	RGF84			229.300	GRBF84
	952.200	RGF84			293.400	GRBF84
10P56	538.600	RGF84			309.500	GRBF84
	551.500	RGF84			329.800	GRBF84
10R02	605.000	RGF84			333.600	GRBF84
10R12	263.500	RGF84			344.800	GRBF84

CH2CHF

	351.000	GRBF84
	352.800	GRBF84
	353.000	GRBF84
	354.500	GRBF84
	356.000	GRBF84
	356.600	GRBF84
	360.900	GRBF84
	361.800	GRBF84
	407.600	GRBF84
	421.800	GRBF84
	429.900	GRBF84
	433.500	GRBF84
	445.200	GRBF84
	446.700	GRBF84
	458.000	TJD86
	467.200	GRBF84
	476.000	GRBF84
	487.500	GRBF84
	489.300	GRBF84
	493.500	GRBF84
	519.600	GRBF84
	583.100	GRBF84
	606.600	GRBF84
	655.000	GRBF84
	660.200	GRBF84
	796.300	GRBF84
	939.000	GRBF84
	949.000	GRBF84
	961.000	TJD86
	981.100	GRBF84
	1170.700	GRBF84
	1234.300	GRBF84
	1264.300	GRBF84

CH2CL2

10P12	195.000	HW82
10P18	208.300	HW82
10P22	231.000	HW82
	298.500	HW82
10P24	235.500	HW82
10P26	254.700	HW82
	294.600	HW82

CH2CLF

9P08	1014.000	ADF84
9P12	176.000	ADF84
9P22	218.000	ADF84
9P26	244.000	ADF84
9P32	344.000	ADF84
9R04	324.000	ADF84
9R10	349.000	ADF84
9R16	284.000	ADF84
9R20	292.000	ADF84
9R22	221.000	ADF84
	308.000	ADF84
9R26	246.000	ADF84
	296.000	ADF84

CH2DOH

9P06	273.004	SPEJ80	1098125.9
9P10	183.621	SPEJ80	1632666.9
	295.397	SPEJ80	1014881.0
9P12	108.818	SPEJ80	2754995.7
	112.532	SPEJ80	2664058.3
	171.000	ZD78	
	172.846	SPEJ80	1734446.4
	322.452	SPEJ80	929726.8
9P14	206.687	SPEJ80	1450463.1
	308.040	SPEJ80	973224.3
9P16	102.023	SPEJ80	2938465.1
9P18	87.100	SPEJ80	
	100.000	SPEJ80	
	167.000	ZD78	
	167.541	SPEJ80	1789365.9
	396.000	ZD78	
	762.500	SPEJ80	
9P20	140.300	SPEJ80	
9P26	468.236	SPEJ80	640259.5
	616.335	SPEJ80	486411.5
9P30	44.000	SPEJ80	
9P32	108.941	SPEJ80	2751872.9
	117.085	SPEJ80	2560467.0
	167.352	SPEJ80	1791384.9
	266.735	SPEJ80	1123932.7
	451.475	SPEJ80	664028.4
9P36	195.496	SPEJ80	1533499.9
	336.246	SPEJ80	891586.3
9P38	42.500	SPEJ80	
	200.000	SPEJ80	
9P40	87.900	SPEJ80	
	387.559	SPEJ80	773539.9
	523.091	SPEJ80	573116.8
9P46	226.297	SPEJ80	1324771.9
	452.400	SPEJ80	
9R08	135.834	SPEJ80	2207058.3
	164.746	SPEJ80	1819720.3
	422.151	SPEJ80	710154.3
9R16	216.800	SPEJ80	
9R18	164.000	ZD78	
9R22	171.800	SPEJ80	
	182.100	SPEJ80	
	218.000	SPEJ80	
9R24	152.700	SPEJ80	
	219.096	SPEJ80	1368315.4
	272.252	SPEJ80	1101159.4
	682.600	SPEJ80	
10P18	238.000	ZD78	
10P26	150.572	SPEJ80	1991028.3
	188.411	SPEJ80	1591161.2
10P28	189.300	SPEJ80	

CH2DOH

	196.100	SPEJ80	
10P30	90.400	SPEJ80	
	162.700	SPEJ80	
10P34	124.432	SPEJ80	2409293.3
	125.000	ZD78	
	248.122	SPEJ80	1208246.0
	249.720	SPEJ80	1200512.7
10P36	149.388	SPEJ80	2006805.2
	224.226	SPEJ80	1337012.5
	427.200	SPEJ80	
10P46	374.086	SPEJ80	801399.6
	509.372	IMEJ86	588553.4
10R16	212.500	SPEJ80	
	363.000	ZD78	
10R32	135.172	SPEJ80	2217863.3
	135.173	SPEJ80	2217849.9
	149.613	SPEJ80	2003788.3
	340.357	SPEJ80	880818.6
10R34	150.816	SPEJ80	1987798.9
	159.218	SPEJ80	1882906.3
	295.639	SPEJ80	1014047.7
	308.296	IMEJ86	972418.7

9P04	289.500	PSE80	1035552.7
	724.920	PSE80	413552.3
9P06	394.701	PSE80	759543.3
	464.412	PSE80	645530.9
9P08	122.466	PSE80	2447974.6
	355.126	PSE80	844185.9
9P10	127.300	IMMSD85	
	127.800	MH84A	
	158.513	PSE80	1891274.3
	182.200	MH84A	
	272.339	PSE80	1100806.7
	382.639	PSE80	783486.0
	657.239	PSE80	456139.1
9P16	105.518	PSE80	2841142.9
9P18	227.657	PSE80	1316860.5
9P20	128.100	MH84A	
	129.100	IMMSD85	
	158.960	PSE80	1885959.3
	210.100	IMMSD85	
	211.400	MH84A	
	293.901	PSE80	1020044.0
9P22	133.998	PSE80	2237296.4
	191.848	PSE80	1562655.9
9P24	109.296	PSE80	2742946.0
	135.269	PSE80	2216263.5
	256.027	PSE80	1170941.0
9P32	281.200	MH84A	
9P34	207.200	MH84A	
9P38	261.729	PSE80	1145430.1
9R06	201.800	DW78C	
	202.465	PSE80	1480712.9
	235.500	DW78C	
	236.592	PSE80	1267131.0
	236.601	PSE80	1267081.5
	432.400	DW78C	
	434.951	PSE80	689255.1
	503.057	PSE80	595941.7
9R12	95.551	PSE80	3137510.6
	193.500	DW78C	
	194.448	PSE80	1541764.7
	417.000	DW78C	
	418.270	PSE80	716743.3
9R14	223.570	IMMSD85	
	326.423	PSE80	918417.0
	337.775	IIMSD86	887551.1
9R18	227.660	IMMSD85	
	246.330	IMMSD85	
9R20	117.000	DW78C	
	117.727	PSE80	2546495.0
	165.900	DW78C	
	166.631	PSE80	1799139.3

9R22	121.700	DW78C	
	122.466	PSE80	2447968.5
	165.800	DW78C	
	166.677	PSE80	1798647.0
	193.904	PSE80	1546083.4
	270.005	PSE80	1110319.9
9R26	250.970	IMMSD85	
	260.000	DW84	
9R28	511.445	PSE80	586167.4
	567.532	PSE80	528239.2
9R32	184.306	PSE80	1626602.6
	196.100	DV88	
	235.654	PSE80	1272171.4
9R34	214.579	PSE80	1397118.6
	230.200	LBG85	
	248.800	LBG85	
	287.667	PSE80	1042150.4
9R36	298.211	PSE80	1005303.3
	381.996	PSE80	784806.0
9R42	230.106	PSE80	1302845.8
	540.986	PSE80	554159.0
9R44	642.600	PSE80	466530.5
	1448.096	PSE80	207025.3
9R46	588.028	PSE80	509827.2
C18O2	115.935	GCK86	2585856.8
	126.545	GCK86	2369056.7
	134.900	PD82B	
	143.186	GCK86	2093728.7
	145.081	GCK86	2066379.1
	153.195	GCK86	1956935.8
	154.160	GCK86	1944679.6
	163.120	GCK86	1837861.1
	166.800	PD82B	
	190.300	PD82B	
	193.173	GCK86	1551938.8
	205.981	GCK86	1455434.8
	208.400	PD82B	
	223.600	PD82B	
	236.599	GCK86	1267091.3
	237.758	GCK86	1260914.2
	243.356	GCK86	1231911.0
	247.679	GCK86	1210408.8
	252.336	GCK86	1188068.7
	262.248	GCK86	1143163.2
	268.062	GCK86	1118369.3
	273.400	PD82B	
	281.053	GCK86	1066675.4
	282.900	PD82B	
	283.783	GCK86	1056414.8
	284.354	GCK86	1054291.8
	287.908	GCK86	1041279.4

CH2F2

289.139	GCK86	1036844.9
290.812	GCK86	1030879.8
298.470	GCK86	1004430.7
300.476	GCK86	997725.4
309.193	GCK86	969596.9
357.901	GCK86	837640.8
360.053	GCK86	832635.0
401.444	GCK86	746784.7
403.710	GCK86	742593.9
439.063	GCK86	682800.4
528.880	GCK86	566843.6
533.573	GCK86	561858.6
587.884	GCK86	509951.3
591.165	GCK86	507121.4
592.759	GCK86	505758.1
677.962	GCK86	442196.7
718.700	PD82B	
1091.637	GCK86	274626.6

CH2NOH

10P10	301.200	DP84
13CO2	264.900	DP84
	278.300	DP84
	291.300	DP84
	653.700	DP84

CH318OH

9P06	215.800	IMPSG89
	294.300	IMPSG89
9P10	165.100	IMPSG89
	230.700	IMPSG89
	359.200	IMPSG89
9P14	182.190	IMPSG89
	214.200	IMPSG89
	482.120	IMPSG89
	653.220	IMPSG89
9P16	49.500	IMPSG89
	193.250	IMPSG89
9P18	77.650	IMPSG89
	179.800	IMPSG89
	465.500	IMPSG89
9P20	206.600	IMPSG89
9P22	65.550	IMPSG89
	93.400	IMPSG89
	104.600	IMPSG89
	151.650	IMPSG89
	242.470	IMPSG89
9P26	119.840	IMPSG89
	222.500	IMPSG89
	262.400	IMPSG89
9P28	99.140	IMPSG89
9P30	34.600	IMPSG89
	43.700	IMPSG89
	123.900	IMPSG89
	134.600	IMPSG89
	149.000	IMPSG89
	218.700	IMPSG89
	221.860	IMPSG89
	284.900	IMPSG89
9P32	114.200	IMPSG89
	142.430	IMPSG89
	181.200	IMPSG89
	251.900	IMPSG89
	327.500	IMPSG89
9P34	92.600	IMPSG89
	123.850	IMPSG89
	364.500	IMPSG89
9P36	115.700	IMPSG89
	153.540	IMPSG89
	506.250	IMPSG89
9P38	438.100	IMPSG89
9P40	115.800	IMPSG89
	143.640	IMPSG89
9P42	191.040	IMPSG89
	307.200	IMPSG89
	505.800	IMPSG89
9P44	229.400	IMPSG89
	277.000	IMPSG89

CH318OH		
9R06	87.650	IMPSG89
9R08	621.700	IMPSG89
9R10	284.150	IMPSG89
9R26	170.180	IMPSG89
9R30	131.690	IMPSG89
9R34	465.700	IMPSG89
9R38	268.300	IMPSG89
10P06	111.600	IMPSG89
	203.800	IMPSG89
	232.650	IMPSG89
	241.750	IMPSG89
10P10	227.000	IMPSG89
	241.500	IMPSG89
10P24	170.100	IMPSG89
	184.800	IMPSG89
10P26	40.000	IMPSG89
	183.360	IMPSG89
	407.500	IMPSG89
10P42	90.970	IMPSG89
	342.800	IMPSG89
10R04	109.300	IMPSG89
	176.450	IMPSG89
	285.250	IMPSG89
10R06	181.600	IMPSG89
	284.500	IMPSG89
10R10	98.650	IMPSG89
10R12	69.900	IMPSG89
10R16	144.180	IMPSG89
	434.950	IMPSG89
10R18	193.550	IMPSG89
	220.270	IMPSG89
	364.300	IMPSG89
	382.880	IMPSG89
10R20	219.800	IMPSG89
	219.900	IMPSG89
	300.600	IMPSG89
	362.650	IMPSG89
	363.860	IMPSG89
	555.750	IMPSG89
10R24	78.200	IMPSG89
10R26	142.800	IMPSG89
	199.900	IMPSG89
10R30	53.600	IMPSG89
	127.770	IMPSG89
	181.100	IMPSG89
10R36	35.000	IMPSG89
	52.700	IMPSG89
	546.800	IMPSG89
10R38	48.400	IMPSG89

CH3BR			
9P28	245.040	CM76	
9P40	585.777	DF78	511785.8
10P04	1572.640	CM76	
10P08	333.150	CM76	
10P10	990.569	DF78	302646.6
	2650.000	MD78	
10P14	749.371	MD78	400058.7
10P16	631.930	CM76	
10P18	1056.852	DF78	283665.4
10P20	311.100	CM76	
10P22	632.050	DF78	474318.0
10P24	531.038	DF78	564540.7
10P26	418.310	CM76	
10P28	564.680	CM76	
	831.267	DF78	360645.1
	1965.652	DF78	152515.5
10P38	545.279	DF78	549796.0
10P40	311.200	CM76	
10R02	414.980	CM76	
10R04	1310.569	IMEJ86	228749.8
10R06	332.860	CM76	
10R10	264.050	CM76	
10R12	311.070	CM76	
	735.820	IMEJ86	407426.4
10R14	715.390	DF78	419061.7
	749.368	DF78	400060.3
	749.390	DF78	400048.7
10R18	380.020	CM76	
10R20	264.350	VHHR86	
	265.000	VHHR86	
	265.800	IMMS82	
	458.662	DF78	653624.5
	600.700	PRP83	
	660.700	CM76	
10R26	422.780	CM76	
10R28	294.280	CM76	
10R32	545.412	DF78	549662.8
10R42	508.480	CM76	
10R46	925.520	CM76	
10R50	311.210	CM76	
10R52	279.810	CM76	
9P18	352.750	CM76	
9P28	407.720	CM76	
9P56	658.530	CM76	

CH3CCH

10P10	427.890	CM71
	757.410	CM71
10P12	488.880	CM71
10P14	647.890	CM71
10P20	798.550	CM71
10P24	563.130	CM71
10P34	649.590	CM71
10P44	1174.870	CM71
9P06	531.080	CM76
9P08	1097.110	CM76
9P18	566.440	CM76
9P20	583.770	CM76
9P40	675.290	CM76
9R12	516.770	CM76
9R38	428.870	CM76

CH3CD2OH

9P40	491.800	BS85
10R20	1010.000	BS85
10R24	581.600	BS85

CH3CF3

10P02	676.700	FR87
10P06	766.600	FR87
	805.800	FR87
10P08	485.400	FR87
10P10	634.700	FR87
	852.500	FR87
10P12	485.800	FR87
10P14	383.200	FR87
10P18	510.700	FR87
10P20	393.300	FR87
10R04	518.800	FR87
10R08	878.100	FR87
10R12	463.000	FR87
10R14	369.100	FR87
	393.300	FR87
	422.000	FR87
	1613.000	FR87
10R16	851.000	FR87
10R18	833.300	FR87
10R20	767.800	FR87
	853.300	FR87
10R26	606.800	FR87
10R28	629.300	FR87
10R30	580.800	FR87
	784.500	FR87
10R32	486.100	FR87
	634.000	FR87
	967.900	FR87
10R36	709.500	FR87
10R38	485.600	FR87
10R40	709.800	FR87
10R46	580.600	FR87
10R48	501.600	FR87
10R50	477.100	FR87
10R52	454.800	FR87
CO2S	782.700	FR87
	968.900	FR87
N2O	388.900	FR87
	410.100	FR87
	411.200	FR87
	454.600	FR87
	471.200	FR87
	510.400	FR87
	519.200	FR87
	558.800	FR87
	633.400	FR87
	709.200	FR87
	783.200	FR87
	878.500	FR87
	969.000	FR87
	973.000	FR87

CH3CF3

1035.600	FR87	
1264.300	FR87	
1324.300	FR87	

CH3CH2BR

10P08	1059.000	BS83	
10P10	896.500	BS83	
10P14	707.800	BS83	
10R10	327.600	BS83	
10R20	838.300	BS83	
10R22	769.800	BS83	
10R30	527.900	BS83	
10R34	453.600	BS83	

CH3CH2CL

10R26	1669.000	DK82	
10R28	447.000	DK82	
	1306.000	DK82	
10R38	698.000	DK82	

CH3CH2F

9P08	1440.000	Knigh81	
9P10	1521.376	IMEJ86	197053.5
	1546.000	Radfo75	
9P18	264.700	WZN73	
9P22	620.400	WZN73	
9P28	1013.000	Radfo75	
9P30	851.900	WZN73	
9P34	404.000	WZN73	
9P36	593.506	RPJM77	505121.4
9R04	519.075	RPJM77	577551.1
9R10	1069.000	Radfo75	
9R16	336.700	WZN73	
	660.000	Knigh81	
9R22	452.000	Knigh81	
9R24	486.000	Radfo75	
	502.262	RPJM77	596884.2
	504.000	Radfo75	
9R30	404.000	Radfo75	
	405.504	RPJM77	739307.5
10P36	206.600	WZN73	
10P40	226.900	WZN73	
9P32	462.920	WZN73	
9P38	540.900	WZN73	
9R12	282.300	WZN73	
9R14	217.100	WZN73	
	376.000	WZN73	
9R18	362.100	WZN73	
9R22	330.200	WZN73	
9R32	378.000	WZN73	

CH3CH2I

10P20	1044.400	BS83	
10P26	660.328	TD86	454005.5
10P30	542.000	BS83	
10P32	504.000	BS83	
10P34	493.000	BS83	
10R04	1049.810	TD86	285568.4
10R06	626.800	BS83	

CH3CH2OH

9P22	449.000	BS85	
9P26	311.900	BS85	
	552.000	BS85	
9P32	388.060	VJE86	772542.0
	396.000	JEJ75	
9P34	575.300	BS85	
9P40	285.300	BS85	
9R04	529.300	BS85	
9R12	620.300	BS85	
9R28	566.100	BS85	

CH3CHDOH

9P24	432.300	BS85
9P32	379.500	BS85
9P46	351.500	BS85
9R12	889.000	BS85

CH3CHF2

10P02	790.800	FBRG85
10P10	632.900	FBRG85
10P14	770.000	BT77
10P18	1201.400	FBRG85
10P20	458.000	HRB73
	533.000	HRB73
	1133.800	FBRG85
10P22	2042.500	FBRG85
	2388.800	FBRG85
10P26	637.500	FBRG85
10P28	582.500	FBRG85
10P30	690.400	FBRG85
	718.000	FBRG85
10P32	1210.700	FBRG85
	1237.100	FBRG85
10P34	698.600	FBRG85
10P40	605.400	FBRG85
	633.000	BT77
	646.500	FBRG85
10P56	822.300	FBRG85
10R02	435.900	FBRG85
10R06	370.800	FBRG85
	1044.800	FBRG85
10R08	939.500	FBRG85
10R18	449.300	FBRG85
10R20	482.200	FBRG85
10R22	397.700	FBRG85
10R28	387.800	FBRG85
CO2H	319.000	FBRG85
	822.300	FBRG85
	877.200	FBRG85
CO2S	443.600	FBRG85
	892.000	FBRG85
	1065.000	FBRG85
N2O	371.300	FBRG85
	421.300	FBRG85
	433.900	FBRG85
	449.500	FBRG85
	464.800	FBRG85
	505.000	FBRG85
	513.400	FBRG85
	518.900	FBRG85
	521.400	FBRG85
	534.200	FBRG85
	569.400	FBRG85
	583.300	FBRG85
	644.500	FBRG85
	671.500	FBRG85
	680.000	FBRG85
	746.400	FBRG85
	761.200	FBRG85

CH3CHF2

	865.500	FBRG85
	937.900	FBRG85
	1005.800	FBRG85
	1042.200	FBRG85
	1292.100	FBRG85
	1405.000	FBRG85
	1503.400	FBRG85

CH3CHO

9R22	328.000	LSB81
9R30	343.000	LSB81
	385.000	LSB81
9R36	509.000	LSB81
9R40	176.000	LSB81
	415.000	LSB81

CH3CL

9P26	1886.695	GCK86	158898.2
9P42	333.935	GCK86	897758.2
9R12	944.019	GCK86	317570.3
	944.028	GCK86	317567.5
10P10	240.980	CM76	
10P20	271.290	CM76	
10P34	261.030	CM76	
10R18	349.387	GCK86	858053.3
10R26	568.810	CM76	
10R34	286.790	CM76	
10R52	511.900	CM76	
9P38	958.250	CM76	
9P48	227.150	CM76	
9P52	870.800	CM76	
9R02	236.250	CM76	
9R14	275.000	CM76	
	281.670	CM76	
9R16	278.570	CM76	
9R36	275.090	CM76	
9R42	461.200	CM76	
CO2S	307.650	CM76	

CH3CN

9P06	494.646	RPJM77	606074.7
9P16	750.000	Knigh81	
	854.585	DF78	350804.8
9P30	589.321	IMEJ86	508708.2
	652.680	CM76	
9R12	387.310	CM76	
9R16	453.397	RPJM77	661213.4
	456.000	Radfo75	
10P10	303.540	CM71	
10P16	380.710	CM71	
10P18	430.482	DF78	696410.9
10P20	372.814	DF78	804134.8
10P24	422.117	IMEJ86	710212.3
10P32	713.720	CM71	
10P46	1814.370	CM71	
9P06	510.160	CM76	
9P08	1016.330	CM76	
9P10	1164.830	CM76	
9P16	346.320	CM76	
9P22	388.390	CM76	
9P26	427.040	CM76	
9P34	281.180	CM76	
9P40	1086.890	CM76	
9P46	386.410	CM76	
9P50	281.980	CM76	
	286.880	CM76	
9R08	561.410	CM76	
	741.620	CM76	
9R14	1014.890	CM76	
9R16	441.150	CM76	
	466.250	CM76	
	480.010	CM76	
9R20	1351.780	CM76	
9R34	704.530	CM76	

CH3COOD

9P32	451.000	DFSY81A
9R06	628.000	DFSY81A
10P18	676.000	DFSY81A
	756.000	DFSY81A
10P20	433.000	DFSY81A
10R12	525.000	DFSY81A
10R18	701.000	DFSY81A
10R20	465.000	DFSY81A
10R22	675.000	DFSY81A
10R28	363.000	DFSY81A

CH3F

9P20	496.070	IIMSD86	604334.7
	496.101	KW76	604297.3
	496.151	IIMSD86	604236.9
10R34	192.780	CM71	
	199.140	CM71	
	251.910	CM71	
9P50	372.680	CM71	
	397.510	CM71	
C18O2	186.828	XKP84	1604647.7
	496.890	DFHAL83	
CO2S	992.000	DW78D	

CH3NC

10P14	2140.000	GB81
10P30	823.000	GB81
10P32	938.000	GB81
10P42	481.000	LSB81
10R04	404.000	DJ80
	454.000	LSB81
10R12	402.000	LSB81
10R18	284.000	LSB81
	288.000	DJ80
10R22	250.000	LSB81
10R24	277.000	LSB81
	280.000	DJ80

CH3I

9P34	508.370	CM76	
9R16	377.450	CM76	
10P08	459.180	CM76	
10P14	517.330	CM76	
10P16	576.170	CM76	
10P18	447.142	KW76	670463.0
	457.250	CM76	
10P22	719.300	CM76	
	964.000	DFBS75	
10P26	542.990	CM76	
	545.000	DFBS75	
10P28	670.990	CM76	
10P32	1253.738	DFBS75	239118.9
	1253.815	DBMF78	239104.2
	1253.822	DBMF78	239102.8
	1253.859	DBMF78	239095.8
10P36	529.280	CM76	
10P38	1063.290	CM76	
	1067.000	DFBS75	
10P42	390.530	CM76	
10R34	578.900	CM76	
9P04	525.320	CM76	
	583.870	CM76	
9P06	639.730	CM76	
9P26	477.870	CM76	
9R14	392.480	CM76	
CO2H	723.080	GRF87	
N2O	302.500	GRF87	
	331.720	GRF87	
	381.600	GRF87	
	1188.000	GRF87	
	1340.000	GRF87	

CH3NH2

9P06	118.000	DFSY81B	
9P08	118.000	PCD73	
	119.000	Lands80D	
	153.000	PCD73	
9P10	150.000	DFSY81B	D?
9P12	178.000	DFSY81B	D?
9P14	68.000	DFSY81B	
	226.000	DFSY81B	D?
9P18	102.000	DFSY81B	
	137.000	DFSY81B	
	377.000	DFSY81B	D?
9P20	130.000	DFSY81B	
9P22	179.000	Lands80D	
	250.000	DFSY81B	
9P24	147.000	PCD73	
	147.845	RPJM77	2027752.6
	148.500	DSF72	
	159.000	PCD73	
	197.940	IMEJ86	1514562.6
	218.000	DSF72	
	218.749	IMEJ86	1370485.0
	243.000	PCD73	
	244.890	DFSY81B	
	250.138	IMEJ86	1198510.1
	251.180	DFSY81B	
9P28	104.000	PCD73	
	105.000	DFSY81B	
9P32	166.000	PCD73	
	219.000	PCD73	
	220.000	Lands80D	
9P34	87.000	DFSY81B	
	203.000	DFSY81B	D?
	245.000	DFSY81B	D?
	387.000	DFSY81B	D?
9P40	109.000	DFSY81B	
	267.000	PCD73	
	600.000	DFSY81B	D?
9P44	115.500	PCD73	
	116.000	Lands80D	
9P46	180.000	Radfo75	
	283.000	Lands80D	
	351.000	Lands80D	
9R04	288.000	Radfo75	
	314.847	RPJM77	952185.0
9R08	146.000	Lands80D	
	194.000	PCD73	
9R12	177.000	PCD73	
	178.000	Lands80D	
	201.000	PCD73	
	208.000	PCD73	
	268.000	PCD73	

CH3NH2

	271.000	Lands80D	
9R14	99.500	PCD73	
	100.000	Lands80D	
	134.000	PCD73	
	139.000	PCD73	
	143.000	PCD73	
	145.000	Lands80D	
	183.000	PCD73	
	185.000	Lands80D	
	281.000	DFSY81B	
9R18	134.000	PCD73	
	164.000	PCD73	
	165.000	Lands80D	
	270.000	Lands80D	
9R20	92.000	DFSY81B	
	198.000	PCD73	
	199.000	DFSY81B	
9R22	168.000	PCD73	
	169.000	Lands80D	
	251.000	Lands80D	
10P12	246.000	Lands80D	
10P24	120.000	Lands80D	
	165.000	Lands80D	
	166.000	DFSY81B	D?
10R06	126.000	PCD73	
	175.000	PCD73	
10R12	128.000	Lands80D	
10R18	349.000	Lands80D	
10R20	142.000	Lands80D	
	221.000	Lands80D	
	347.000	PCD73	
	349.937	IMEJ86	856703.7
10R22	141.000	PCD73	
	142.000	DFSY81B	
10R32	176.000	PCD73	
	178.000	Lands80D	
10R36	147.000	Lands80D	
10R40	254.000	Lands80D	

In the above table the symbol D? indicates that the molecule may have been partially deuterated – i.e. CH_3NHD or CH_3ND_2.

CH3NO2		
9P06	376.000	DFSY81A
	552.000	DFSY81A
9P08	340.000	DFSY81A
	973.000	DFSY81A
9P14	470.000	DFSY81A
9P28	514.000	DFSY81A
9R04	414.000	DFSY81A
9R06	487.000	DFSY81A
9R08	472.000	DFSY81A
9R10	841.000	DFSY81A
9R12	487.000	DFSY81A
9R14	675.000	DFSY81A
	717.000	DFSY81A
9R16	530.000	DFSY81A
	598.000	DFSY81A
9R18	620.000	DFSY81A
	646.000	DFSY81A
9R20	656.000	DFSY81A
9R22	426.000	DFSY81A
9R26	1001.000	DFSY81A
9R28	524.000	DFSY81A
	594.000	DFSY81A
9R30	634.000	DFSY81A
	673.000	DFSY81A
	697.000	DFSY81A
9R32	780.000	DFSY81A
9R34	450.000	DFSY81A
	1070.000	DFSY81A
9R36	398.000	DFSY81A
9R40	778.000	DFSY81A
9R42	845.000	DFSY81A
10P16	311.000	DFSY81A
10P18	318.000	DFSY81A
10P20	344.000	DFSY81A
	351.000	DFSY81A
10P22	344.000	DFSY81A
	378.000	DFSY81A
10P24	424.000	DFSY81A
10P26	454.000	DFSY81A
10P28	489.000	DFSY81A
10P30	550.000	DFSY81A
	564.000	DFSY81A
10P32	631.000	DFSY81A
10P34	735.000	DFSY81A
10P36	869.000	DFSY81A
10P40	809.000	DFSY81A

CH3OCH3		
9R22	209.300	BSKK82
	338.900	BSKK82
9R26	497.400	BSKK82
9R28	530.700	BSKK82
9R32	220.300	BSKK82
9R34	645.500	BSKK82
9R40	304.300	BSKK82
10P08	441.300	BSKK82
10P12	495.000	PCD73
	520.000	PCD73
	526.300	BSKK82
10P16	564.700	BSKK82
10P20	375.000	PCD73
	496.500	BSKK82
	934.200	BSKK82
10P34	378.200	BSKK82
	461.000	PCD73
	480.000	PCD73
	492.000	PCD73
10P52	511.900	BSKK82

9P02	279.400	FK86	
	392.000	FK86	
	715.300	FK86	
9P06	134.700	BP77	
	182.100	NH80	
	229.100	BP77	
	417.100	BP77	
	487.600	FK86	
	510.000	KFK82	
	515.800	FK86	
	560.000	FK86	
	759.600	FK86	
9P10	133.000	YYSF82	
	134.000	LD79	
	416.700	FK86	
9P22	128.000	DSF74A	
9P26	100.800	LD79	
	101.440	VHHR86	
	117.227	RPJM77	2557365.4
	1325.000	FK86	
	1671.000	FK86	
9P30	81.900	NH80	
	89.600	NH80	
	103.125	RPJM77	2907088.9
	104.000	KYHH75	
	145.662	BKM78	2058141.8
	168.100	DSF74A	
	320.000	DSF74A	
	352.500	DSF74A	
	420.200	FK86	
9P32	80.180	VHHR86	
	88.720	VHHR86	
	108.720	VHHR86	
	110.000	YYSF82	
	110.700	NH80	
	113.800	NH80	
	141.000	YYSF82	
	145.600	NH80	
	178.000	YYSF82	
	178.820	VHHR86	
	279.400	DSF74A	
	498.000	DSF74A	
9P52	46.600	FK86	
	445.400	FK86	
9R02	106.000	Lands80C	
	114.960	GSRF80	
	186.400	FK86	
	372.400	FK86	
9R04	212.900	PD85	
	332.600	PD85	
	351.300	FK86	

	550.000	FK86	
	613.500	FK86	
	797.500	FK86	
	820.000	FK86	
	1134.000	FK86	
9R06	69.500	LD79	
	220.100	PD85	
	221.000	Lands80C	
	275.100	PD85	
	282.000	PD85	
9R08	46.700	BP77	
	47.650	VHHR86	
	57.240	PD85	
	289.600	PD85	
	294.811	BKM78	1016897.2
	305.726	BKM78	980591.6
	307.000	KYHH75	
	825.000	FK86	
9R14	215.372	BKM78	1391972.1
	233.400	VHHR86	
	234.000	NH80	
	238.000	LD79	
9R16	57.340	PD85	
	69.400	YYSF82	
	70.300	LD79	
	294.300	PD85	
9R22	167.000	YYSF82	
	169.250	VHHR86	
9R26	106.000	Lands80C	
	177.000	YYSF82	
	182.000	Lands80C	
9R28	186.000	Lands80C	
9R34	151.560	GSRF80	
10P10	134.000	HP79	
10P18	280.000	NH80	
	283.200	FK86	
10P24	136.000	HP79	
10P38	164.200	FK86	
10P40	173.500	FK86	
10P46	182.200	FK86	
	784.400	FK86	
10R14	172.600	PD85	
10R22	77.930	PD85	
	136.000	YYSF82	
	137.170	GSRF80	
	225.200	PD85	
	351.300	FK86	
10R34	223.700	FK86	
	238.000	NH80	
10R36	131.200	FK86	
10R38	249.700	FK86	

CH3OD		
10R42	141.000	Lands80C
10R44	110.000	KYHH75
	236.000	Lands80C
	241.000	NH80
	734.600	FK86
13CO2	74.380	PD85
	86.700	DVPA81
	91.080	PD85
	217.900	PD85
	310.000	DVPA81
	320.700	PD85
	917.000	DVPA81
C18O2	55.590	PD85
	172.700	PD85
	209.600	PD85
CO2H	140.400	FK86
	221.500	FK86
	226.300	FK86
	235.900	FK86
	243.500	FK86
	294.700	FK86
	307.500	FK86
CO2S	113.500	FK86
	306.500	FK86
N2O	139.850	GSRF80

CH3OH			
9P04	214.300	PS87	
9P06	418.790	Henni86	
9P10	214.350	Henni78	
	218.220	Henni78	
	232.080	Henni86	
	289.700	Henni82	
9P12	46.450	MH84B	
	163.574	VWE87	1832768.6
	164.200	IIMMS81	
	206.785	PEJS80	1449778.0
	211.315	PEJS80	1418701.0
	261.500	IIMMS81	
	290.620	Henni78	
	448.455	VWE87	668500.1
	450.400	IIMMS81	
	781.000	DFSY81B	
9P14	37.000	Knigh81	
	117.960	PEJS80	2541485.6
	164.508	PEJS80	1822362.7
	164.564	Knigh81	1821735.5
	301.994	PEJS80	992708.9
	386.339	PEJS80	775982.4
	416.522	PEJS80	719751.1
9P16	44.240	Henni86	
	48.700	MH84B	
	164.600	PEJS80	1821335.2
	223.500	CBB70B	
	369.114	PEJWG75	812195.4
	570.569	PEJWG75	525427.5
	1223.660	PEJS80	244996.6
9P18	304.800	Strum84	
	676.000	DFSY81B	
9P20	51.207	TOH88	
	53.831	TOH88	
9P22	47.800	VHHR86	
	48.100	SBW84	
	213.462	PEJS80	1404427.0
	214.800	DSF74A	
	346.488	PEJS80	865233.1
	658.900	PS87	
9P24	92.544	PEJS80	3239461.6
	133.120	PEJS80	2252054.2
	164.697	PEJS80	1820261.5
	165.600	DSF74A	
	311.200	DSF74A	
	470.000	DFSY81B	
	602.487	PEJS80	497591.6
	614.285	PEJS80	488034.7
	694.189	PEJS80	431859.8
9P26	43.450	MH84B	
	82.700	MH84B	

	516.000	DFSY81B	
	973.000	DFSY81B	
9P28	312.000	DFSY81B	
	416.710	Henni86	
9P30	210.000	DFSY81B	
	313.880	Henni86	
9P32	37.500	HRB73	
	37.854	PEJS80	7919660.2
	41.700	HRB73	
	42.159	PEJS80	7110981.4
	240.000	DFSY81B	
	270.000	IMS80	
	275.000	DFSY81B	
	372.000	DFSY81B	
	418.000	DFSY81B	
9P34	39.924	PEJS80	7509036.2
	40.200	HRB73	
	42.300	Henni82	
	43.400	HRB73	
	44.307	TOH88	
	51.240	TOH88	
	63.370	PEJS80	4730860.6
	65.600	HRB73	
	70.512	PEJS80	4251674.0
	80.300	HRB73	
	180.676	PEJS80	1659278.6
	185.500	PEJS80	1616128.4
	186.319	PEJS80	1609026.7
	190.726	PEJS80	1571849.7
	205.600	IIMMS81	
	208.300	IIMMS81	
	237.600	CBB70B	
	253.553	PEJS80	1182366.2
	254.041	PEJS80	1180092.5
	255.500	Strum84	
	263.683	PEJS80	1136942.0
	264.536	PEJS80	1133277.0
	292.500	CBB70B	
	303.000	DFSY81B	
	363.000	DFSY81B	
	622.000	DFSY81B	
	699.423	PEJWG75	428628.5
9P36	99.280	IIMMS81	
	110.716	VWE87	2707749.3
	118.834	PEJS80	2522781.6
	135.710	IIMMS81	
	162.218	VWE87	1848083.8
	170.576	RPJM77	1757526.3
	202.400	CBB70B	
	332.000	DFSY81B	
	392.069	PEJWG75	764642.6
	418.083	PEJS80	717065.0
9P38	193.142	PEJS80	1552190.1
	198.664	PEJS80	1509040.2
	278.805	PEJS80	1075277.1
	292.141	PEJS80	1026189.3
	624.430	PEJS80	480105.7
9P40	55.370	PEJS80	5414344.1
	60.173	PEJS80	4982153.1
	73.306	PEJS80	4089579.6
	85.601	PEJS80	3502210.2
	697.000	DFSY81B	
9P44	112.946	VWE87	2654310.7
	196.564	VWE87	1525164.1
	265.600	PS87	
9R02	94.700	SSKYM81	
	105.100	SSKYM81	
	152.000	Lands80C	
	176.000	Lands80C	
	261.000	Lands80C	
9R08	77.406	PEJS80	3873005.1
	86.239	PEJS80	3476282.5
	113.732	PEJS80	2635958.0
	225.516	PEJS80	1329362.9
	259.000	DFSY81B	
	461.000	DFSY81B	
9R10	96.000	Radfo75	
	96.522	PEJWG75	3105936.8
	164.783	PEJS80	1819314.0
	189.942	IIMSD86	1578339.2
	190.650	HIMS82	
	224.530	HIMS82	
	232.000	Radfo75	
	232.939	PEJWG75	1286999.5
	285.000	DFSY81B	
9R12	430.000	DFSY81B	
	448.080	Henni86	
9R14	100.806	PEJS80	2973940.6
	194.063	PEJS80	1544818.7
	209.930	PEJS80	1428057.6
	319.000	DFSY81B	
9R16	33.400	SMNM83	
	36.666	Henni86	
	41.910	VHHR86	
	56.730	Henni86	
	66.249	Henni86	
	191.683	Henni86	
	310.000	DFSY81B	
	419.000	DFSY81B	
9R18	61.613	PEJS80	4865709.8
	67.495	PEJS80	4441675.2
	186.042	PEJS80	1611421.9

	251.432	PEJS80	1192338.3
	280.934	PEJS80	1067127.2
9R20	174.300	PS87	
9R22	48.630	Henni86	
	48.740	SH84	
	232.788	PEJS80	1287832.2
9R24	56.150	SH84	
	461.700	PS87	
9R26	151.254	PEJS80	1982050.6
	159.676	PEJS80	1877508.5
	290.000	DFSY81B	
	346.000	DFSY81B	
	461.000	DFSY81B	
	781.000	DFSY81B	
9R28	157.600	Henni86C	
	195.000	PS87	
	289.000	DFSY81B	
9R34	136.800	PS87	
9R?	281.500	DSF74A	
10P10	84.005	Henni86	
	121.270	Henni86	
	161.530	Henni86	
	270.700	Henni86	
	467.850	Henni86	
10P12	42.400	Henni86	
	368.000	DFSY81B	
10P16	75.821	Henni86	
	84.913	Henni86	
	88.819	Henni86	
	99.861	VWE87	3002087.5
	123.640	Henni86	
	337.040	Henni86	
	479.150	Henni86	
	566.750	Henni86	
10P18	75.932	Henni86	
	100.010	Henni86	
	296.480	Henni86	
	523.120	Henni86	
	663.670	Henni86	
10R02	178.000	Lands80C	
10R04	179.728	PEJS80	1668035.0
	191.200	WZN73	
	211.263	PEJS80	1419049.3
	493.541	VWE87	607431.2
10R06	181.100	PS87	
10R08	151.000	DFSY81B	
	262.000	DFSY81B	
10R10	125.100	MH84B	
	191.620	PEJS80	1564518.7
	292.000	Radfo75	
	293.822	PEJS80	1020321.1

	588.700	PS87	
	881.300	PS87	
10R16	62.966	PEJS80	4761182.4
	69.680	PEJS80	4302444.9
	77.905	PEJS80	3848185.5
	564.000	DFSY81B	
	695.350	PEJS80	431139.0
10R20	145.050	Strum84	
	209.030	Strum84	
10R22	262.700	HL82	
	622.000	DFSY81B	
10R26	329.000	DFSY81B	
10R28	65.300	MH84B	
10R32	30.500	SBW84	
	100.166	VWE87	2992957.0
	145.252	VWE87	2063941.1
	242.847	VWE87	1234490.4
	362.000	HL82	
	390.000	DFSY81B	
10R34	40.030	TOH88	
	43.470	Henni78	
	48.760	TOH88	
	63.006	TOH88	
	92.664	PEJS80	3235253.6
	129.550	PEJS80	2314111.3
	130.000	KYHH75	
	163.010	WZN73	
	223.800	TOH88	
	242.473	PEJS80	1236396.8
	249.000	Radfo75	
	250.781	PEJWG75	1195433.9
	267.443	PEJS80	1120957.7
10R36	43.100	IMS80	
	53.500	IMS80	
	53.861	PEJS80	5566052.7
	233.000	DFSY81B	
10R38	163.034	PEJWG75	1838839.3
	164.000	DTS74	
	213.900	HL82	
	246.000	DTS74	
	251.140	PEJWG75	1193727.3
	251.400	HL82	
	254.000	Radfo75	
	261.700	HL82	
	469.023	PEJWG75	639184.6
10R40	97.519	PEJS80	3074210.0
	98.000	TYY75	
	167.587	PEJS80	1788876.6
	244.000	DFSY81B	
	470.900	PS87	
10R44	120.902	VWE87	2479622.2

CH3OH

	162.670	Henni86	
	231.300	HL82	
	251.000	TYY75	
	251.912	VWE87	1190069.1
	284.330	Henni86	
	381.820	Henni86	
10R46	53.000	IMMS81B	
	65.000	IMMS81B	
	274.245	VWE87	1093154.7
10R48	97.800	Henni86	
	149.800	Henni86	
	164.000	IMS80	
	286.155	VWE87	1047657.6
13C18O2	41.330	PD84	
	52.270	PD84	
	63.880	PD84	
	66.210	PD84	
	69.950	PD84	
	78.390	PD84	
	89.310	PD84	
	89.980	PD84	
	95.100	PD84	
	102.600	PD84	
	116.000	PD84	
	164.400	PD84	
	187.200	PD84	
	254.700	PD84	
	273.000	PD84	
	279.800	PD84	
	313.500	PD84	
	351.000	PD84	
	368.900	PD84	
	390.200	PD84	
	443.800	PD84	
	552.600	PD84	
	569.700	PD84	
	973.000	PD84	
13CO2	19.520	PD82A	
	49.980	PD84	
	51.850	PD82A	
	52.480	PD82A	
	56.230	PD82A	
	60.350	PD82A	
	77.840	PD84	
	85.830	PD82A	
	97.290	PD84	
	119.700	PD82A	
	124.700	PD82A	
	126.600	PD82A	
	132.900	PD82A	
	145.000	PD82A	

	148.000	PD82A
	177.000	PD82A
	200.900	PD84
	212.200	PD82A
	297.700	PD82A
	351.100	PD82A
	419.300	PD82A
	448.300	PD82A
	674.800	PD82A
9P04	50.699	TOH88
9P06	117.630	Henni86
	167.810	Henni86
	183.900	HL82
9P08	65.544	TOH88
	77.487	TOH88
	96.812	TOH88
	181.100	HL82
9P10	46.164	Henni86
	50.224	Henni86
	68.698	Henni86
	74.384	Henni86
	156.510	TOH88
	208.950	TOH88
9P12	152.670	TOH88
	257.800	HL82
9P14	80.843	Henni86
9P16	35.860	Henni83
	35.968	Henni86
	41.034	Henni86
	41.871	Henni86
	48.363	Henni86
	113.450	Henni86
	306.300	HL82
	627.340	WZN73
	1217.000	TTMYY74
9P18	205.200	HL82
9P22	654.920	Henni86
9P24	43.970	Henni83
	47.800	Henni83
9P26	25.270	Henni86
	41.171	Henni86
	42.953	Henni86
	81.903	Henni86
	507.480	Henni86
9P28	223.840	Henni86
	447.080	Henni86
9P30	41.500	TOH88
	52.171	TH86
	63.948	TOH88
	152.420	TH86
	200.210	Henni86

CH3OH

9P34	50.015	TH86			460.440	Henni86
	77.341	TH86			781.000	Henni86
	85.825	TH86	9R28		119.800	Henni86
	164.090	TH86			135.000	Henni86
	177.510	TH86			194.300	Henni86
9P36	64.397	TOH88	9R30		181.580	TOH88
	104.850	TH86			194.260	TOH88
9P40	98.862	TOH88	9R32		131.560	TOH88
	172.020	TOH88			183.680	TOH88
	365.830	TOH88	C18O2		41.060	PD82A
9R04	178.410	TOH88			60.580	PD82A
	235.100	HL82			73.200	PD82A
	261.200	TOH88			75.060	PD82A
9R06	53.988	TOH88			79.980	PD82A
	67.224	TOH88			93.960	PD84
	90.303	TOH88			95.250	PD82A
	190.270	TOH88			102.300	PD82A
	235.800	HL82			108.600	PD82A
	242.310	TOH88			109.400	PD82A
9R08	60.086	TOH88			117.400	PD82A
	62.171	TOH88			123.900	PD82A
	232.930	WZN73			131.600	PD82A
	234.610	TOH88			149.800	PD82A
	254.230	TOH88			167.100	PD82A
9R10	136.120	Henni86			176.500	PD82A
	242.200	HL82			193.500	PD82A
9R12	189.190	Henni86			205.100	PD82A
	327.770	Henni86			211.000	PD82A
9R14	130.100	Henni86			216.500	PD82A
	216.800	HL82			224.700	PD82A
	218.500	Henni86			233.400	PD82A
9R16	33.522	Henni86			240.400	PD82A
	188.900	Henni83			247.400	PD82A
9R18	33.540	Henni83			262.400	PD82A
	36.690	Henni83			271.500	PD82A
	39.785	TOH88			272.900	PD82A
	48.784	TOH88			312.100	PD82A
	63.681	TOH88			348.300	PD82A
	216.100	TOH88			356.500	PD84
9R20	172.000	Henni86			552.000	PD82A
	180.600	Henni86	CO2S		80.600	WGS77
	186.300	HL82			159.200	WGS77
	248.620	TOH88			171.300	WGS77
	366.420	TOH88			390.100	WGS77
9R22	67.430	Henni86			486.100	WGS77
	84.908	TOH88				
9R24	117.000	Henni86				
	176.000	Henni86				
	240.290	Henni86				
	289.170	Henni86				
	347.640	Henni86				

CH3SH

9P12	403.000	Lands80B
9P16	341.000	Lands80B
	384.000	Lands80B
9P18	127.000	Lands80B
9P22	379.000	Lands80B
9P24	128.000	Lands80B
9P30	298.000	Lands80B
9P38	262.000	Lands80B
	319.000	Lands80B
9P44	224.000	Lands80B
	316.000	Lands80B
	456.000	Lands80B
9R18	185.000	Lands80B
	205.000	Lands80B
9R28	351.000	Lands80B
9R30	161.000	Lands80B
9R34	116.000	Lands80B
10R16	324.000	Lands80B
10R24	124.000	Lands80B
10R34	117.000	Lands80B
	147.000	Lands80B
	234.000	Lands80B
	370.000	Lands80B

CHCL2F

9P08	495.963	VWPE83	604465.0
9P16	375.980	VWPE83	797362.5
9P18	365.725	VWPE83	819720.5
9P20	340.300	VWPE83	880965.6
9R04	832.757	VWPE83	359999.9
	905.428	VWPE83	331105.8
9R06	467.515	VWPE83	641246.9
	530.854	VWPE83	564736.1
9R08	549.258	VWPE83	545813.2
9R10	547.529	VWPE83	547537.6
9R12	580.869	VWPE83	516110.2
9R30	661.153	VWPE83	453438.9
9R34	470.386	VWPE83	637332.6
9R36	492.040	VWPE83	609284.6
9R40	561.028	VWPE83	534362.8

CHCLF2			
9P28	301.000	DFP78	
9R08	324.000	DFP78	
9R10	396.000	DFP78	
	747.050	DFP78	401301.7
9R12	326.000	DFP78	
9R14	360.606	DFP78	831356.9
9R16	335.467	DFP78	893656.9
	387.000	DFP78	
	842.125	DFP78	355995.2
9R18	385.687	DFP78	777294.8
	386.966	DFP78	774726.1
9R22	372.870	DFP78	804012.9
9R24	298.049	DFP78	1005850.0
	306.053	DFP78	979544.5
9R26	427.807	DFP78	700766.2
9R28	328.960	DFP78	911332.7
9R30	366.273	DFP78	818494.6
	380.000	DFP78	
	415.075	DFP78	722260.5
	481.452	DFP78	622683.6
9R32	337.094	DFP78	889343.5
	487.144	DFP78	615408.5
9R34	534.430	DFP78	560957.0
	556.097	DFP78	539101.3
	562.450	DFP78	533011.8
9R36	388.000	DFP78	
	414.351	DFP78	723522.9
9R38	899.384	DFP78	333330.9
9R40	370.000	DFP78	
	432.244	DFP78	693572.9
	444.000	DFP78	
	592.441	DFP78	506029.4
	617.656	DFP78	485371.3
9R44	467.700	TD80	
10P06	432.000	DFP78	
10P12	590.000	DFP78	
10P14	682.175	DFP78	439465.6
10P18	665.885	DFP78	450216.4
10P34	345.000	DFP78	
10R14	615.329	DFP78	487206.6
10R16	533.137	DFP78	562317.4
10R24	591.130	DFP78	507151.1
10R30	476.000	DFP78	
10R34	433.438	DFP78	691662.4
10R40	382.766	DFP78	783226.7

CHD2F			
9P10	231.000	Tobin84	
9P16	232.000	Tobin84	
9P34	163.300	Tobin84	
9P42	193.400	Tobin84	
9P44	301.000	Tobin84	
9R06	375.407	Tobin84	798579.5
9R16	260.000	Tobin84	
9R20	285.100	Tobin84	
9R30	292.700	Tobin84	
10P28	384.319	Tobin84	780061.5
	406.878	Tobin84	736812.0
10P46	984.795	Tobin84	304421.1
10P48	204.000	Tobin84	
10R20	862.544	Tobin84	347568.0
	986.070	Tobin84	304027.5
10R26	691.250	Tobin84	433695.8
	768.012	Tobin84	390348.8
10R38	435.427	Tobin84	688503.0

CHD2OH

9P04	246.800	FPVSF89
	512.800	FPVSF89
9P06	482.900	FPVSF89
9P08	196.800	FPVSF89
9P16	226.800	FPVSF89
9P18	104.600	FPVSF89
9P20	246.100	FPVSF89
	254.300	FPVSF89
	346.000	ZD78
	501.900	FPVSF89
9P22	484.400	FPVSF89
9P24	203.100	FPVSF89
9P26	128.500	FPVSF89
	204.700	FPVSF89
	255.300	FPVSF89
	437.400	FPVSF89
9P28	404.600	FPVSF89
9P30	385.400	FPVSF89
	518.000	ZD78
9P34	137.000	FPVSF89
	606.700	FPVSF89
	607.300	FPVSF89
9P36	249.700	FPVSF89
9P38	123.900	FPVSF89
	187.500	FPVSF89
9R04	165.100	FPVSF89
9R10	221.200	FPVSF89
9R14	280.800	FPVSF89
	317.000	FPVSF89
	598.300	FPVSF89
9R16	217.900	FPVSF89
9R18	165.000	ZD78
9R20	249.600	FPVSF89
9R24	164.400	FPVSF89
9R26	135.000	FPVSF89
	144.800	FPVSF89
	202.600	FPVSF89
9R30	120.900	FPVSF89
	132.200	FPVSF89
9R32	145.300	FPVSF89
	179.800	FPVSF89
9R34	279.000	FPVSF89
	290.200	FPVSF89
9R38	109.300	FPVSF89
	111.400	FPVSF89
	117.300	FPVSF89
	172.400	FPVSF89
10P04	57.900	FPVSF89
10P08	123.800	FPVSF89
	152.600	FPVSF89
10P10	517.800	FPVSF89

CHD2OH

10P12	171.100	FPVSF89
10P14	103.000	FPVSF89
10P16	103.000	FPVSF89
10P18	212.900	FPVSF89
	238.000	ZD78
	355.000	ZD78
10P20	291.300	FPVSF89
	427.100	FPVSF89
10P28	74.800	FPVSF89
	124.400	FPVSF89
	558.800	FPVSF89
10P40	125.400	FPVSF89
	142.900	FPVSF89
10R02	344.900	FPVSF89
10R04	74.100	FPVSF89
10R06	105.000	FPVSF89
10R08	305.600	FPVSF89
	452.500	FPVSF89
10R10	228.700	FPVSF89
	270.000	FPVSF89
10R14	83.700	FPVSF89
10R16	179.000	ZD78
	363.000	ZD78
10R18	55.600	FPVSF89
	127.400	FPVSF89
	259.900	FPVSF89
10R20	172.100	FPVSF89
	260.000	ZD78
10R22	288.300	FPVSF89
10R24	107.400	FPVSF89
10R26	41.800	FPVSF89
	45.100	FPVSF89
	136.200	FPVSF89
10R28	227.500	FPVSF89
10R32	83.900	FPVSF89
	93.000	FPVSF89
10R34	111.900	FPVSF89
10R36	278.400	FPVSF89
10R38	168.000	ZD78
	426.000	ZD78
10R40	80.000	FPVSF89
	111.600	FPVSF89

CHFCHF

9P16	411.100	BH82	
	433.800	BH82	
9P20	228.100	BH82	
	546.800	BH82	
9P28	286.500	DRH80	
	422.500	BH82	
9P30	360.500	BH82	
9P34	310.800	DRH80	
	543.200	BH82	
9P36	213.300	BH82	
9P38	687.200	BH82	
9P40	557.000	BH82	
9P42	583.700	BH82	
9R04	190.000	AD80	
9R08	185.000	AD80	
10P06	326.600	BH82	
	549.500	BH82	
10P08	219.300	BH82	
10P10	161.800	BH82	
10P14	272.100	BH82	
10P20	241.500	BH82	
10P26	220.700	BH82	
	339.000	BH82	
10R04	196.300	BH82	
	198.000	AD80	
10R06	437.600	BH82	
10R14	242.600	DRH80	
	424.000	BH82	
10R16	376.700	BH82	
	705.000	BH82	
10R18	219.500	BH82	
	284.600	BH82	
10R20	260.100	DRH80	
	389.300	BH82	
10R24	262.000	AD80	
	289.800	BH82	
10R26	232.800	BH82	
10R30	307.500	DRH80	
	442.100	BH82	
10R34	310.000	AD80	
10R38	231.100	BH82	
	360.000	BH82	
10R40	386.500	BH82	

CLO2

9R12	300.000	DSF75B	
9R14	775.000	DSF75B	
	914.721	DSF75B	327742.0
	914.735	DSF75B	327736.8
	914.755	DSF75B	327729.7
	914.780	DSF75B	327720.8
9R18	255.000	DSF75B	
9R20	207.000	DSF75B	
9R22	380.000	DSF75B	
9R24	264.000	DSF75B	
9R26	337.000	DSF75B	
9R36	509.859	DSF75B	587991.1
	525.000	DSF75B	
9R40	340.000	DSF75B	
10P14	1310.748	DSF75B	228718.6
10P16	949.685	DSF75B	315675.7
	1134.113	DSF75B	264340.9
10P20	409.000	DSF75B	
	1827.424	DSF75B	164051.9
10R08	233.000	DSF75B	
	418.000	DSF75B	
10R20	285.000	DSF75B	
10R24	264.000	DSF75B	
	459.886	DSF75B	651884.5
10R30	247.000	DSF75B	
10R32	215.000	DSF75B	
C18O2	176.000	DJ80	
	196.000	DJ80	
	204.000	DJ80	
	216.000	DJ80	

CO2

ELEC	13.151	HK66	
	13.537	HK66	
	16.580	HK66	
	17.024	HK66	
	17.372	HK66	

COF2

10P06	402.915	TD86	744058.0
	1650.312	TD86	181658.1
10P10	1079.380	GRF88	
	1184.380	GRF88	
10P16	538.415	TD86	556805.5
10P22	437.000	TW82	
	485.270	GRF88	
	505.829	TD86	592676.0
10P24	478.072	TD86	627086.5
	488.110	GRF88	
10P32	357.000	TW82	
	358.111	DPK87	837149.1
	1191.563	TD86	251596.0
10P36	440.000	TW82	
	444.745	TD86	674077.8
10P38	312.910	GRF88	
10P40	297.090	GRF88	
10R08	339.000	TW82	
	516.382	TD86	580562.9
	1891.062	TD86	158531.3
	1900.000	Tobin82	
10R10	509.440	GRF88	
10R12	572.510	GRF88	
10R14	837.270	GRF88	
	1135.070	TD86	264118.1
10R16	384.916	DPK87	778852.3
	839.400	GRF88	
10R18	640.350	GRF88	
10R22	527.000	TW82	
	699.636	TD86	428497.6
10R32	393.330	GRF88	
10R38	390.780	TD86	767165.1
10R40	379.242	TD86	790504.6
10R50	305.240	GRF88	
10R52	304.350	GRF88	
10R54	301.370	GRF88	
CO2S	369.620	GRF88	
N2O	335.850	GRF88	
	345.500	GRF88	
	354.630	GRF88	
	379.590	GRF88	
	424.130	GRF88	
	430.910	GRF88	
	539.100	GRF88	
	552.940	GRF88	
	601.670	GRF88	
	650.700	GRF88	
	665.700	GRF88	
	765.420	GRF88	
	799.170	GRF88	
	817.500	GRF88	

COF2

867.270	GRF88

CS2

ELEC	11.482	Patel65
	11.489	Patel65
	11.503	Patel65
	11.510	Patel65
	11.517	Patel65
	11.524	Patel65
	11.531	Patel65
	11.538	Patel65
	11.545	Patel65
	11.596	Patel65

D2CO

9P32	319.268	DDG80	939000.3
	733.574	DDSB77A	408673.8
	733.597	DDG80	408661.1
9P34	320.000	Lands80A	
	341.000	Lands80A	
9R14	233.126	DDG80	1285968.5
	245.000	Lands80A	
9R24	243.848	DDG80	1229421.8
	245.000	DDSB77A	
	256.000	Lands80A	
9R32	752.681	DDSB77A	398299.6
10P08	278.399	DF78	1076842.8
	294.000	Lands80A	
10P16	244.000	Lands80A	
10P22	324.000	Lands80A	
	346.000	Lands80A	
10P24	324.423	DF78	924078.4
10P28	364.484	DG81	822512.2
10R32	737.113	DDG80	406711.7

D2O			
CO2S	94.500	DW78D	
	112.600	DW78D	
ELEC	33.896	MC64	
	35.090	MC64	
	36.319	MC64	
	36.524	MC64	
	37.791	MC64	
	40.994	MC64	
	56.845	MC64	
	71.965	MC64	
	72.429	MC64	
	72.748	PEJWG75	4120984.3
	73.337	MC64	
	74.545	MC64	
	76.305	MC64	
	84.111	MC64	
	84.279	HSJ69	3557143.0
	107.720	PEJWG75	2783066.6
	171.670	BPT69	

DCN			
ELEC	181.788	Maki68	
	189.949	HJ68A	1578278.7
	190.008	HJ68B	1577789.0
	194.703	HJ68B	1539745.0
	194.764	HJ68B	1539257.0
	204.387	HJ68B	1466787.0

DCOOD			
9P12	843.237	DSF76	355525.8
	998.514	DSF76	300238.6
	1070.231	DSF76	280119.5
9P16	276.000	DSF76	
	935.009	DSF76	320630.4
9P38	1281.649	DSF76	233911.6
9R04	367.000	DSF76	
9R08	335.709	DSF76	893013.6
9R16	283.000	DSF76	
10P04	737.000	DSF76	
	795.000	DSF76	
	812.000	DSF76	
10P08	491.891	DSF76	609469.8
	508.791	DSF76	589225.0
	1160.000	DSF76	
10P10	726.920	DSF76	412414.5
	761.762	DSF76	393551.5
10P12	469.000	DSF76	
10P14	443.000	DSF76	
10P18	425.000	DSF76	
	592.000	DSF76	
10P20	561.294	DSF76	534109.6
	927.981	DSF76	323058.7
10P26	779.874	DSF76	384411.2
	877.548	DSF76	341625.1
	936.602	DSF76	320085.1
10P30	457.341	DSF76	655511.9
	666.000	DSF76	
10P34	514.951	DSF76	582177.0
	526.486	DSF76	569421.9
	527.215	DSF76	568634.6
10P44	414.000	DSF76	
10R04	479.000	DSF76	
10R06	298.000	DSF76	
10R10	395.149	DSF76	758682.5
	452.000	DSF76	
10R12	380.565	DSF76	787755.5
	389.907	DSF76	768882.0
10R14	352.000	DSF76	
	415.000	DSF76	
10R18	1009.409	DSF76	296997.9
10R20	218.000	DSF76	
	396.000	DSF76	
	789.420	DSF76	379762.8
	835.000	DSF76	
10R22	397.000	DSF76	
10R24	304.083	DSF76	985889.7
	310.000	DSF76	
	645.000	DSF76	
10R26	567.868	DSF76	527926.0
	591.616	DSF76	506735.1

DCOOD

10R28	325.000	DSF76
	508.000	DSF76
10R30	323.000	DSF76
10R32	265.000	DSF76
10R36	241.000	DSF76
10R40	350.000	DSF76

DCOOH

10P06	710.000	DSF76	
10P08	639.128	DSF76	469064.7
10P14	466.546	DSF76	642578.4
	479.904	DSF76	624692.6
10P16	433.000	DSF76	
10P20	362.000	DSF76	
10P22	328.457	DSF76	912729.7
10P30	272.000	DSF76	
10R12	365.000	DSF76	
	1047.579	DSF76	286176.6
10R14	433.235	DSF76	691985.3
10R16	342.000	DSF76	
10R20	265.000	DSF76	
10R24	312.000	DSF76	
	1237.966	DSF76	242165.4
10R28	971.806	DSF76	308489.9
10R30	647.348	DSF76	463108.3
10R34	713.106	DSF76	420404.0
	752.748	DSF76	398263.8
10R36	697.455	DSF76	429837.6

DFCO

9P12	450.000	JDL83
9P20	750.000	JDL83
9R12	124.000	JDL83
	164.000	JDL83
	384.000	JDL83
9R16	144.000	JDL83
10P16	514.000	JDL83
10P20	384.000	JDL83
13C18O2	906.000	JDL83
C18O2	198.000	JDL83
	354.000	JDL83
	358.000	JDL83
	569.000	JDL83
	608.000	JDL83
	664.000	JDL83
	788.000	JDL83
	1005.000	JDL83
	2216.000	JDL83

FCN

9R28	988.000	TJD86
C18O2	308.000	DJ80

H13COOH

9P06	313.797	DG82	955370.3
	313.797	DG82	955368.1
9P12	788.919	DDSB77B	380004.0
	838.369	DG82	357589.7
9P14	255.000	DG82	
	1491.846	DG82	200953.9
9P16	258.425	DG82	1160071.8
9P20	548.843	DG82	546225.3
9P24	381.615	DG82	785587.2
	536.096	DG82	559214.1
9P26	382.357	DG82	784063.1
9P30	477.963	DG82	627229.1
9P32	393.485	DG82	761888.8
9R26	448.534	DG82	668383.6
	464.627	DG82	645231.8
9R30	1030.378	DDSB77B	290953.8
	1116.483	DG82	268514.8
9R32	572.330	DG82	523810.4
10R28	310.000	DG82	
10R32	891.087	DG82	336434.4
10R46	480.000	DG82	

H2CO

ELEC	101.900	HM76
	119.600	HM76
	122.800	HM76
	125.900	HM76
	155.100	HM76
	157.600	HM76
	159.500	HM76
	163.800	HM76
	170.200	HM76
	184.400	HM76

H2O

ELEC	11.830	TP68	
	11.960	TP68	
	16.931	MC64	
	23.359	MC64	
	26.666	MC64	
	27.971	BBEK73	10718068.7
	28.054	MC64	
	28.273	MC64	
	28.356	MC64	
	32.929	MC64	
	33.033	MC64	
	35.000	MC64	
	35.841	MC64	
	36.619	MC64	
	37.859	MC64	
	38.094	MC64	
	39.698	MC64	
	40.629	MC64	
	45.523	MC64	
	47.251	MC64	
	47.463	DHJRS69	
	47.693	MC64	
	48.677	MC64	
	53.906	MC64	
	55.077	MC64	
	57.660	MC64	
	67.177	MC64	
	73.402	MC64	
	78.443	EWME70	3821775.0
	79.091	PEJWG75	3790474.5
	89.775	MC64	
	115.420	MC64	
	118.591	FSPB67	2527952.8
	120.080	MC64	
	220.230	PFS68	1361282.6
	791.060	MF66	

H2S

ELEC	33.470	HC69
	33.640	HC69
	49.620	HC69
	52.400	HC69
	56.840	HC69
	60.290	HC69
	61.500	HC69
	73.520	HC69
	80.500	HC69
	83.430	HC69
	87.470	HC69
	92.000	HC69
	96.380	HC69
	103.300	HC69
	108.800	HC69
	116.800	HC69
	126.200	HC69
	129.100	HC69
	130.800	HC69
	135.500	HC69
	140.600	HC69
	162.400	HC69
	192.900	HC69
	225.300	HC69

HBR

ELEC	19.399	AY70
	19.988	AY70
	20.360	AY70
	20.896	AY70
	20.949	AY70
	21.501	AY70
	21.546	AY70
	22.136	AY70
	22.226	AY70
	22.855	AY70
	23.436	AY70
	29.786	AY70
	30.445	AY70
	30.948	AY70
	31.368	AY70
	31.849	AY70
	32.469	AY70
	32.799	AY70
	33.409	AY70
	40.526	AY70

HCCCH2F

9P18	1547.000	TJD86
9P24	623.000	TJD86
9P32	1006.000	TJD86

HCCCHO

10P14	516.000	DJ80
10P18	148.000	DJ80
10P22	156.000	DJ80
10P26	336.000	DJ80

HCCF

9R18	1028.000	DJ80
C18O2	590.000	DJ80

HCL

ELEC	13.872	Deuts67C
	14.099	Deuts67C
	14.343	Deuts67C
	16.213	Deuts67C
	16.609	Deuts67C
	16.644	AY70
	16.765	AY70
	17.034	Deuts67C
	17.125	AY70
	17.492	Deuts67C
	17.575	AY70
	17.987	Deuts67C
	17.997	AY70
	18.035	AY70
	18.522	Deuts67C
	18.555	AY70
	18.593	AY70
	19.122	AY70
	19.145	AY70
	19.183	AY70
	19.700	Deuts67C
	19.783	AY70
	19.821	AY70
	20.346	Deuts67C
	20.411	Deuts67C
	20.999	Deuts67C
	21.047	Deuts67C
	21.156	Deuts67C
	21.813	Deuts67C
	21.971	Deuts67C
	22.651	Deuts67C
	22.864	Deuts67C
	23.571	Deuts67C
	23.849	Deuts67C
	24.318	Deuts67C
	24.583	Deuts67C
	24.618	Deuts67C
	24.937	Deuts67C
	25.704	Deuts67C
	26.146	Deuts67C
	26.247	AY70
	27.508	AY70

HCN				HCOOD			
ELEC	12.850	TP68		9P10	395.000	DSF76	
	71.899	MCW68		9P12	813.757	DSF76	368405.3
	73.101	MCW68			826.000	DSF76	
	76.093	MCW68		9P14	477.000	DSF76	
	77.001	MCW68		9P16	395.000	DSF76	
	81.554	MCW68		9P18	582.554	DSF76	514617.8
	96.401	MCW68		9P22	657.000	DSF76	
	98.693	MCW68		9P28	340.000	DSF76	
	101.257	MCW68			447.000	DSF76	
	110.240	MCW68		9P30	1541.750	DSF76	194449.4
	112.066	MCW68		9P34	374.000	DSF76	
	113.311	MCW68		9P36	351.000	DSF76	
	116.132	MCW68		9P38	355.000	DSF76	
	126.164	Maki68			361.000	DSF76	
	128.629	MCW68		9P40	411.000	DSF76	
	130.839	Maki68			531.000	DSF76	
	134.933	Maki68		9P44	498.000	DSF76	
	138.768	MCW68		9R06	304.000	DSF76	
	165.150	MCW68		9R16	393.000	DSF76	
	201.059	MCW68		9R22	417.000	DSF76	
	211.001	MCW68		9R30	594.000	DSF76	
	222.949	MCW68		9R38	473.000	DSF76	
	284.000	HJ67		9R40	668.000	DSF76	
	309.714	HJ67	967965.8	10P06	430.438	DSF76	696482.3
	310.887	HJRFS67	964312.3	10P12	450.980	DSF76	664757.9
	335.183	HJ67	894414.2	10P14	567.107	DSF76	528635.2
	336.558	HJRFS67	890759.5		590.000	DSF76	
	372.528	HJ67	804750.9	10P16	461.261	DSF76	649941.0
					472.000	DSF76	
				10P24	429.690	DSF76	697695.1
				10P26	372.000	DSF76	
				10P28	353.000	DSF76	
				10P30	356.000	DSF76	
				10P32	292.000	DSF76	
				10P36	819.000	DSF76	
				10R04	388.000	DSF76	
				10R06	353.000	DSF76	
				10R08	347.000	DSF76	
				10R10	324.000	DSF76	
					630.166	DSF76	475735.6
				10R12	395.712	DSF76	757601.9
					660.000	DSF76	
					692.000	DSF76	
				10R14	240.000	DSF76	
					926.209	DSF76	323677.0
				10R16	358.000	DSF76	
					433.000	DSF76	
				10R20	1161.676	DSF76	258068.8
				10R22	352.000	DSF76	
					398.000	DSF76	
				10R24	1730.833	DSF76	173207.0

HCOOD			
10R26	450.000	DSF76	
	689.998	DSF76	434483.0
10R28	369.968	DSF76	810320.5
10R30	326.000	DSF76	
10R32	919.936	DSF76	325884.2
	986.313	DSF76	303952.8
10R36	372.000	DSF76	
	695.672	DSF76	430939.4
	733.000	DSF76	
10R38	391.689	DSF76	765384.6
	1157.318	DSF76	259040.7
10R40	493.156	DSF76	607905.7
10R42	727.949	DSF76	411831.6

HCOOH			
9P16	437.451	DSF76	685316.6
	515.170	DSF76	581929.7
	533.678	DSF76	561747.5
9P20	254.800	WZN73	
9P26	405.750	WZN73	
9P28	1213.362	DSF76	247075.8
9P30	278.610	WZN73	
9P38	580.801	DSF76	516170.7
9R04	302.278	RPJM77	991777.8
	309.230	WZN73	
9R06	705.000	DSF76	
9R08	420.391	DSF76	713127.6
9R12	336.000	DDSB77B	
9R14	336.300	DSF76	
	393.500	DSF74D	
9R16	446.873	DSF76	670867.2
9R18	393.631	KW76	761608.3
	405.585	RPJM77	739161.0
	418.300	SJFWM87	
9R20	432.631	KW76	692951.4
	432.667	DSF76	692895.0
	441.810	DH78A	
	445.900	DSF76	672331.8
	460.000	SJFWM87	
9R22	133.900	Lands80A	
	196.500	Lands80A	
	418.613	RPJM77	716156.8
	432.109	DDSB77B	693788.5
	446.505	DDSB77B	671419.5
	580.387	DDSB77B	516538.7
9R24	418.510	WZN73	
	744.050	DDSB77B	402919.6
	761.000	Radfo75	
9R28	513.002	DDSB77B	584388.2
	513.016	DSF76	584372.9
	533.701	DSF76	561724.0
9R30	669.531	DDSB77B	447765.0
9R32	786.942	DSF76	380958.8
9R34	359.810	WZN73	
9R36	789.840	DSF76	379561.2
9R38	458.523	DDSB77B	653822.2
9R40	742.573	DSF76	403721.5
	786.162	DDSB77B	381336.9
9R42	396.000	DDSB77B	
10P14	445.210	WZN73	
10R14	930.000	Knigh81	
10R22	309.000	PCD73	
	311.554	EWTR79	962250.0
	319.480	WZN73	
10R24	320.000	DSF76	
10R32	930.000	Knigh81	

HCOOH

10R42	404.100	WZN73	
9P08	302.080	WZN73	
9P14	493.280	WZN73	
9P16	435.000	PCD73	
	518.830	WZN73	
9P18	334.820	WZN73	
9P38	577.000	PCD73	
9P42	492.000	WZN73	
9R10	460.510	WZN73	
9R14	334.910	WZN73	
	342.740	WZN73	
9R16	388.000	PCD73	
	401.000	PCD73	
	413.000	PCD73	
9R18	368.000	PCD73	
	392.000	PCD73	
	403.000	PCD73	
	421.000	WZN73	
	428.000	PCD73	
	441.000	PCD73	
	496.000	PCD73	
9R20	414.000	PCD73	
	428.000	PCD73	
9R22	414.000	PCD73	
	419.550	WZN73	
	433.100	WZN73	
9R26	512.000	PCD73	
	530.000	PCD73	
9R28	530.000	PCD73	
9R32	229.390	WZN73	
9R36	458.430	WZN73	

HDCO

9P08	194.352	DDG80	1542524.6
	196.000	DDSB77A	
9P16	405.486	DDG80	739340.3
9R22	152.000	DDG80	
9R26	195.000	DDSB77A	
10P30	331.088	DDG80	905477.0
10R34	155.000	DDG80	

HE

ELEC	95.763	TM76
	216.120	TM76
	216.300	KS80

HF

ELEC	10.198	Deuts67B
	10.458	Deuts67B
	10.582	Deuts67B
	10.744	Deuts67B
	10.812	Deuts67B
	11.057	Deuts67B
	11.403	Deuts67B
	11.541	Deuts67B
	11.785	Deuts67B
	11.863	AY70
	12.208	Deuts67B
	12.262	Deuts67B
	12.678	Deuts67B
	12.701	Deuts67B
	13.188	Deuts67B
	13.201	Deuts67B
	13.221	Deuts67B
	13.728	Deuts67B
	13.784	Deuts67B
	14.288	Deuts67B
	14.441	Deuts67B
	15.016	Deuts67B
	16.022	Deuts67B
	16.444	AY70
	16.655	AY70
	16.975	AY70
	17.095	AY70
	17.325	AY70
	17.645	AY70
	18.085	AY70
	18.801	Deuts67B
	19.113	Deuts67B
	20.134	Deuts67B
	20.351	Deuts67B
	20.939	Deuts67B
	21.699	Deuts67B
	21.789	Deuts67B

HFCO				NE		
9P16	260.000	JD81		ELEC	10.060	FMPG64
9P18	654.000	JD81			10.981	Patel68
9P24	120.000	JDL83			11.865	FMPG64
9R06	196.000	JDL83			12.820	FMPG64
9R08	120.000	JDL83			13.735	FMPG64
9R20	196.000	JD81			13.757	FMPG64
9R22	128.000	JD81			14.930	FMPG64
9R28	280.000	JD81			16.630	FMPG64
9R32	432.000	JD81			16.668	Patel68
9R36	220.000	JD81			16.893	Patel68
C18O2	258.000	JD81			16.947	Patel68
	280.000	JD81			17.156	FMPG64
	282.000	JD81			17.189	FMPG64
	306.000	JD81			17.802	FMPG64
					17.841	Patel68
					17.888	Patel68
ND2ND2					18.396	Patel68
					20.480	Patel68
9P14	293.000	SDEF85			21.752	Patel68
	434.000	SDEF85			22.836	Patel68
9P20	296.000	SDEF85			25.423	Patel68
9P22	249.000	SDEF85			28.064	FMPG64
	641.000	SDEF85			31.550	FMPG64
9P36	159.500	SDEF85			31.928	PFMG64B
10P10	658.500	SDEF85			32.020	FMPG64
10P16	115.000	SDEF85			32.520	FMPG64
10P18	134.000	SDEF85			32.830	FMPG64
10P22	244.000	SDEF85			34.550	FMPG64
	389.000	SDEF85			34.679	PFMG64B
10P30	311.000	SDEF85			35.602	PFMG64B
10P32	286.000	SDEF85			37.231	PFMG64B
	296.000	SDEF85			41.741	PFMG64B
	354.500	SDEF85			50.700	PFMG64A
10P34	587.500	SDEF85			52.390	PFMG64A
10P38	278.000	SDEF85			53.486	PFMG64B
	285.500	SDEF85			54.019	PFMG64B
	533.000	SDEF85			54.117	PFMG64B
	641.000	SDEF85			55.680	PFMG64A
10R08	290.000	SDEF85			57.355	PFMG64B
10R12	275.000	SDEF85			68.329	Patel68
	552.000	SDEF85			72.150	PFMG64A
	699.000	SDEF85			85.047	Patel68
10R14	217.000	SDEF85			86.900	PFMG64A
10R18	386.500	SDEF85			88.460	PFMG64A
10R24	301.000	SDEF85			89.930	PFMG64A
10R30	724.000	SDEF85			93.020	PFMG64A
10R36	252.000	SDEF85			106.020	PFMG64A
10R38	285.000	SDEF85			124.400	PFMG64A
10R40	454.000	SDEF85			126.100	PFMG64A
					132.800	PFMG64A

NH2NH2			
9P12	331.669	RPJM77	903889.4
	333.000	DSF74C	
	527.873	RPJM77	567925.4
9P20	311.075	RPJM77	963731.4
	483.500	DSF74C	
9P30	331.500	DSF74C	
9R08	250.500	DSF74C	
9R18	368.862	IMEJ86	812750.0
9R22	327.000	DSF74C	
10P06	181.926	RPJM77	1647877.4
	246.500	DSF74C	
10P12	721.000	DSF74C	
10P16	461.072	RPJM77	650207.7
10P18	271.500	DSF74C	
	372.500	DSF74C	
10P22	1007.000	DSF74C	
10P24	192.907	RPJM77	1554076.0
	193.500	DSF74C	
	336.000	DSF74C	
	435.772	RPJM77	687957.4
10P28	262.000	DSF74C	
10P32	795.000	DSF74C	
10R08	233.916	RPJM77	1281625.8
	235.000	DSF74C	
	533.655	IMEJ86	561772.0
10R12	301.275	RPJM77	995077.8
	373.000	DSF74C	
10R18	264.000	DSF74C	
10R20	264.801	RPJM77	1132140.6
10R24	802.400	DSF74C	
10R28	265.000	DSF74C	
10R34	234.000	DSF74C	
10R38	734.162	Knigh81	408346.7
13C18O2	195.000	JTT82	
	267.000	JTT82	
	863.000	JTT82	
13CO2	219.000	JTT82	
	289.000	JTT82	
	945.000	JTT82	
C18O2	705.000	JTT82	
N2O	200.000	JTT82	
	237.000	JTT82	
	320.000	JTT82	
	371.000	JTT82	
	487.000	JTT82	
	575.000	JTT82	

NH2OH		
9P12	290.000	TJD86
9R24	277.000	TJD86
13CO2	292.000	TJD86
C18O2	545.000	TJD86
	659.000	TJD86

9P40	88.000	GDEFJ83	(D)		12.682	SRD86	14
9R30	10.338	KR86	14		12.697	KR86	14
	10.342	KR86	14		12.720	KR86	14
	10.359	KR86	14		12.811	SRD86	14
	10.367	KR86	14		12.849	SRD86	14
	10.507	KR86	14		12.972	SRD86	14
	10.546	KR86	14		13.024	KR86	14
	10.718	KR86	14		13.051	KR86	14
	10.732	KR86	14		13.146	SRD86	14
	10.737	KR86	14		13.270	SRD86	14
	10.744	SRD86	14		13.415	KR86	14
	10.754	KR86	14		13.453	KR86	14
	10.762	KR86	14		13.578	KR86	14
	10.767	KR86	14		13.826	KR86	14
	10.784	SRD86	14		67.200	RGF79	14
	10.855	KR86	14		89.000	GDEFJ83	
	11.011	SRD86	14		147.040	WZN73	14
	11.209	SRD86	14		148.000	GDEFJ83	
	11.212	KR86	14		389.000	GDEFJ83	
	11.260	KR86	14	9R40	87.000	Lands80D	14D3
	11.263	KR86	14	9R42	375.000	GDEFJ83	
	11.460	KR86	14	10P02	234.400	Wille81	14
	11.471	KR86	14	10P04	90.000	GDEFJ83	(D)
	11.521	KR86	14	10P40	86.000	Lands80D	14D
	11.521	SRD86	14		113.000	Lands80D	14D
	11.712	SRD86	14		972.000	GDEFJ83	(D)
	11.716	KR86	14	10R02	370.400	WHH80	14
	11.727	KR86	14	10R06	291.000	Wille81	14
	11.746	SRD86	14	10R14	77.000	Lands80D	14D
	11.794	KR86	14		124.000	Lands80D	14D
	11.798	SRD86	14		694.000	GDEFJ83	(D)
	11.979	KR86	14	10R26	108.000	Lands80D	14D
	11.990	KR86	14		695.000	GDEFJ83	(D)
	12.010	SRD86	14	10R30	77.000	Lands80D	14D
	12.039	KR86	14		337.000	GDEFJ83	(D)
	12.079	SRD86	14	10R34	52.000	GDEFJ83	(D)
	12.080	KR86	14		465.000	GDEFJ83	(D)
	12.100	KR86	14	10R36	58.000	GDEFJ83	(D)
	12.245	SRD86	14		259.000	GDEFJ83	(D)
	12.249	KR86	14	10R42	10.789	SRD86	15
	12.261	KR86	14		11.257	SRD86	15
	12.281	SRD86	14		11.586	SRD86	15
	12.311	KR86	14		11.763	SRD86	15
	12.350	KR86	14		11.798	SRD86	15
	12.384	SRD86	14		11.866	SRD86	15
	12.403	KR86	14		12.063	SRD86	15
	12.528	KR86	14		12.148	SRD86	15
	12.540	KR86	14		12.299	SRD86	15
	12.561	SRD86	14		12.336	SRD86	15
	12.591	KR86	14		12.616	SRD86	15
	12.631	KR86	14		12.739	SRD86	15

NH3

	12.905	SRD86	15
	12.967	SRD86	15
	13.030	SRD86	15
	373.400	FWSGW82	15
13CO2	51.370	DP84	15
	66.870	DP84	14
	78.500	DVPA81	15
	110.000	DVPA81	15
	112.300	DVPA81	15
	124.600	DVPA81	
	152.700	WDVP80	15
	287.700	MPD83	15
	308.600	DP84	14
	388.500	DVPA81	15
C18O2	93.690	DP84	15
	102.900	DP84	14
CO2S	11.520	ZRGB80	14
	12.977	SRD86	15
	87.400	DW78D	14
	93.600	FWSGW82	15
	111.900	DW78A	15
	112.300	FWSGW82	14
	218.600	FWSGW82	15
	273.400	FWSGW82	15
	290.200	DW78D	14
	524.900	FWSGW82	15
N2O	42.600	Wille81	14
	81.497	KMNT79 3678577.0	14
	263.400	CBB70A	
ELEC	14.780	AW69	
	15.040	AW69	
	15.080	AW69	
	15.410	AW69	
	15.470	AW69	
	18.210	AW69	
	21.460	AW69	
	22.540	AW69	
	22.710	AW69	
	23.680	AW69	
	23.860	AW69	
	24.920	AW69	
	25.120	AW69	
	26.270	AW69	
	30.690	AW69	
	31.470	AW69	
	31.920	AW69	
	32.130	AW69	

In the preceding table the symbols in the righthand column indicate the molecular species as follows:

$14 - {}^{14}NH_3$
$15 - {}^{15}NH_3$
$14D3 - {}^{14}ND_3$
$14D - {}^{14}NH_2D$
(D) – deuterated state uncertain, i.e. NH3(D) in Table A.
no symbol – isotope not specified

O

ELEC	10.400	Wille71

O3

9P06	149.200	DDM83	
9P30	171.500	WZN73	
	217.830	DDM83	1376271.1
	313.600	DDM83	
9P40	163.610	WZN73	
9R32	489.038	DDM83	613025.0
9P14	121.000	WZN73	

OCS

9R08	378.400	Lands80C
ELEC	123.000	HC69
	132.000	HC69

PH3

9P12	135.940	SCLB81
9R14	104.000	SCLB81
10P42	194.000	SCLB81
10R34	83.770	SCLB81

SIH2F2		
9P20	175.500	DFS85
10P12	494.000	DFS85
10P22	330.000	DFS85
10R14	343.000	DFS85
10R18	494.000	DFS85
10R20	471.000	DFS85
10R22	613.000	DFS85
10R28	195.500	DFS85
13CO2	184.500	DFS85
	192.000	DFS85
	263.000	DFS85
	352.500	DFS85
	443.000	DFS85
	1053.000	DFS85
C18O2	169.000	DFS85
	190.500	DFS85
	193.000	DFS85
	261.500	DFS85
	317.500	DFS85
	355.000	DFS85
	375.500	DFS85

SIHF3		
9P32	345.000	DFS85
9P34	361.500	DFS85
10R12	334.000	DFS85
10R14	301.000	DFS85
10R16	465.000	DFS85
10R20	498.500	DFS85
10R24	488.000	DFS85
10R28	455.500	DFS85
10R30	436.500	DFS85
10R32	322.500	DFS85
13CO2	330.000	DFS85
	355.500	DFS85
	412.000	DFS85
	439.000	DFS85
	488.000	DFS85
C18O2	149.000	DFS85
	487.000	DFS85
	523.500	DFS85

SIH3F		
10P22	330.000	DS82
10R06	236.000	DS82
10R10	187.000	DS82
10R14	343.500	DS82
10R16	369.000	DS82
10R20	516.000	DFS85
10R22	340.000	DS82
10R26	264.500	DS82
10R30	1014.000	DS82
10R32	1286.000	DFS85
10R36	221.000	DS82
10R38	280.500	DS82
13CO2	622.000	DFS85
	689.000	DFS85
	1056.000	DFS85
	1058.000	DFS85

SO2		
9R14	139.600	CD81A
	146.200	SL84
	205.300	SL84
	208.000	BS81
9R18	128.100	SL84
	142.100	CD81A
	149.700	SL84
	182.000	BS81
9R26	165.200	SL84
9R28	142.100	SL84
	159.500	BS81
	160.000	CD81A
	169.600	BS84
	171.400	BS84
	258.000	BS81
	282.100	BS84
	684.800	BS81
9R40	180.000	SL84
	193.100	SL84
	312.100	BS81
	349.100	SL84
ELEC	139.800	HHCS71
	140.780	KS80
	140.880	KS80
	142.000	HHCS71
	150.000	HHCS71
	151.190	KS80
	151.310	KS80
	192.710	KS80
	206.400	HHCS71
	215.330	KS80

SO2(I)		
9P10	134.200	BS84
	192.000	BS84
	279.900	BS84
9P16	148.200	CD81B
	215.300	BS84
9P32	218.200	BS84
9R16	570.300	BS84
	1570.200	BS84
9R18	166.800	BS84
	208.800	BS84
	471.800	CD81B
9R20	185.100	BS84
	194.500	BS84
	298.900	BS84
9R22	184.100	BS84
9R24	174.700	BS84
	505.000	BS84
9R30	221.200	BS84
	232.900	BS84
	525.300	BS84

TRIOX					XE		
9P10	710.000	DWC81			ELEC	11.297	FMPG64
9P16	500.000	DWC81				12.266	Patel68
9P18	389.000	DWC81				12.917	Patel68
9P20	279.000	DWC81				18.514	FMPG64
9P26	1581.705	DWC81	189537.5			75.578	PP65
9P30	749.372	DWC81	400058.4				
9P32	813.654	DWC81	368452.2				
9R10	558.577	DWC81	536707.3				
9R12	661.000	DWC81					
9R14	789.000	DWC81					
9R16	694.428	DWC81	431711.4				
9R20	889.466	DWC81	337047.8				
9R22	459.428	DWC81	652533.9				
9R24	948.925	DDSB77A	315928.6				
9R26	497.000	DWC81					
	889.716	DWC81	336952.9				
9R30	384.869	DWC81	778946.7				
9R32	711.752	DWC81	421203.7				
10P08	952.000	DWC81					
10P12	381.000	DWC81					
10P18	501.164	DWC81	598192.4				
	750.000	DDSB77A					
	1055.000	DWC81					
10P20	733.000	DWC81					
	750.606	DWC81	399400.6				
	771.038	DWC81	388816.9				
10P22	419.839	DWC81	714065.8				
	467.000	DWC81					
10P30	491.376	DWC81	610108.3				
10P32	813.000	DWC81					
10P34	679.766	DWC81	441023.2				
10P36	607.714	DWC81	493311.8				
	656.000	DWC81					
10P38	593.000	DWC81					
10P40	509.890	DWC81	587955.6				
	512.000	DDSB77A					
10P44	433.000	DDSB77A					
10R06	948.250	DWC81	316153.4				
10R10	366.000	DWC81					
	1292.743	DWC81	231904.1				
10R12	404.000	DWC81					
	1579.903	DWC81	189753.7				
10R22	619.000	DDSB77A					
	695.000	DWC81					
	727.000	DWC81					
	782.000	DWC81					
	1053.477	DWC81	284574.4				
	1185.079	DWC81	252972.6				
10R26	393.000	DWC81					
10R30	2031.281	DWC81	147587.9				

9. Table C. **Partnerless Lines**

The 988 lines in the following table are:

(i) all pumped by regular CO_2 laser lines;
(ii) all "partnerless" – that is, each line is the only one corresponding to that particular molecule and pump transition.

The list is ordered by wavelength. It should be of interest to experimentalists who require a monochromatic source near a given wavelength and who might not be in a position to easily filter out weak emission at undesirable wavelengths, or for whom even tiny amounts of extraneous emission could be disastrous (for example in the testing of heterodyne systems). The lines in this list are (almost) certain to be the only ones to lase under the given conditions. I would certainly be glad to hear of any exceptions!

Three things should be noted:

(i) in order not to eliminate useful lines the species belonging to "families" with uncertain isotopic identification (see Chapter 8) have been regarded as *distinct* in compiling this list. Table B is arranged in such a way as to make it easy to check for lines which may lase under the same conditions as those of the desired line.
(ii) some caution is called for also because of the uncertainties in identifying pump transitions, especially R-branch transitions. For this reason too it is advisable to consult Table B, and possibly the relevant footnotes in Table A.
(iii) as with Table B, the * symbols indicating long-pulsed lines, and the footnotes, have been removed.

The list can be readily supplemented, in a given wavelength range, by making use of the 646 lines pumped by hot, sequence, isotopic CO_2 and N_2O lasers. Most of these are indeed partnerless, which can be verified after using Tables A and B to find the appropriate references. Whether such a supplementation is useful or not would depend upon the experimental techniques available. The list here presented will hopefully satisfy most requirements for monochromatic sources.

λ [μm]	Molecule	Pump	Reference	Freq. [MHz]
16.000	CF4	9R12	Telle83	
34.790	13CH3OH	10R22	HP78	
43.700	CD3OH	10R48	CIMPS87	
44.000	CH2DOH	9P30	SPEJ80	
48.400	CH318OH	10R38	IMPSG89	
52.200	13CD3OH	9R18	IEPSV84	
56.870	CD3OH	9P26	CIMPS87	
57.900	CHD2OH	10P04	FPVSF89	
60.600	CD3OD	10P04	FK86	
65.300	CH3OH	10R28	MH84B	
65.400	13CD3OH	9R10	IEPSV84	
67.800	13CD3OH	10R12	IEPSV84	
68.000	CH3NH2	9P14	DFSY81B	
68.800	CD3OH	10R46	SEJZS87	
69.900	CH318OH	10R12	IMPSG89	
70.000	13CH3OH	9P42	IMSD85	
71.700	CD3OH	10R10	CIMPS87	
73.467	13CD3OH	10R20	IEPSV84	4080637.2
74.100	CHD2OH	10R04	FPVSF89	
75.275	13CD3OD	9R24	VE85	3982631.1
75.500	13CD3OD	9R10	VE85	
77.000	14NH2D	10R30	Lands80D	
78.200	CH318OH	10R24	IMPSG89	
78.780	CD3OH	9R42	CIMPS87	
79.600	CD3OD	10R42	PVSEP85	
80.300	13CH3OH	10P34	PS87	
80.500	CD3OD	10R20	VSPE81	
82.100	13CD3OD	9R34	VE85	
82.600	CD3OD	9R18	PVSEP85	
83.600	CD3OD	10P30	PVSEP85	
83.700	CHD2OH	10R14	FPVSF89	
83.770	PH3	10R34	SCLB81	
84.400	13CD3OD	10P12	VE85	
84.500	CD3OH	10P06	PFS86	
86.300	CD3OH	10P56	CIMPS87	
87.000	CH3NH2	9P34	DFSY81B	
87.000	14ND3	9R40	Lands80D	
87.650	CH318OH	9R06	IMPSG89	
87.900	13CH3OH	9P08	HP78	
88.000	NH3(D)	9P40	GDEFJ83	
89.000	13CH3OH	9P44	IMSD85	

λ [μm]	Molecule	Pump	Reference	Freq. [MHz]
90.000	NH3(D)	10P04	GDEFJ83	
98.000	13CH3OH	9P28	PS87	
98.500	13CD3OH	9R34	IEPSV84	
98.650	CH318OH	10R10	IMPSG89	
99.140	CH318OH	9P28	IMPSG89	
101.300	13CH3OH	9R40	PS87	
102.023	CH2DOH	9P16	SPEJ80	2938465.1
102.200	CD3OD	9R24	PVSEP85	
103.000	CHD2OH	10P14	FPVSF89	
103.000	CD3OH	10P14	PFS86	
103.000	CHD2OH	10P16	FPVSF89	
103.000	13CH3OH	9P24	IMSD85	
104.000	PH3	9R14	SCLB81	
104.600	CHD2OH	9P18	FPVSF89	
105.000	CHD2OH	10R06	FPVSF89	
105.518	CH2F2	9P16	PSE80	2841142.9
106.400	13CH2F2	9R14	STPVE85	
107.400	CHD2OH	10R24	FPVSF89	
108.000	14NH2D	10R26	Lands80D	
109.100	C2H4O2H2	9R16	PCD73	
110.900	13CH3OH	10R46	PS87	
111.900	CHD2OH	10R34	FPVSF89	
112.000	13CH2F2	9P44	STPVE85	
115.000	ND2ND2	10P16	SDEF85	
116.000	CH3SH	9R34	Lands80B	
118.000	C2H4O2H2	10P30	PCD73	
119.400	13CD3OH	9R14	IEPSV84	
120.000	HFCO	9R08	JDL83	
120.000	HFCO	9P24	JDL83	
121.000	O3	9P14	WZN73	
124.000	CH3SH	10R24	Lands80B	
127.000	CH3SH	9P18	Lands80B	
127.300	CD3OH	10P04	PFS86	
128.000	HFCO	9R22	JD81	
128.000	CH3SH	9P24	Lands80B	
128.000	CH3OD	9P22	DSF74A	
128.000	CH3NH2	10R12	Lands80D	
129.200	13CD3OD	9P38	VE85	
129.600	CD3OD	9P42	PVSEP85	
130.000	CH3NH2	9P20	DFSY81B	
130.500	CD3OD	10R06	VSPE81	

λ [μm]	Molecule	Pump	Reference	Freq. [MHz]
131.200	CH3OD	10R36	FK86	
131.500	CD3OD	10R32	PVSEP85	
131.690	CH318OH	9R30	IMPSG89	
132.000	C2H4O2H2	9P24	PCD73	
133.700	13CH3OH	10P12	PS87	
134.000	ND2ND2	10P18	SDEF85	
134.000	CH3OD	10P10	HP79	
135.940	PH3	9P12	SCLB81	
136.000	CH3OD	10P24	HP79	
136.800	CH3OH	9R34	PS87	
140.000	CD3OH	9R10	PFS86	
140.300	CH2DOH	9P20	SPEJ80	
140.405	13CH2F2	9R38	STPVE85	2135193.0
140.900	13CH3OH	10P30	PS87	
141.000	CH3OD	10R42	Lands80C	
141.700	CD3OD	9R12	PVSEP85	
144.000	DFCO	9R16	JDL83	
145.563	13CD3OH	10R24	IEPSV84	2059531.6
147.000	CH3NH2	10R36	Lands80D	
148.000	HCCCHO	10P18	DJ80	
148.617	13CD3OD	10R30	VE85	2017218.5
150.000	CH3NH2?	9P10	DFSY81B	
150.200	13CD3OH	9R30	IEPSV84	
150.300	13CD3OD	9R32	VE85	
151.000	13CD3OD	9P24	VE85	
151.560	CH3OD	9R34	GSRF80	
152.000	HDCO	9R22	DDG80	
152.300	CD3OD	9P44	PVSEP85	
153.700	CD3OH	10P46	CIMPS87	
155.000	HDCO	10R34	DDG80	
155.000	13CH3OH	10P08	PS87	
155.600	CD3F	10P10	TSW79B	
156.000	13CD3OH	9P38	IMSD85	
156.000	HCCCHO	10P22	DJ80	
159.500	ND2ND2	9P36	SDEF85	
161.000	CH3SH	9R30	Lands80B	
161.300	CD3OD	10P08	PVSEP85	
161.800	CHFCHF	10P10	BH82	
163.300	CHD2F	9P34	Tobin84	
163.610	O3	9P40	WZN73	
164.000	C2H4O2H2	9P16	PCD73	

λ [μm]	Molecule	Pump	Reference	Freq. [MHz]
164.000	CH2DOH	9R18	ZD78	
164.200	CH3OD	10P38	FK86	
164.400	CHD2OH	9R24	FPVSF89	
164.656	13CH2F2	9P12	STPVE85	1820715.0
165.000	CHD2OH	9R18	ZD78	
165.100	CHD2OH	9R04	FPVSF89	
165.200	SO2	9R26	SL84	
165.300	CD3OD	10P14	PVSEP85	
166.000	CH3NH2?	10P24	DFSY81B	
170.180	CH318OH	9R26	IMPSG89	
171.000	C2H4O2H2	9R08	PCD73	
171.100	CHD2OH	10P12	FPVSF89	
172.000	CD3OD	9P46	PVSEP85	
172.600	CH3OD	10R14	PD85	
172.800	CD3F	10R06	TSW79B	
173.500	CH3OD	10P40	FK86	
173.637	13CD3OD	10R20	VE85	1726548.5
175.100	13CD3OD	9R22	VE85	
175.500	SIH2F2	9P20	DFS85	
176.000	CH2CLF	9P12	ADF84	
177.600	13CD3OH	9R40	IEPSV84	
178.000	CH3OH	10R02	Lands80C	
178.000	CH3NH2?	9P12	DFSY81B	
178.600	CD3OD	9R02	PVSEP85	
180.600	13CH2F2	9P34	STPVE85	
181.100	CH3OH	10R06	PS87	
183.400	13CD3F	9P36	TF81	
184.000	CD3OH	9R08	DSF75A	
184.100	SO2(l)	9R22	BS84	
185.000	CHFCHF	9R08	AD80	
186.000	CH3OD	9R28	Lands80C	
186.043	13CH2F2	9P38	STPVE85	1611414.0
187.000	SIH3F	10R10	DS82	
189.200	CD3OD	10P42	FK86	
190.000	CHFCHF	9R04	AD80	
190.300	13CH3OH	9P04	PS87	
191.500	CD2CL2	10P18	SDFY86	
193.400	CHD2F	9P42	Tobin84	
194.000	PH3	10P42	SCLB81	
195.000	HDCO	9R26	DDSB77A	
195.500	SIH2F2	10R28	DFS85	

λ [μm]	Molecule	Pump	Reference	Freq. [MHz]
196.000	HFCO	9R20	JD81	
196.000	HFCO	9R06	JDL83	
196.800	CHD2OH	9P08	FPVSF89	
197.388	13CH2F2	9P20	STPVE85	1518795.0
200.000	CD3F	9P28	TSW79B	
201.500	CD3F	10P12	TSW79B	
203.100	CHD2OH	9P24	FPVSF89	
203.500	CD3OH	10P16	PFS86	
204.000	CHD2F	10P48	Tobin84	
206.000	CD3F	9P16	TSW79B	
206.600	CH3CH2F	10P36	WZN73	
206.600	CH318OH	9P20	IMPSG89	
207.000	CLO2	9R20	DSF75B	
207.200	CH2F2	9P34	MH84A	
208.300	CH2CL2	10P18	HW82	
208.300	CD3OD	10R08	PVSEP85	
209.100	13CD3F	9P24	TF81	
213.300	CHFCHF	9P36	BH82	
215.000	CLO2	10R3?	DSF75B	
216.500	13CH3OH	9R32	PS87	
216.800	CH2DOH	9R16	SPEJ80	
216.900	CD3OD	10P20	PVSEP85	
217.000	ND2ND2	10R14	SDEF85	
217.900	CHD2OH	9R16	FPVSF89	
218.000	CH2CLF	9P22	ADF84	
218.200	SO2(I)	9P32	BS84	
219.300	CHFCHF	10P08	BH82	
220.000	HFCO	9R36	JD81	
220.300	CH3OCH3	9R32	BSKK82	
221.000	SIH3F	10R36	DS82	
221.200	CHD2OH	9R10	FPVSF89	
224.000	CD3CL	9R28	DFBS75	
225.000	CD3OH	9R46	PFS86	
226.000	CH3NH2?	9P14	DFSY81B	
226.800	CHD2OH	9P16	FPVSF89	
226.900	CH3CH2F	10P40	WZN73	
227.000	13CD3OD	9R28	VE85	
227.150	CH3CL	9P48	CM76	
227.500	CHD2OH	10R28	FPVSF89	
227.657	CH2F2	9P18	PSE80	1316860.5
231.000	CHD2F	9P10	Tobin84	

λ [μm]	Molecule	Pump	Reference	Freq. [MHz]
232.000	CHD2F	9P16	Tobin84	
232.800	CHFCHF	10R26	BH82	
233.685	CD2F2	9P10	VPE81	1282892.0
234.000	NH2NH2	10R34	DSF74C	
234.400	14NH3	10P02	Wille81	
234.800	13CH2F2	9R18	STPVE85	
235.500	CH2CL2	10P24	HW82	
235.700	CD3OD	10P06	PVSEP85	
236.000	SIH3F	10R06	DS82	
236.100	CD3OD	9P38	PVSEP85	
236.250	CH3CL	9R02	CM76	
237.500	CD2CL2	10P26	SDFY86	
238.000	CH2DOH	10P18	ZD78	
240.100	13CH3OH	10R06	PS87	
240.980	CH3CL	10P10	CM76	
241.000	DCOOD	10R36	DSF76	
241.500	CHFCHF	10P20	BH82	
244.000	CH2CLF	9P26	ADF84	
244.000	D2CO	10P16	Lands80A	
245.000	CD3CL	9P32	DFBS75	
246.000	CH3NH2	10P12	Lands80D	
246.000	CD3CL	10R14	DFBS75	
247.000	CLO2	10R30	DSF75B	
247.400	13CH3OH	10P10	PS87	
247.500	CD3F	9R10	TSW79B	
248.606	13CH2F2	9R26	STPVE85	1205896.0
249.000	CD3CL	9P38	DFBS75	
249.600	CHD2OH	9R20	FPVSF89	
249.700	CH3OD	10R38	FK86	
249.700	CHD2OH	9P36	FPVSF89	
249.900	13CD3F	10R34	TF81	
250.000	CH3NC	10R22	LSB81	
250.500	NH2NH2	9R08	DSF74C	
252.000	ND2ND2	10R36	SDEF85	
254.000	CD2CL2	10R36	ZD78	
254.000	CH3NH2	10R40	Lands80D	
254.800	HCOOH	9P20	WZN73	
255.000	CLO2	9R18	DSF75B	
255.000	CD3OD	9P48	PVSEP85	
256.400	CD3OD	9P08	PVSEP85	
258.300	CD3OH	9P46	CIMPS87	

λ [μm]	Molecule	Pump	Reference	Freq. [MHz]
258.425	H13COOH	9P16	DG82	1160071.8
260.000	CHD2F	9R16	Tobin84	
260.000	HFCO	9P16	JD81	
261.030	CH3CL	10P34	CM76	
261.729	CH2F2	9P38	PSE80	1145430.1
262.000	NH2NH2	10P28	DSF74C	
262.100	CD3OD	10P48	PVSEP85	
263.200	CD3OD	9R42	FK86	
263.500	CH2CHF	10R12	RGF84	
264.000	CLO2	9R24	DSF75B	
264.000	NH2NH2	10R18	DSF74C	
264.050	CH3BR	10R10	CM76	
264.500	SIH3F	10R26	DS82	
264.700	CH3CH2F	9P18	WZN73	
264.801	NH2NH2	10R20	RPJM77	1132140.6
265.000	DCOOH	10R20	DSF76	
265.000	CD3F	9P52	TSW79A	
265.000	DCOOD	10R32	DSF76	
265.000	NH2NH2	10R28	DSF74C	
266.000	CDF3	10P14	TF80	
266.900	CDF3	9P44	TF80	
267.823	CD2F2	9P18	VPE81	1119368.0
268.300	CH318OH	9R38	IMPSG89	
271.290	CH3CL	10P20	CM76	
272.000	CD3I	9P12	DFBS75	
272.000	DCOOH	10P30	DSF76	
272.100	CHFCHF	10P14	BH82	
272.500	CD3OD	9P32	PVSEP85	
272.958	13CD3OD	10R26	VE85	1098307.9
275.090	CH3CL	9R36	CM76	
277.000	NH2OH	9R24	TJD86	
278.400	CHD2OH	10R36	FPVSF89	
278.570	CH3CL	9R16	CM76	
278.610	HCOOH	9P30	WZN73	
279.000	TRIOX	9P20	DWC81	
279.810	CH3BR	10R52	CM76	
280.000	HFCO	9R28	JD81	
280.500	SIH3F	10R38	DS82	
280.800	13CD3F	10P20	TF81	
281.180	CH3CN	9P34	CM76	
281.200	CH2F2	9P32	MH84A	

λ [μm]	Molecule	Pump	Reference	Freq. [MHz]
281.500	CH3OH	9R?	DSF74A	
281.600	CH2CHF	10R44	RGF84	
282.300	CH3CH2F	9R12	WZN73	
283.000	DCOOD	9R16	DSF76	
283.750	CD3OH	9P42	CIMPS87	
284.000	CH2CLF	9R16	ADF84	
284.150	CH318OH	9R10	IMPSG89	
285.000	ND2ND2	10R38	SDEF85	
285.000	CLO2	10R20	DSF75B	
285.100	CHD2F	9R20	Tobin84	
285.300	CH3CH2OH	9P40	BS85	
286.300	CDF3	10R48	TF80	
286.790	CH3CL	10R34	CM76	
286.800	CDF3	10R04	TF80	
287.950	CD3OH	9P50	CIMPS87	
288.000	CD3CL	9P16	DFBS75	
288.000	CD3CL	10R18	DFBS75	
288.300	CHD2OH	10R22	FPVSF89	
289.400	CD2CL2	10R08	BF86	
290.000	CD3BR	10P18	Lands80A	
290.000	NH2OH	9P12	TJD86	
290.000	ND2ND2	10R08	SDEF85	
291.000	14NH3	10R06	Wille81	
292.000	CH2CLF	9R20	ADF84	
292.000	HCOOD	10P32	DSF76	
292.700	CHD2F	9R30	Tobin84	
293.648	CD3CL	9P24	DFBS75	1020924.7
293.800	CH2CF2	10R08	HW84	
294.280	CH3BR	10R28	CM76	
296.000	ND2ND2	9P20	SDEF85	
297.090	COF2	10P40	GRF88	
298.000	DCOOD	10R06	DSF76	
298.000	CH3SH	9P30	Lands80B	
298.000	CH2CHF	10P08	TW82	
299.500	13CD3F	9P48	TF81	
300.000	CLO2	9R12	DSF75B	
300.233	13CH2F2	9R36	STPVE85	998532.1
300.246	13CH2F2	10R38	STPVE85	998487.9
301.000	CD3I	9R26	DFBS75	
301.000	CHCLF2	9P28	DFP78	
301.000	ND2ND2	10R24	SDEF85	

λ [μm]	Molecule	Pump	Reference	Freq. [MHz]
301.000	SIHF3	10R14	DFS85	
301.000	CHD2F	9P44	Tobin84	
301.200	CH2NOH	10P10	DP84	
301.370	COF2	10R54	GRF88	
301.654	13CH2F2	10R04	STPVE85	993829.9
302.080	HCOOH	9P08	WZN73	
303.540	CH3CN	10P10	CM71	
303.800	CD2F2	10R48	VPE81	
304.000	HCOOD	9R06	DSF76	
304.300	CH3OCH3	9R40	BSKK82	
304.350	COF2	10R52	GRF88	
305.240	COF2	10R50	GRF88	
308.000	CD3OD	10R52	FK86	
310.000	H13COOH	10R28	DG82	
310.000	CHFCHF	10R34	AD80	
311.000	ND2ND2	10P30	SDEF85	
311.000	CH3NO2	10P16	DFSY81A	
311.100	CH3BR	10P20	CM76	
311.200	CH3BR	10P40	CM76	
311.210	CH3BR	10R50	CM76	
312.276	13CH2F2	9P24	STPVE85	960024.5
312.910	COF2	10P38	GRF88	
316.000	CD3OD	9R26	YKYSF81	
318.000	CH3NO2	10P18	DFSY81A	
318.000	CD3CL	10R28	DFBS75	
320.000	HCOOH	10R24	DSF76	
320.597	CD2F2	10R44	VPE81	935107.5
322.500	SIHF3	10R32	DFS85	
322.800	CH2CHF	10R26	RGF84	
323.000	DCOOD	10R30	DSF76	
323.300	CD3F	10P08	TSW79B	
324.000	CHCLF2	9R08	DFP78	
324.000	CH3SH	10R16	Lands80B	
324.000	CH2CLF	9R04	ADF84	
324.140	13CD3OD	10R14	VE85	924885.8
324.423	D2CO	10P24	DF78	924078.4
325.900	13CD3F	9R22	TF81	
326.000	CHCLF2	9R12	DFP78	
326.000	HCOOD	10R30	DSF76	
327.000	NH2NH2	9R22	DSF74C	
327.600	CH3CH2BR	10R10	BS83	

λ [μm]	Molecule	Pump	Reference	Freq. [MHz]
328.000	CH3CHO	9R22	LSB81	
328.457	DCOOH	10P22	DSF76	912729.7
328.900	13CH3OH	9P38	PS87	
328.960	CHCLF2	9R28	DFP78	911332.7
329.000	CH3OH	10R26	DFSY81B	
329.500	CH2CF2	10R36	HW84	
329.900	CD3OD	10P16	PVSEP85	
330.000	SIH2F2	10P22	DFS85	
330.000	SIH3F	10P22	DS82	
330.019	CDF3	10R08	TLD83	908408.6
330.100	CH2CHF	10R22	RGF84	
330.991	CD2F2	9P34	VPE81	905742.6
331.088	HDCO	10P30	DDG80	905477.0
331.500	NH2NH2	9P30	DSF74C	
331.700	CD3OD	10P22	FK86	
332.860	CH3BR	10R06	CM76	
333.150	CH3BR	10P08	CM76	
333.935	CH3CL	9P42	GCK86	897758.2
334.000	SIHF3	10R12	DFS85	
334.820	HCOOH	9P18	WZN73	
335.000	CH2CHF	10P06	TW82	
335.709	DCOOD	9R08	DSF76	893013.6
336.000	HCOOH	9R12	DDSB77B	
336.000	HCCCHO	10P26	DJ80	
336.380	CD3F	10P50	TD86	891231.4
337.000	CLO2	9R26	DSF75B	
337.000	NH3(D)	10R30	GDEFJ83	
339.900	13CH3OH	10R36	PS87	
340.000	SIH3F	10R22	DS82	
340.000	13CH3OH	10R32	IMSD85	
340.000	CLO2	9R40	DSF75B	
340.300	CHCL2F	9P20	VWPE83	880965.6
341.000	CD3BR	9P36	Lands80A	
342.000	DCOOH	10R16	DSF76	
342.127	CD2F2	9P44	VPE81	876261.3
343.000	SIH2F2	10R14	DFS85	
343.500	SIH3F	10R14	DS82	
344.000	C2H4O2H2	9P22	PCD73	
344.000	CH2CLF	9P32	ADF84	
344.900	CHD2OH	10R02	FPVSF89	
345.000	SIHF3	9P32	DFS85	

λ [μm]	Molecule	Pump	Reference	Freq. [MHz]
345.000	CHCLF2	10P34	DFP78	
345.800	CDF3	9R22	TF80	
347.000	HCOOD	10R08	DSF76	
348.899	CD3F	9P34	TD86	859252.2
349.000	CH2CLF	9R10	ADF84	
349.000	CH3NH2	10R18	Lands80D	
349.387	CH3CL	10R18	GCK86	858053.3
350.000	DCOOD	10R40	DSF76	
351.000	CH3SH	9R28	Lands80B	
351.000	HCOOD	9P36	DSF76	
351.500	CH3CHDOH	9P46	BS85	
352.750	CH3BR	9P18	CM76	
353.000	HCOOD	10P28	DSF76	
353.000	HCOOD	10R06	DSF76	
353.100	13CD3OD	9R26	VE85	
353.800	CD3OD	9P12	PVSEP85	
355.550	13CH3I	10P30	GRF87	
356.000	HCOOD	10P30	DSF76	
358.000	CH2CF2	10R18	HW84	
359.810	HCOOH	9R34	WZN73	
360.500	CHFCHF	9P30	BH82	
360.606	CHCLF2	9R14	DFP78	831356.9
361.231	CDF3	10P24	TLD83	829918.3
361.500	SIHF3	9P34	DFS85	
362.000	DCOOH	10P20	DSF76	
362.100	CH3CH2F	9R18	WZN73	
362.423	CDF3	9R24	TLD83	827188.4
362.800	CH2CHF	10R28	RGF84	
363.000	CH3COOD	10R28	DFSY81A	
364.484	D2CO	10P28	DG81	822512.2
365.725	CHCL2F	9P18	VWPE83	819720.5
366.625	CD3BR	9P14	DF78	817708.3
367.000	DCOOD	9R04	DSF76	
368.448	CD3F	10R48	TD86	813662.6
368.862	NH2NH2	9R18	IMEJ86	812750.0
369.000	SIH3F	10R16	DS82	
369.700	CD3OH	10P38	CIMPS87	
369.968	HCOOD	10R28	DSF76	810320.5
370.400	14NH3	10R02	WHH80	
372.000	HCOOD	10P26	DSF76	
372.814	CH3CN	10P20	DF78	804134.8

λ [μm]	Molecule	Pump	Reference	Freq. [MHz]
372.870	CHCLF2	9R22	DFP78	804012.9
373.000	CD2CL2	10P10	SDFY86	
373.400	CD3OD	9P20	PVSEP85	
374.000	HCOOD	9P34	DSF76	
375.000	NH3	9R42	GDEFJ83	
375.407	CHD2F	9R06	Tobin84	798579.5
375.980	CHCL2F	9P16	VWPE83	797362.5
376.800	13CD3F	10P38	TF81	
377.000	CH3NH2?	9P18	DFSY81B	
377.400	CH2CHF	10P14	CD81B	
377.450	CH3I	9R16	CM76	
378.000	CH3CH2F	9R32	WZN73	
378.400	OCS	9R08	Lands80C	
379.000	CH3SH	9P22	Lands80B	
379.242	COF2	10R40	TD86	790504.6
379.500	CH3CHDOH	9P32	BS85	
380.000	CLO2	9R22	DSF75B	
380.020	CH3BR	10R18	CM76	
380.710	CH3CN	10P16	CM71	
381.000	TRIOX	10P12	DWC81	
382.357	H13COOH	9P26	DG82	784063.1
382.766	CHCLF2	10R40	DFP78	783226.7
383.200	CH3CF3	10P14	FR87	
383.285	CD3CL	9R34	DFBS75	782166.1
384.000	DFCO	10P20	JDL83	
384.869	TRIOX	9R30	DWC81	778946.7
385.800	CH2CHCN	10R22	GRBKF85	
386.410	CH3CN	9P46	CM76	
386.500	CHFCHF	10R40	BH82	
386.500	ND2ND2	10R18	SDEF85	
387.310	CH3CN	9R12	CM76	
387.800	CH3CHF2	10R28	FBRG85	
388.000	HCOOD	10R04	DSF76	
388.273	CDF3	10R32	TLD83	772117.0
388.390	CH3CN	9P22	CM76	
388.652	CDF3	10R42	TLD83	771365.4
389.000	TRIOX	9P18	DWC81	
390.000	CD3I	9P26	DFBS75	
390.530	CH3I	10P42	CM76	
390.780	COF2	10R38	TD86	767165.1
392.480	CH3I	9R14	CM76	

λ [μm]	Molecule	Pump	Reference	Freq. [MHz]
393.000	HCOOD	9R16	DSF76	
393.000	TRIOX	10R26	DWC81	
393.000	CD2F2	10R08	VPE81	
393.300	CH3CF3	10P20	FR87	
393.330	COF2	10R32	GRF88	
393.485	H13COOH	9P32	DG82	761888.8
395.000	HCOOD	9P16	DSF76	
395.000	HCOOD	9P10	DSF76	
396.000	CH2CHBR	9P34	DESF76	
396.000	HCOOH	9R42	DDSB77B	
397.000	DCOOD	10R22	DSF76	
397.700	CH3CHF2	10R22	FBRG85	
398.000	CH3NO2	9R36	DFSY81A	
399.800	13CD3OH	9P34	IEPSV84	
401.250	CH2CHCN	9P42	GRBKF85	
401.300	CH2CF2	10P20	HW84	
402.000	CH3NC	10R12	LSB81	
403.777	13CH2F2	9P32	STPVE85	742470.4
404.000	CH3CH2F	9P34	WZN73	
404.100	HCOOH	10R42	WZN73	
404.600	CHD2OH	9P28	FPVSF89	
405.486	HDCO	9P16	DDG80	739340.3
405.750	HCOOH	9P26	WZN73	
405.950	CH2CHCN	9R06	GRBKF85	
414.000	CH3NO2	9R04	DFSY81A	
414.000	DCOOD	10P44	DSF76	
414.980	CH3BR	10R02	CM76	
416.000	CH2CHBR	10R22	DESF76	
417.000	HCOOD	9R22	DSF76	
417.244	CD2F2	9P20	VPE81	718505.6
418.200	CD3OD	10R36	FK86	
418.310	CH3BR	10P26	CM76	
420.000	CH2CHF	10P40	TW82	
420.391	HCOOH	9R08	DSF76	713127.6
420.980	CDF3	10R46	TLD83	712130.6
421.000	CH2CHCL	10P04	LSB81	
422.117	CH3CN	10P24	IMEJ86	710212.3
422.780	CH3BR	10R26	CM76	
423.000	CH2CHF	10R46	RGF84	
423.354	CH2CHCL	10R30	DSFE74	708137.1
424.000	CH3NO2	10P24	DFSY81A	

λ [μm]	Molecule	Pump	Reference	Freq. [MHz]
424.550	CD3I	10P02	GRF87	
425.650	CH2CHCN	10P18	GRBKF85	
426.000	CH3NO2	9R22	DFSY81A	
427.040	CH3CN	9P26	CM76	
427.807	CHCLF2	9R26	DFP78	700766.2
428.000	CD3BR	10R02	Lands80A	
428.870	CH3CCH	9R38	CM76	
429.690	HCOOD	10P24	DSF76	697695.1
430.100	CH2CF2	10P36	HW84	
430.438	HCOOD	10P06	DSF76	696482.3
430.482	CH3CN	10P18	DF78	696410.9
432.000	HFCO	9R32	JD81	
432.000	CHCLF2	10P06	DFP78	
432.300	CH3CHDOH	9P24	BS85	
433.000	TRIOX	10P44	DDSB77A	
433.000	CH3COOD	10P20	DFSY81A	
433.000	DCOOH	10P16	DSF76	
433.104	CD3I	9P28	DFBS75	692195.5
433.235	DCOOH	10R14	DSF76	691985.3
433.438	CHCLF2	10R34	DFP78	691662.4
435.427	CHD2F	10R38	Tobin84	688503.0
435.900	CH3CHF2	10R02	FBRG85	
436.500	SIHF3	10R30	DFS85	
437.600	CHFCHF	10R06	BH82	
438.000	CH2CHCL	10P42	FBM84	
438.100	CH318OH	9P38	IMPSG89	
440.884	CD2F2	10P32	VPE81	679979.8
441.300	CH3OCH3	10P08	BSKK82	
443.000	DCOOD	10P14	DSF76	
443.265	CD3CL	9P10	DFBS75	676328.5
444.386	CD3I	9R32	DFBS75	674621.3
445.210	HCOOH	10P14	WZN73	
449.000	CH3CH2OH	9P22	BS85	
449.300	CH3CHF2	10R18	FBRG85	
450.000	DFCO	9P12	JDL83	
450.980	HCOOD	10P12	DSF76	664757.9
451.000	CH3COOD	9P32	DFSY81A	
452.400	13CH3OH	9R36	PS87	
453.570	CH2CHCN	10R34	GRBKF85	
453.600	CH3CH2BR	10R34	BS83	
453.800	CH2CHBR	9P36	BGRF84	

λ [μm]	Molecule	Pump	Reference	Freq. [MHz]
454.000	ND2ND2	10R40	SDEF85	
454.000	CH3NO2	10P26	DFSY81A	
454.800	CH3CF3	10R52	FR87	
455.073	CD3CN	9P08	DF78	658778.6
455.500	SIHF3	10R28	DFS85	
458.523	HCOOH	9R38	DDSB77B	653822.2
459.180	CH3I	10P08	CM76	
459.400	CH2CHCL	10P52	FBM84	
459.400	CDF3	10R18	TSD82	
459.428	TRIOX	9R22	DWC81	652533.9
460.510	HCOOH	9R10	WZN73	
460.562	CD3I	9R12	DFBS75	650927.5
461.072	NH2NH2	10P16	RPJM77	650207.7
461.200	CH3CL	9R42	CM76	
462.920	CH3CH2F	9P32	WZN73	
463.000	CH3CF3	10R12	FR87	
464.400	CH2CHCN	10P36	GRBKF85	
464.700	13CD3OD	10R08	VE85	
465.000	SIHF3	10R16	DFS85	
465.000	CH3COOD	10R20	DFSY81A	
465.700	CH318OH	9R34	IMPSG89	
466.643	CD3BR	9R26	DF78	642445.1
467.700	CHCLF2	9R44	TD80	
469.000	DCOOD	10P12	DSF76	
469.000	CD2CL2	10R04	ZD78	
470.000	CH3NO2	9P14	DFSY81A	
470.065	13CD3F	10P34	TD86	637768.5
470.386	CHCL2F	9R34	VWPE83	637332.6
471.000	SIH2F2	10R20	DFS85	
472.000	CH3NO2	9R08	DFSY81A	
473.000	HCOOD	9R38	DSF76	
474.600	CH2CHCL	10P48	FBM84	
475.100	CH2CF2	10P32	HW84	
476.000	CHCLF2	10R30	DFP78	
477.000	HCOOD	9P14	DSF76	
477.100	CH3CF3	10R50	FR87	
477.870	CH3I	9P26	CM76	
477.963	H13COOH	9P30	DG82	627229.1
479.000	DCOOD	10R04	DSF76	
480.000	H13COOH	10R46	DG82	
481.000	CH3NC	10P42	LSB81	

λ [μm]	Molecule	Pump	Reference	Freq. [MHz]
482.200	CH3CHF2	10R20	FBRG85	
482.900	CHD2OH	9P06	FPVSF89	
484.400	CHD2OH	9P22	FPVSF89	
485.400	CH3CF3	10P08	FR87	
485.600	CH3CF3	10R38	FR87	
485.800	CH3CF3	10P12	FR87	
486.100	CH2CF2	10R30	HW84	
487.000	CH3NO2	9R06	DFSY81A	
487.000	CH2CHCL	9P10	Radfo75	
487.000	CH3NO2	9R12	DFSY81A	
487.226	CD3I	9P10	DFBS75	615304.6
488.000	SIHF3	10R24	DFS85	
488.880	CH3CCH	10P12	CM71	
489.000	CH3NO2	10P28	DFSY81A	
489.038	O3	9R32	DDM83	613025.0
490.391	CD3I	9R22	DFBS75	611333.6
491.376	TRIOX	10P30	DWC81	610108.3
491.800	CH3CD2OH	9P40	BS85	
492.000	HCOOH	9P42	WZN73	
492.040	CHCL2F	9R36	VWPE83	609284.6
493.000	CH3CH2I	10P34	BS83	
493.000	CD2CL2	10P28	SDFY86	
493.156	HCOOD	10R40	DSF76	607905.7
493.280	HCOOH	9P14	WZN73	
494.000	SIH2F2	10P12	DFS85	
494.000	SIH2F2	10R18	DFS85	
497.000	CD2CL2	10R28	SDFY86	
497.400	CH3OCH3	9R26	BSKK82	
498.000	HCOOD	9P44	DSF76	
498.500	SIHF3	10R20	DFS85	
500.000	TRIOX	9P16	DWC81	
500.577	CD2F2	10R24	VPE81	598893.7
501.600	CH3CF3	10R48	FR87	
503.000	CH2CHCN	9R12	Radfo75	
504.000	CH3CH2I	10P32	BS83	
506.000	CH2CHBR	10R38	DESF76	
508.370	CH3I	9P34	CM76	
508.480	CH3BR	10R42	CM76	
509.000	CH3CHO	9R36	LSB81	
509.160	CH2CHCN	10R36	GRBKF85	
509.440	COF2	10R10	GRF88	

λ [μm]	Molecule	Pump	Reference	Freq. [MHz]
510.700	CH3CF3	10P18	FR87	
511.900	CH3CL	10R52	CM76	
511.900	CH3OCH3	10P52	BSKK82	
514.000	DFCO	10P16	JDL83	
514.000	CH3NO2	9P28	DFSY81A	
515.800	CH2CHBR	10P04	BGRF84	
516.000	HCCCHO	10P14	DJ80	
516.000	SIH3F	10R20	DFS85	
516.253	CD3CN	9P30	DF78	580708.2
516.770	CH3CCH	9R12	CM76	
517.330	CH3I	10P14	CM76	
517.800	CHD2OH	10P10	FPVSF89	
518.600	CH2CHF	10P52	RGF84	
518.800	CH3CF3	10R04	FR87	
519.000	CH2CHCL	10P34	Radfo75	
519.075	CH3CH2F	9R04	RPJM77	577551.1
521.237	CDF3	10R24	TLD83	575156.1
523.406	CD3I	10P38	DFBS75	572772.1
525.000	CH3COOD	10R12	DFSY81A	
525.560	CH2CHCN	10P30	GRBKF85	
527.700	CH2CF2	10P40	HW84	
527.900	CH3CH2BR	10R30	BS83	
528.497	CH2CHBR	10R40	DESF76	567255.3
529.280	CH3I	10P36	CM76	
529.300	CH3CH2OH	9R04	BS85	
529.880	CD3CN	9R04	DF78	565774.2
530.132	CD3BR	10R10	DF78	565505.1
530.533	CH2CHCL	9P16	DSFE74	565077.8
530.700	CH3OCH3	9R28	BSKK82	
531.038	CH3BR	10P24	DF78	564540.7
531.080	CH3CCH	9P06	CM76	
531.363	13CH2F2	9P08	STPVE85	564195.3
533.137	CHCLF2	10R16	DFP78	562317.4
538.000	CH2CHCL	10R04	Radfo75	
538.415	COF2	10P16	TD86	556805.5
540.000	CD3I	9R06	DFBS75	
540.900	CH3CH2F	9P38	WZN73	
542.000	CH3CH2I	10P30	BS83	
545.279	CH3BR	10P38	DF78	549796.0
545.412	CH3BR	10R32	DF78	549662.8
545.500	CH2CHCL	10P18	FBM84	

λ [μm]	Molecule	Pump	Reference	Freq. [MHz]
547.529	CHCL2F	9R10	VWPE83	547537.6
548.700	CD2F2	10R28	TSD82	
548.843	H13COOH	9P20	DG82	546225.3
549.258	CHCL2F	9R08	VWPE83	545813.2
549.686	CH2CHCN	10P14	DSFE74	545388.2
551.100	13CH2F2	9R06	STPVE85	
552.400	CD3OD	9R10	PVSEP85	
553.300	CH2CHBR	9P40	BGRF84	
555.900	CH2CF2	10R12	FGRD84	
556.803	CD3BR	9R30	DF78	538417.8
557.000	CHFCHF	9P40	BH82	
558.577	TRIOX	9R10	DWC81	536707.3
560.803	CDF3	10R40	TLD83	534577.4
561.028	CHCL2F	9R40	VWPE83	534362.8
563.000	CH2CHF	10R36	TW82	
563.130	CH3CCH	10P24	CM71	
564.700	CH3OCH3	10P16	BSKK82	
566.100	CH3CH2OH	9R28	BS85	
566.440	CH3CCH	9P18	CM76	
567.800	CD3OD	9R40	PVSEP85	
568.810	CH3CL	10R26	CM76	
572.330	H13COOH	9R32	DG82	523810.4
572.510	COF2	10R12	GRF88	
575.300	CH3CH2OH	9P34	BS85	
576.170	CH3I	10P16	CM76	
578.900	CH3I	10R34	CM76	
580.600	CH3CF3	10R46	FR87	
580.800	CH2CHCL	10R50	FBM84	
580.869	CHCL2F	9R12	VWPE83	516110.2
581.600	CH3CD2OH	10R24	BS85	
581.984	CDF3	10R28	TLD83	515121.1
582.500	CH3CHF2	10P28	FBRG85	
582.500	CH2CHF	10R32	RGF84	
582.554	HCOOD	9P18	DSF76	514617.8
583.700	CHFCHF	9P42	BH82	
583.770	CH3CCH	9P20	CM76	
585.777	CH3BR	9P40	DF78	511785.8
587.500	ND2ND2	10P34	SDEF85	
588.028	CH2F2	9R46	PSE80	509827.2
590.000	CHCLF2	10P12	DFP78	
591.130	CHCLF2	10R24	DFP78	507151.1

λ [μm]	Molecule	Pump	Reference	Freq. [MHz]
593.000	TRIOX	10P38	DWC81	
593.100	CD3OH	10R06	PFS86	
593.506	CH3CH2F	9P36	RPJM77	505121.4
594.000	HCOOD	9R30	DSF76	
594.729	CH2CHBR	10P32	DESF76	504082.8
600.000	CH3NH2?	9P40	DFSY81B	
601.897	CH2CHCL	10P38	RPJM77	498079.1
602.000	CH2CF2	10R10	HW84	
605.000	CH2CHF	10R02	RGF84	
606.000	CH2CHCL	10P14	FBM84	
606.700	CH2CHBR	10R46	BGRF84	
606.700	CH2CHCL	10P44	FBM84	
606.800	CH3CF3	10R26	FR87	
613.000	SIH2F2	10R22	DFS85	
614.300	CF2CL2	10P32	LPMD81	
615.329	CHCLF2	10R14	DFP78	487206.6
618.446	CH2CHBR	10R30	DESF76	484751.1
619.300	CH2CHBR	10R28	BGRF84	
620.000	CH2CHCL	9P46	FBM84	
620.300	CH3CH2OH	9R12	BS85	
620.400	CH3CH2F	9P22	WZN73	
621.700	CH318OH	9R08	IMPSG89	
622.300	CH2CHCL	10R34	FBM84	
623.000	HCCCH2F	9P24	TJD86	
626.800	CH3CH2I	10R06	BS83	
628.000	CH3COOD	9R06	DFSY81A	
629.300	CH3CF3	10R28	FR87	
631.000	CH3NO2	10P32	DFSY81A	
631.000	CD2CL2	10R18	ZD78	
631.930	CH3BR	10P16	CM76	
632.050	CH3BR	10P22	DF78	474318.0
632.900	CH3CHF2	10P10	FBRG85	
634.471	CH2CHCL	10P20	RPJM77	472507.8
637.500	CH3CHF2	10P26	FBRG85	
638.000	CH2CHCL	10P06	Radfo75	
639.128	DCOOH	10P08	DSF76	469064.7
639.730	CH3I	9P06	CM76	
640.350	COF2	10R18	GRF88	
640.700	CH2CHBR	10P56	BGRF84	
641.430	CH2CHCN	9R22	GRBKF85	
644.000	CD3I	10P16	DFBS75	

λ [μm]	Molecule	Pump	Reference	Freq. [MHz]
645.500	CH3OCH3	9R34	BSKK82	
647.348	DCOOH	10R30	DSF76	463108.3
647.890	CH3CCH	10P14	CM71	
649.590	CH3CCH	10P34	CM71	
649.600	CD2CL2	10P04	BF86	
654.000	HFCO	9P18	JD81	
655.400	CH2CHCL	10R40	FBM84	
656.000	CH3NO2	9R20	DFSY81A	
657.000	HCOOD	9P22	DSF76	
657.590	CH2CHCN	9P44	GRBKF85	
657.938	CDF3	10P12	TLD83	455654.7
658.152	CDF3	10P06	TLD83	455506.2
658.500	ND2ND2	10P10	SDEF85	
658.530	CH3BR	9P56	CM76	
660.328	CH3CH2I	10P26	TD86	454005.5
660.582	CD3I	10P46	DFBS75	453830.6
660.882	13CH3BR	10R20	IMEJ86	453624.6
661.000	TRIOX	9R12	DWC81	
661.153	CHCL2F	9R30	VWPE83	453438.9
663.000	CD3OD	9P22	PVSEP85	
665.885	CHCLF2	10P18	DFP78	450216.4
667.232	CD3I	10P10	DFBS75	449307.5
668.000	HCOOD	9R40	DSF76	
669.531	HCOOH	9R30	DDSB77B	447765.0
670.790	CH2CHCN	10R08	GRBKF85	
670.990	CH3I	10P28	CM76	
671.150	CH2CHCN	10R10	GRBKF85	
675.000	CH3COOD	10R22	DFSY81A	
675.200	CH2CF2	10P44	HW84	
675.290	CH3CCH	9P40	CM76	
676.700	CH3CF3	10P02	FR87	
679.766	TRIOX	10P34	DWC81	441023.2
682.175	CHCLF2	10P14	DFP78	439465.6
687.200	CHFCHF	9P38	BH82	
690.000	13CD3I	10P10	DH78A	
690.000	CD2CL2	10R24	SDFY86	
691.119	CD3I	9R20	DFBS75	433778.2
692.025	CD3BR	9P26	DF78	433210.6
693.800	CH2CHBR	10P50	BGRF84	
694.000	NH3(D)	10R14	GDEFJ83	
694.428	TRIOX	9R16	DWC81	431711.4

λ [μm]	Molecule	Pump	Reference	Freq. [MHz]
695.000	NH3(D)	10R26	GDEFJ83	
697.455	DCOOH	10R36	DSF76	429837.6
698.000	CH3CH2CL	10R38	DK82	
698.600	CH3CHF2	10P34	FBRG85	
701.000	CH3COOD	10R18	DFSY81A	
704.530	CH3CN	9R34	CM76	
704.600	CD2CL2	10P12	BF86	
705.000	HCOOH	9R06	DSF76	
705.300	CH2CHBR	10R52	BGRF84	
706.600	CD3OH	9P10	PFS86	
707.800	CH3CH2BR	10P14	BS83	
709.500	CH3CF3	10R36	FR87	
709.800	CH3CF3	10R40	FR87	
710.000	DCOOH	10P06	DSF76	
711.752	TRIOX	9R32	DWC81	421203.7
712.000	CH2CHBR	10R10	DESF76	
712.760	CH2CHCN	9R24	GRBKF85	
713.720	CH3CN	10P32	CM71	
721.000	NH2NH2	10P12	DSF74C	
723.000	CD2CL2	10R26	SDFY86	
724.000	ND2ND2	10R30	SDEF85	
724.140	CH2CHBR	10P14	DESF76	413998.0
727.949	HCOOD	10R42	DSF76	411831.6
728.900	CH2CF2	10R06	HW84	
734.162	NH2NH2	10R38	Knigh81	408346.7
734.262	CD3I	9P22	DFBS75	408290.6
735.000	CH3NO2	10P34	DFSY81A	
737.113	D2CO	10R32	DDG80	406711.7
745.000	CD3I	10P08	DFBS75	
749.371	CH3BR	10P14	MD78	400058.7
749.372	TRIOX	9P30	DWC81	400058.4
750.000	DFCO	9P20	JDL83	
750.380	CH2CHCN	9P26	GRBKF85	
751.400	CF2CL2	10P30	LPMD81	
752.681	D2CO	9R32	DDSB77A	398299.6
769.800	CH3CH2BR	10R22	BS83	
770.000	CH3CHF2	10P14	BT77	
770.400	CH2CHCL	10R48	FBM84	
775.000	CH2CHCN	10R42	Radfo75	
778.000	CH3NO2	9R40	DFSY81A	
780.000	CH3NO2	9R32	DFSY81A	

λ [μm]	Molecule	Pump	Reference	Freq. [MHz]
780.133	CH2CHBR	10R14	DESF76	384283.8
788.482	CD3I	10P12	DFBS75	380214.9
789.000	TRIOX	9R14	DWC81	
790.800	CH3CHF2	10P02	FBRG85	
792.000	CD3CL	9P28	DH78A	
793.000	CH2CHCN	10R40	Radfo75	
795.000	NH2NH2	10P32	DSF74C	
798.550	CH3CCH	10P20	CM71	
802.400	NH2NH2	10R24	DSF74C	
806.000	13CD3I	10P12	DH78A	
809.000	CH3NO2	10P40	DFSY81A	
813.000	TRIOX	10P32	DWC81	
813.654	TRIOX	9P32	DWC81	368452.2
819.000	HCOOD	10P36	DSF76	
820.000	13CH3I	10P48	GRF87	
822.300	CH3CHF2	10P56	FBRG85	
823.000	CH3NC	10P30	GB81	
830.450	CH2CHCN	10R26	GRBKF85	
832.700	CH2CHBR	10R42	BGRF84	
833.300	CH3CF3	10R18	FR87	
838.300	CH3CH2BR	10R20	BS83	
841.000	CH3NO2	9R10	DFSY81A	
845.000	CH3NO2	9R42	DFSY81A	
851.000	CH3CF3	10R16	FR87	
851.900	CH3CH2F	9P30	WZN73	
863.100	CH2CHCL	10R16	FBM84	
867.200	CH2CF2	10P50	FGRD84	
869.000	CH3NO2	10P36	DFSY81A	
870.800	CH3CL	9P52	CM76	
878.100	CH3CF3	10R08	FR87	
883.000	CF3BR	9R10	PLM79	
883.598	CD3CL	9P34	DFBS75	339286.0
886.300	CH2CHBR	10P42	BGRF84	
889.000	CH3CHDOH	9R12	BS85	
889.466	TRIOX	9R20	DWC81	337047.8
891.087	H13COOH	10R32	DG82	336434.4
895.000	CD3I	10P30	DFBS75	
896.500	CH3CH2BR	10P10	BS83	
899.384	CHCLF2	9R38	DFP78	333330.9
925.520	CH3BR	10R46	CM76	
930.000	HCOOH	10R14	Knigh81	

λ [μm]	Molecule	Pump	Reference	Freq. [MHz]
930.000	HCOOH	10R32	Knigh81	
934.223	CH2CHBR	9P28	DESF76	320900.3
935.000	CH2CHCL	10P46	Radfo75	
938.000	CH3NC	10P32	GB81	
939.500	CH3CHF2	10R08	FBRG85	
948.250	TRIOX	10R06	DWC81	316153.4
948.925	TRIOX	9R24	DDSB77A	315928.6
952.000	TRIOX	10P08	DWC81	
958.250	CH3CL	9P38	CM76	
971.806	DCOOH	10R28	DSF76	308489.9
972.000	NH3(D)	10P40	GDEFJ83	
984.795	CHD2F	10P46	Tobin84	304421.1
985.859	CH2CHBR	10R02	DESF76	304092.7
988.000	FCN	9R28	TJD86	
990.630	CH2CHBR	10R04	DESF76	302628.0
994.900	CH2CHBR	10R06	BGRF84	
995.000	CH2CHCL	10R26	Radfo75	
1001.000	CH3NO2	9R26	DFSY81A	
1005.000	CH2CHCL	9P44	FBM84	
1006.000	HCCCH2F	9P32	TJD86	
1007.000	NH2NH2	10P22	DSF74C	
1009.409	DCOOD	10R18	DSF76	296997.9
1010.000	CH3CD2OH	10R20	BS85	
1013.000	CH3CH2F	9P28	Radfo75	
1014.000	SIH3F	10R30	DS82	
1014.890	CH3CN	9R14	CM76	
1016.009	CH2CHCN	10P32	DSFE74	295068.8
1016.330	CH3CN	9P08	CM76	
1017.800	CH2CHCN	10R24	GRBKF85	
1026.680	CH2CHCL	10R38	DSFE74	292001.8
1028.000	HCCF	9R18	DJ80	
1041.000	CH2CHCL	10R36	Radfo75	
1043.000	CF3BR	9R12	LPM81	
1044.400	CH3CH2I	10P20	BS83	
1049.810	CH3CH2I	10R04	TD86	285568.4
1056.852	CH3BR	10P18	DF78	283665.4
1059.000	CH3CH2BR	10P08	BS83	
1069.000	CH3CH2F	9R10	Radfo75	
1080.600	CH2CHBR	10P48	BGRF84	
1083.000	CF3BR	9R28	LPM81	
1086.890	CH3CN	9P40	CM76	

λ [μm]	Molecule	Pump	Reference	Freq. [MHz]
1097.110	CH3CCH	9P08	CM76	
1137.500	CD3I	10P04	GRF87	
1161.676	HCOOD	10R20	DSF76	258068.8
1164.830	CH3CN	9P10	CM76	
1165.700	CH2CHF	10P46	RGF84	
1167.600	CH2CHF	10P28	RGF84	
1174.870	CH3CCH	10P44	CM71	
1197.100	CH2CHCN	10R04	GRBKF85	
1201.400	CH3CHF2	10P18	FBRG85	
1202.200	CH2CHCN	9R34	GRBKF85	
1213.362	HCOOH	9P28	DSF76	247075.8
1221.893	13CH3F	9P32	TLD83	245350.7
1239.480	CD3CL	9P12	DFBS75	241869.6
1247.594	CH2CHBR	10R12	DESF76	240296.5
1250.000	CH2CHBR	10P06	BGRF84	
1255.700	CH2CHF	10P34	RGF84	
1260.561	CDF3	10R16	TLD83	237824.7
1281.649	DCOOD	9P38	DSF76	233911.6
1286.000	SIH3F	10R32	DFS85	
1296.400	CH2CHBR	10P12	BGRF84	
1310.569	CH3BR	10R04	IMEJ86	228749.8
1310.748	CLO2	10P14	DSF75B	228718.6
1351.780	CH3CN	9R20	CM76	
1377.000	CDF3	10R22	TSD82	
1406.000	CH2CHCL	10R12	FBM84	
1432.500	CH2CHCN	10P50	GRBKF85	
1440.000	CH3CH2F	9P08	Knigh81	
1450.000	CD3F	9R36	TSW79A	
1526.000	CF3BR	9R16	LPM81	
1541.750	HCOOD	9P30	DSF76	194449.4
1547.000	HCCCH2F	9P18	TJD86	
1549.505	CD3I	9R10	DFBS75	193476.3
1555.000	CH2CHCN	10R28	GRBKF85	
1556.000	CF3BR	9R38	LPM81	
1572.640	CH3BR	10P04	CM76	
1581.705	TRIOX	9P26	DWC81	189537.5
1669.000	CH3CH2CL	10R26	DK82	
1687.000	CF3BR	9R40	LPM81	
1730.833	HCOOD	10R24	DSF76	173207.0
1814.370	CH3CN	10P46	CM71	
1895.000	CF3BR	9R20	LPM81	

λ [μm]	Molecule	Pump	Reference	Freq. [MHz]
1990.757	CD3CL	9P14	DFBS75	150592.2
2031.281	TRIOX	10R30	DWC81	147587.9
2140.000	CH3NC	10P14	GB81	
2206.000	CH2CHCL	9P32	FBM84	

10. Molecule Codes

The coding is self-explanatory, being based on a simplified "structural formula" such as CH3BR for ethyl bromide. Only capital letters are used. Thus we have CH3CL = CH_3Cl, 18O3 = $^{18}O_3$, etc. Isotopic variants of hydrogen, carbon and oxygen are distinguished in the codes, the others may be presumed to be their "natural" variants unless noted otherwise. Isotopic variants of N are given when known, but this is not always the case. In the cases of chlorine and bromine (here CL and BR) no attempt has been made to keep track of the isotopic variants, which are often not given in the literature. I chose to show organic compounds as explicitly as possible (e.g. CH2CF2 rather than "C2H2F2").

A few cases occur in which lines differ only by a slight change in molecule, for example the deuteration of a single hydrogen. Since laser action has been observed on less than 1% pure samples, the suspicion arises that such pairs are in fact identical. They have been tagged in Table A by the symbol mm# where # is an integer common to both lines.

In the following table, which is sorted numero-alphabetically by molecule code, the column headings are:

1. codes as used in Tables A, B and C;
2. the number of lines in for this molecule in Table A or B;
3. chemical name(s) – this consists of the names recommeded by the International Union of Chemistry followed in brackets by other names, many of which are of course much more common.

The list of codes is followed by an alphabetical listing of names to facilitate the location of the appropriate code.

10.1 Listing by Code

Code	Number of Lines	Names
13CD3F	10	trideuterofluoromethane-^{13}C
13CD3I	8	trideuteroiodomethane-^{13}C
13CD3OD	34	deuteroxytrideuteromethane-^{13}C
13CD3OH	38	trideuteromethanol-^{13}C
13CH2F2	65	difluoromethane-^{13}C
13CH3BR	1	bromomethane-^{13}C
13CH3F	1	fluoromethane-^{13}C
13CH3I	14	iodomethane-^{13}C
13CH3OH	99	methanol-^{13}C
14ND3	1	trideuteroammonia
14NH2D	6	deuteroammonia
14NH3	78	ammonia-^{14}N
15NH3	30	ammonia-^{15}N
18O3	1	ozone-^{18}O
AR	5	argon
BCL3	8	boron chloride
C2H4O2H2	42	($C_2H_4(OH)_2$) dihydroxyethane (ethylene glycol)
CD2CL2	101	dideuterodichloromethane
CD2F2	57	dideuterodifluoromethane
CD3BR	17	trideuterobromomethane
CD3CL	20	trideuterochloromethane
CD3CN	3	trideuterocyanomethane
CD3F	14	trideuterofluromethane
CD3I	70	trideuteroiodomethane
CD3OD	185	deuteroxytrideuteromethanol
CD3OH	366	trideuteromethanol
CDF3	35	deuterotrifluoromethane
CF2CL2	11	dichlorodifluoromethane (fluorocarbon)
CF3BR	10	bromotrifluoromethane
CF4	1	tetrafluoromethane (carbon tetrafluoride)
CH2CF2	125	1,1-difluoroethene (1,1-difluoroethylene)
CH2CHBR	118	bromoethene (vinyl bromide)
CH2CHCL	94	chloroethene (vinyl chloride)
CH2CHCN	123	(acrylonitrile, vinyl cyanide)
CH2CHF	135	fluoroethene (vinyl fluoride)
CH2CL2	7	dichloromethane
CH2CLF	13	chlorofluoromethane
CH2DOH	73	deuteromethanol

Code	Number of Lines	Names
CH2F2	120	difluoromethane
CH2NOH	5	(formaldoxime)
CH318OH	100	methanol-^{18}O
CH3BR	44	bromomethane (methyl bromide)
CH3CCH	15	(methylacetylene)
CH3CD2OH	3	1,1-dideuteroethanol
CH3CF3	54	1,1,1-trifluoroethane (methyl fluoroform)
CH3CH2BR	8	bromoethane (ethyl bromide)
CH3CH2CL	4	chloroethane (ethyl chloride)
CH3CH2F	29	fluoroethane (ethyl fluoride)
CH3CH2I	7	iodoethane (ethyl iodide)
CH3CH2OH	10	ethanol
CH3CHDOH	4	1-deuteroethanol
CH3CHF2	58	1,1-difluoroethane
CH3CHO	6	ethanal (acetaldehyde)
CH3CL	21	chloromethane (methyl chloride)
CH3CN	34	ethanenitrile (acetonitrile, cyanomethane, methyl cyanide)
CH3COOD	10	deuteroxyethanoic acid
CH3F	11	fluoromethane (methyl fluoride)
CH3I	32	iodomethane (methyl iodide)
CH3NC	12	(methyl isocyanide)
CH3NH2	80	aminomethane (methylamine)
CH3NH2?	9	(methylamine, possibly partially deuterated)
CH3NO2	46	nitromethane
CH3OCH3	20	methoxymethane (dimethyl ether)
CH3OD	127	deuteroxymethanol
CH3OH	453	methanol (methyl alcohol)
CH3SH	23	(methyl mercaptan)
CHCL2F	15	dichlorofluoromethane
CHCLF2	45	chlorodifluoromethane
CHD2F	18	dideuterofluoromethane
CHD2OH	93	dideuteromethanol
CHFCHF	43	cis 1,2-difluoroethene (cis 1,2-difluoroethylene)
CLO2	30	chlorine dioxide
CO2	5	carbon dioxide
COF2	52	carbonyl fluoride
CS2	10	carbon disulphide
D2CO	19	dideuteromethanal (dideuteroformaldehyde)
D2O	19	(deuterium oxide, heavy water)

Code	Number of Lines	Names
DCN	6	deuterohydrocyanic acid (deuterium cyanide)
DCOOD	57	dideuteromethanoic acid (dideutero formic acid)
DCOOH	20	deuteromethanoic acid (deutero formic acid)
DFCO	18	(DCOF) deuterated formyl fluoride
FCN	2	cyanogen fluoride
H13COOH	21	(formic acid-^{13}C)
H2CO	10	methanal (formaldehyde)
H2O	36	water
H2S	24	hydrogen sulphide
HBR	20	hydrogen bromide (hydrobromic acid)
HCCCH2F	3	($FCH_2C:CH$) 3-fluoropropyne (propargyl fluoride)
HCCCHO	4	(HC:CCHO) propynal
HCCF	2	(fluoroacetylene)
HCL	42	hydrogen chloride (hydrochloric acid)
HCN	28	hydrocyanic acid (hydrogen cyanide)
HCOOD	64	deuteroxymethanoic acid (deuteroxy formic acid)
HCOOH	82	methanoic acid (formic acid)
HDCO	7	deuteromethanal (deuteroformaldehyde)
HE	3	helium
HF	37	hydrogen fluoride (hydrofluoric acid)
HFCO	14	(HCOF, FCHO) (fluoroformaldehyde, formyl fluoride)
ND2ND2	31	(N_2D_4) dideuterohydrazine
NE	50	neon
NH2NH2	46	(N_2H_4) hydrazine
NH2OH	5	hydroxylamine
NH3	24	ammonia, N isotope unidentified
NH3(D)	10	ammonia with deuterating agent
O	1	oxygen
O3	7	ozone
OCS	3	(COS) carbonyl sulphide
PH3	4	(phosphine)
SIH2F2	21	(SiH_2F_2) difluorosilane
SIH3F	16	(SiH_3F) fluorosilane
SIHF3	18	($SiHF_3$) trifluorosilane (silyl fluoride)
SO2	31	sulphur dioxide
SO2(I)	20	isotopically substituted sulphur dioxide
TRIOX	50	((H_2CO)$_3$) (trioxane)
XE	5	xenon

10.2 Alphabetical Listing of Molecule Names

Name	Code
acetaldehyde	CH3CHO
acetonitrile	CH3CN
acrylonitrile	CH2CHCN
aminomethane	CH3NH2
ammonia with deuterating agent	NH3(D)
ammonia, N isotope unidentified	NH3
ammonia-^{14}N	14NH3
ammonia-^{15}N	15NH3
argon	AR
boron chloride	BCL3
bromoethane	CH3CH2BR
bromoethene	CH2CHBR
bromomethane	CH3BR
bromomethane-^{13}C	13CH3BR
bromotrifluoromethane	CF3BR
carbon dioxide	CO2
carbon tetrafluoride	CF4
carbonyl fluoride	COF2
carbonyl sulphide	OCS
chlorine dioxide	CLO2
carbon disulphide	CS2
chlorodifluoromethane	CHCLF2
chloroethane	CH3CH2CL
chloroethene	CH2CHCL
chlorofluoromethane	CH2CLF
chloromethane	CH3CL
cyanogen fluoride	FCN
cyanomethane	CH3CN
deuterated formyl fluoride	DFCO
deuterium cyanide	DCN
deuterium oxide	D2O
deutero formic acid	DCOOH
deuteroammonia	14NH2D
deuteroethanol 1,-	CH3CHDOH
deuteroformaldehyde	HDCO
deuterohydrocyanic acid	DCN
deuteromethanal	HDCO
deuteromethanoic acid	DCOOH
deuteromethanol	CH2DOH

Name	Code
deuterotrifluoromethane	CDF3
deuteroxy formic acid	HCOOD
deuteroxyethanoic acid	CH3COOD
deuteroxymethanoic acid	HCOOD
deuteroxymethanol	CH3OD
deuteroxytrideuteromethane-[13]C	13CD3OD
deuteroxytrideuteromethanol	CD3OD
dichlorodifluoromethane	CF2CL2
dichlorofluoromethane	CHCL2F
dichloromethane	CH2CL2
dideutero formic acid	DCOOD
dideuterodichloromethane	CD2CL2
dideuterodifluoromethane	CD2F2
dideuteroethanol 1,1-	CH3CD2OH
dideuterofluoromethane	CHD2F
dideuteroformaldehyde	D2CO
dideuterohydrazine	ND2ND2
dideuteromethanal	D2CO
dideuteromethanoic acid	DCOOD
dideuteromethanol	CHD2OH
difluoroethane 1,1-	CH3CHF2
difluoroethene cis 1,2-	CHFCHF
difluoroethene 1,1-	CH2CF2
difluoroethylene cis 1,2-	CHFCHF
difluoroethylene 1,1-	CH2CF2
difluoromethane	CH2F2
difluoromethane-[13]C	13CH2F2
difluorosilane	SIH2F2
dihydroxyethane	C2H4O2H2
dimethyl ether	CH3OCH3
ethanal	CH3CHO
ethanenitrile	CH3CN
ethanol	CH3CH2OH
ethyl bromide	CH3CH2BR
ethyl chloride	CH3CH2CL
ethyl fluoride	CH3CH2F
ethyl iodide	CH3CH2I
ethylene glycol	C2H4O2H2
fluoroacetylene	HCCF
fluorocarbon	CF2CL2
fluoroethane	CH3CH2F

Name	Code
fluoroethene	CH2CHF
fluoroformaldehyde	HFCO
fluoromethane	CH3F
fluoromethane-[13]C	13CH3F
fluoropropyne 3-	HCCCH2F
fluorosilane	SIH3F
formaldehyde	H2CO
formaldoxime	CH2NOH
formic acid	HCOOH
formic acid-[13]C	H13COOH
formyl fluoride	HFCO
heavy water	D2O
helium	HE
hydrazine	NH2NH2
hydrobromic acid	HBR
hydrochloric acid	HCL
hydrocyanic acid	HCN
hydrofluoric acid	HF
hydrogen bromide	HBR
hydrogen chloride	HCL
hydrogen cyanide	HCN
hydrogen fluoride	HF
hydrogen sulphide	H2S
hydroxylamine	NH2OH
iodoethane	CH3CH2I
iodomethane	CH3I
iodomethane-[13]C	13CH3I
methanal	H2CO
methanoic acid	HCOOH
methanol	CH3OH
methanol-[13]C	13CH3OH
methanol-[18]O	CH318OH
methoxymethane	CH3OCH3
methyl alcohol	CH3OH
methyl bromide	CH3BR
methyl chloride	CH3CL
methyl cyanide	CH3CN
methyl fluoride	CH3F
methyl fluoroform	CH3CF3
methyl iodide	CH3I
methyl isocyanide	CH3NC

Name	Code
methyl mercaptan	CH3SH
methylacetylene	CH3CCH
methylamine	CH3NH2
methylamine, possibly partially deuterated	CH3NH2?
neon	NE
nitromethane	CH3NO2
oxygen	O
ozone	O3
ozone-^{18}O	18O3
phosphine	PH3
propargyl fluoride	HCCCH2F
propynal	HCCCHO
silyl fluoride	SIHF3
sulphur dioxide	SO2
sulphur dioxide, with isotopic substitutions	SO2(I)
tetrafluoromethane	CF4
trideuteroammonia	14ND3
trideuterobromomethane	CD3BR
trideuterochloromethane	CD3CL
trideuterocyanomethane	CD3CN
trideuterofluoromethane-^{13}C	13CD3F
trideuterofluromethane	CD3F
trideuteroiodomethane	CD3I
trideuteroiodomethane-^{13}C	13CD3I
trideuteromethanol	CD3OH
trideuteromethanol-^{13}C	13CD3OH
trifluoroethane 1,1,1-	CH3CF3
trifluorosilane	SIHF3
trioxane	TRIOX
vinyl bromide	CH2CHBR
vinyl chloride	CH2CHCL
vinyl cyanide	CH2CHCN
vinyl fluoride	CH2CHF
water	H2O
xenon	XE

11. Reference Database

This is actually rather more than just a list of references since it shows all the papers I have consulted to establish the accuracy of data, including review articles and other papers which did not contribute data to the final listing, usually because the lines concerned were not cw according to the working definition (see Chapter 6) or because the data were later removed according to the priorities indicated (also in Chapter 6).

The reference code, of up to eight characters, indicates the author(s) and year of publication. In the case of single-author papers the surname is truncted to five characters (e.g. Henni78) while for multiple-author papers up to five initials are used (e.g. CIMPS87). Suffixes A, B etc. are used to remove ambiguities and do not connote any precedence. For the reasons stated above, not every reference in the database will appear in Table A.

Notes about a paper which are too global to have been included in Table A are included here.

ACB84	E. Arimondo, M. Ciocca, G. Baldacchini: Reviews of Infrared and Millimeter Waves 2, K.J. Button, M. Inguscio and F. Strumia (eds.) 467–476 (review CH3CCH)
AD80	K.B. Amos, J.A. Davis: IEEE J. QE-**16**, 574–575
ADF84	J.R. Anacona, P.B. Davies, A.H. Ferguson: IEEE J. QE-**20**, 829–830
Arimo84	E. Arimondo: Reviews of Infrared and Millimeter Waves 2, K.J. Button, M. Inguscio and F. Strumia (eds.) 81–104 (review CH3I, CD3I, 13CD3I)
AW69	D.P. Akitt, C.F. Wittig: J. Appl. Phys. **40**, 902–903 Long-pulsed lines (pulse length around 0.004 ms)
AY70	D.P. Akitt, J.T. Yardley: IEEE.J.QE-**6**, 113–116, Long-pulsed lines (pulse length 0.003 to 0.012 ms). The laser material used was a boron trihalide but emission was attributed to the corresponding hydrogen halide. For an overview of measurements and assignments of diatomic species see Polla71.
BBEK73	T.G. Blaney, C.C. Bradley, G.J. Edwards, D.J.E. Knight: Phys. Lett. **43** A, 471–472
BCKEP80	T.G. Blaney, N.R. Cross, D.J.E. Knight, G.J. Edwards, P.R. Pearce: J. Phys. D: Appl. Phys. **13**, 1365–1370
BDGBI77	E. Bava, A. De Marchi, A. Godone, R. Benedetti, M. Inguscio, P. Minguzzi, F. Strumia, M. Tonelli: Opt. Commun. **21**, 46–48
BDMF77	O.I. Baskakov, S.F. Dyubko, M.V. Moskienko, L.D. Fesenko: Sov. J. QE-**7**, 445–449
BEG78	R. Beck, W. Englisch, K. Gürs: *Table of Laser Lines in Gases and Vapors*, 2nd edn. (Springer-Verlag, Berlin)
BF86	P. Belland, M. Fourrier: Int. J. Infrared Mmwaves **7**, 1251–1256

BGRF84   P. Belland, C. Gastaud, M. Redon, M. Fourrier: Appl. Phys. B **34**, 175–177
BH82   A.S. Bennett, H. Herman: IEEE J. QE-**18**, 323–325
BKM78   T.G. Blaney, D.J.E. Knight, E.K. Murray Lloyd: Opt. Commun. **25**, 176–178
BP77   B.L. Bean, S. Perkowitz: Opt. Lett. **1**, 202–204
         Lines measured at 372.7 μm (CH2CF2) and 296.2 μm (CH3OD) seem to
         be identical with the 375.545 μm and 294.811 μm entries respectively even
         though these differences exceed the stated measurement accuracy
BPT69   W.S. Bennedict, M.A. Pollack, W.J. Tomlinson: IEEE J. QE-**5**, 108–124
         We entered only those lines listed as cw; in other cases the pulse length was
         not stated. In some cases this paper contained more accurate remeasure-
         ments of previously reported lines
BS81   V.A. Bugaev, E.P. Shliteris: Sov. J. QE-**11**, 742–744
         Quasi-cw techniques employed although not necessarily for all lines listed
BS83   V.A. Bugaev, E.P. Shliteris: Sov. J. QE-**13**, 150–154
         Quasi-cw techniques employed although not necessarily for all lines listed
BS84   V.A. Bugaev, E.P. Shliteris: Sov. J. QE. **14**, 1331–1336, The experimenters
         used various isotopic variants of SO2 but since identification was incomplete
         we have denoted all such species as SO2(I)
BS85   V.A. Bugaev, E.P. Shliteris: Sov. J. QE-**15**, 547–550
BSKK82   V.A. Bugaev, E.P. Shliteris, Yu.F. Klement'ev, V.A. Kudryashova: Sov. J.
         QE-**12**, 304–308
BT77   G. Busse, R. Thurmaier: Appl. Phys. Lett. **31**, 194–195
CB70A   T.Y. Chang, T.J. Bridges: Opt. Commun. **1**, 423–426
CB70B   T.Y. Chang, T.J. Bridges: Proc. Symp. Millimeter Waves, Brooklyn Poly-
         technic, 93–98
CBB70A   T.Y. Chang, T.J. Bridges, E.G. Burkhardt: Appl. Phys. Lett. **17**, 357–358
CBB70B   T.Y. Chang, T.J. Bridges, E.G. Burkhardt: Appl. Phys. Lett. **17**, 249–251
CD81A   A.R. Calloway, E.J. Danielewicz: IEEE J. QE-**17**, 579–581
CD81B   A.R. Calloway, E.J. Danielewicz: Int. J. Infrared Mmwaves **2**, 933–942
Chang84 T.Y. Chang: Reviews of Infrared and Millimeter Waves 2, K.J. Button, M.
         Inguscio and F. Strumia (eds.) 1–28 (review CH3F, 13CH3F)
CIMPS87 G. Carelli, N. Ioli, A. Moretti, D. Pereira, F. Strumia: Appl. Phys. B **44**,
         111–117
         Quasi-cw techniques employed for many lines. Accuracy may be less than
         that claimed; see footnote 78 to Table A
CIMPS88 G. Carelli, N. Ioli, A. Moretti, D. Pereira, F. Strumia, R. Densing: Appl.
         Phys. B **45**, 97–100
CM71   T.Y. Chang, J.D. McGee: Appl. Phys. Lett. **19**, 103–105
         Long-pulsed lines (pulse length around 0.15 ms)
CM76   T.Y. Chang, J.D. McGee: IEEE J. QE-**12**, 62–65
         Note that this reference is frequently given incorrectly as IEEE J. QE-**9**,
         62–65.
         Long-pulsed lines (pulse length around 0.15 ms)
         In the case of CH3BR, isotopic identification can be found in DB76
Colem84 P.D. Coleman: Reviews of Infrared and Millimeter Waves 2, K.J. Button,
         M. Inguscio and F. Strumia (eds.) 383–427 (review PH3)
         This review contains only TEA-pulsed lines. In particular the results in
         SCLB81 are not included
CR87   A. Chakrabarti, J. Reid: Rev. Sci. Instrum. **58**, 1413–1416
Danie84 E.J. Danielewicz: Reviews of Infrared and Millimeter Waves 2, K.J. Button,
         M. Inguscio and F. Strumia (eds.) 223–250 (review CH2CF2)
DB76   J.-C. Deroche, C. Betrencourt-Stirnemann: J. Mol. Phys. **32**, 921–930
DB85   J.-C. Deroche, E.K. Benichou: Opt. Commun. **54**, 23–26
DBGD86   J.-C. Deroche, E.K. Benichou, G. Guelachvili, J. Demaison: Int. J. Infrared
         Mmwaves **7**, 1653–1675
DBMF78   S.Y. Dyubko, O.I. Baskakov, M.V. Moskienko, L.D. Fesenko: 3rd Int. Conf.
         Submm. Waves & Appl., Univ. Surrey 1978 (Guildford) 68–69
DC77   E.J. Danielewicz, P.D. Coleman: IEEE J. QE-**13**, 485–490

DD87   J. Dupre-Maquaire, J. Dupre: Int. J. Infrared Mmwaves **8**, 317–332
DDG80   D. Dangoisse, B. Duterage, P. Glorieux: IEEE J. QE-**16**, 296–300
DDM83   D. Dangoisse, J.C. Depannemaecker, N. Monnanteuil: Int. J. Infrared Mmwaves **4**, 913–918
DDSB76   D. Dangoisse, A. Deldalle, J.P. Splingard, J. Bellet: C.R. Acad. Sc. Paris **283** (B), 115–118
            The nine HCOOH frequencies in this reference can also be found in DDSB77B
DDSB77A   D. Dangoisse, A. Deldalle, J.P. Splingard, J. Bellet: IEEE J. QE-**13**, 730–731
DDSB77B   A. Deldalle, D. Dangoisse, J.P. Splingard, J. Bellet: Opt. Commun. **22**, 333–336
            For those lines for which frequencies were not measured wavelength accuracy is claimed to be only ± 2 μm
Deroc78   J.-C. Deroche: J. Mol. Spect. **69**, 19–24
DESF76   S.F. Dyubko, M.N. Efimenko, V.A. Svitch, L.D. Fesenko: Sov. J. QE-**6**, 600–601
Deuts67A   T.F. Deutsch: Appl. Phys. Lett. **10**, 234–236
Deuts67B   T.F. Deutsch: Appl. Phys. Lett. **11**, 18–20, Long-pulsed lines (pulse length 0.001 to 0.005 ms). For an overview of measurements and assignments of diatomic species see Polla71.
Deuts67C   T.F. Deutsch: IEEE. J. QE-**3**, 419–421, See Deuts67A for experimental details. For an overview of measurements and assignments of diatomic species see Polla71.
DF78   S.Y. Dyubko, L.D. Fesenko, A.H. Ferguson: 3rd Int. Conf. Submm. Waves & Appl., Univ. Surrey 1978 (Guildford) 70–73
            In the case of obscure references (their 5, 6) this paper was used as the source of data
DFBS75   S.F. Dyubko, L.D. Fesenko, O.I. Baskakov, V.A. Svich: Prikl. Spektrosk. (USSR) **23**, 317–320
DFHAL83   P.B. Davies, A.H. Ferguson, P.A. Hamilton, T.L. Amyes, I.M.R. van Laere: Int. J. Infrared Mmwaves **4**, 1029–1036
DFP78   S.F. Dyubko, L.D. Fesenko, B.I. Polevoyn: EKON-78 Conference Abstracts, Poznan, 19–20
DFS85   P.B. Davies, A.H. Ferguson, D.P. Stern: Infrared Phys. **25**, 87–90
            Some of the lines reported as new appeared in DS82
DFSY81A   S.F. Dyubko, L.D. Fesenko, A.S. Shevyrev, V.I. Yartsev: Sov. J. QE-**11**, 1247–1248
DFSY81B   S.F. Dyubko, L.D. Fesenko, A.S. Shevyrev, V.I. Yartsev: Sov. J. QE-**11**, 1248–1249
DG81   D. Dangoisse, P. Glorieux: Int. Report, Lab. de Spectroscopie Hertz, Lille
DG82   D. Dangoisse, P. Glorieux: J. Mol. Spec. **92**, 283–297
DG84A   J.-C. Deroche, G. Graner: Reviews of Infrared and Millimeter Waves 2, K.J. Button, M. Inguscio and F. Strumia (eds.) 35–42 [review CH3CL (35/37)], (^{35}Cl and ^{37}Cl)
DG84B   D. Dangoisse, P. Glorieux: Reviews of Infrared and Millimeter Waves 2, K.J. Button, M. Inguscio and F. Strumia (eds.) 429–466 (review HCOOH)
DGFRH79   E.J. Danielewicz, T.A. Galantowicz, F.B. Foote, R.D. Reel, D.T. Hodges: Opt. Lett. **4**, 280–282
DH78A   G. Duxbury, H. Herman: J. Phys. B **11**, 935–949
            The fifth and fourteenth wavelengths in their Table 3 are incorrectly given although the inverse wavelengths are correct. A line at 869 μm (10R18) was removed; VSPE81 remeasured the wavelength and found it to agree with the 858.254 μm entry
DH78B   G. Duxbury, H. Herman: Infrared Phys. **18**, 461–463
            The 697.8 μm line appears to have been communicated as a private communication; data were therefore taken from this paper
DHJRS69   V. Daneu, L.O. Hocker, A. Javan, D. Ramachandra Rao, A. Szoeke: Phys. Lett. **29** A, 319–320
DJ80   P.B. Davies, H. Jones: Appl. Phys. **22**, 53–55
DK82   N.G. Douglas, P.A. Krug: IEEE J. QE-**18**, 1409–1410

Corrects the data given in JEJ75

DP84   G. Duxbury, J.C. Petersen: Appl. Phys. B **35**, 127–129
DPK87   I.H. Davis, K.I. Pharaoh, D.J.E. Knight: Int. J. Infrared Mmwaves **8**, 765–769
DRH80   E.J. Danielewicz, R.D. Reel, D.T. Hodges: IEEE J. QE-**16**, 402–405
DS82   P.B. Davies, D.P. Stern: Int. J. Infrared Mmwaves **3**, 909–915
DSF72   S.F. Dyubko, V.A. Svich, L.D. Fesenko: JETP Lett. **16**, 418–419
DSF74A   S.F. Dyubko, V.A. Svich, L.D. Fesenko: Sov. Phys. Tech. Phys. **18**, 1121
DSF74B   S.F. Dyubko, V.A. Svich, L.D. Fesenko: Opt. Spectrosc. **37**, 118
             The three lines were eliminated as they were more accurately reported in DFBS75. Moreover the pump identifications should read 9R16, 10P18 and 10P32. See also Grane75
DSF74C   S.F. Dyubko, V.A. Svich, L.D. Fesenko: Zh. Prikl. Spektrosk. (USSR) **20**, 718–719
             The authors state that the R-branch transitions were difficult to identify correctly and may be in error
DSF74D   S.F. Dyubko, V.A. Svich, L.D. Fesenko: Sov. J. QE-**3**, 446
             Most of the lines were eliminated as they were more accurately reported in DSF76
DSF75A   S.F. Dyubko, V.A. Svich, L.D. Fesenko: Izv. Vuz. Radiofiz. (USSR) **18**, 1434–1437
             Several corrections were made on the basis of data given in PFS86
DSF75B   S.F. Dyubko, V.A. Svich, L.D. Fesenko: Sov. Tech. Phys. Lett. **1** (6) 192–193
DSF76   S.F. Dyubko, V.A. Svich, L.D. Fesenko: Sov. Phys. Tech. Phys. **20**, 1536–1538
             Many of these data replace those from DSF74D but the older paper contains polarization measurements
DSFE74   S.F. Dyubko, V.A. Svich, L.D. Fesenko, M.N. Efimenko: Russian Proc. Symp. Submm. Mm. Waves Atmos.
             Data reproduced in DF78. Several pump assignments adjusted to agree with Duxbu84B
DSJ85   P.B. Davies, D.P. Stern, H. Jones: Spectrochimica Acta 41A, 367–370
             Comparison between Table 1 and Table 4 shows that the 10R22 assignment should be 10P22, in agreement with DS82, for the 330 μm SIH3F line
DSV68   S.F. Dyubko, V.A. Svich, R.A. Valitov: JETP Lett. **7**, 320
DTS74   Yu.S. Domnin, V.M. Tatarenkov, P.S. Shumyatskii: Sov. J. QE-**4**, 401–402
Duxbu84A   G. Duxbury: Reviews of Infrared and Millimeter Waves 2, K.J. Button, M. Inguscio and F. Strumia (eds.) 29–34 (review CD3F)
Duxbu84B   G. Duxbury: Reviews of Infrared and Millimeter Waves 2, K.J. Button, M. Inguscio and F. Strumia (eds.) 261–276 (review CH2CF2)
             This review is used as the reference for several lines for which the given references are either obscure or apparently incorrect (e.g. several references to GD81). References to BP77 and DH78B appear to be incorrect: the correct reference is SWTRD80. The fourth frequency in Table 1 of Duxbu84B is incorrect. The correct value was given by DSFE74 and DF78
DV88   B.W. Davis, A. Vass: Int. J. Infrared Mmwaves **9**, 279–293
DVPA81   B.W. Davis, A. Vass, C.R. Pidgeon, G.R. Allan: Opt. Commun. **37**, 303–305
DW78A   E.J. Danielewicz, C.O. Weiss: IEEE J. QE-**14**, 222–223
DW78B   E.J. Danielewicz, C.O. Weiss: IEEE J. QE-**14**, 458–459
DW78C   E.J. Danielewicz, C.O. Weiss: IEEE J. QE-**14**, 705–707
DW78D   E.J. Danielewicz, C.O. Weiss: Opt. Commun. **27**, 98–100
DW84   D. Devoy, B. Walker: J. Phys. E. **17**, 1132–1134
DWC81   D. Dangoisse, J. Wascat, J.M. Colmont: Int. J. Infrared Mmwaves **2**, 1177–1191
DWDB79   D. Dangoisse, E. Willemot, A. Deldalle, J. Bellet: Opt. Commun. **28**, 111–116
Egger83   D.F. Eggers: Infrared Phys. **23**, 233–234
EWME70   K.M. Evenson, J.S. Wells, L.M. Matarrese, L.B. Elwell: Appl. Phys. Lett. **16**, 159–161

EWTR79   P.J. Epton, W.L. Wilson, Jr., F.K. Tittel, T.A. Rabson: Appl. Opt. **18**, 1704–1705

FBGR84   M. Fourrier, P. Belland, C. Gastaud, R. Redon, Proceedings Third Conference on Infrared Physics (CIRP3), 803–805. All but six of the lines mentioned were reported in either BGRF84, FBM84, GRBF84 or RGF84. No details available on those six.

FBM84   M. Fourrier, P. Belland, D. Mangili: IEEE J. QE-**20**, 85–87

FBRG85   M. Fourrier, P. Belland, M. Redon, C. Gastaud: IEEE J. QE-**21**, 21–24

FCTP74   H.R. Fetterman, B.J. Clifton, P.E. Tannenwald, C.D. Parker: Appl. Phys. Lett. **24**, 70–72

FGRD84   M. Fourrier, C. Gastaud, M. Redon, J.C. Deroche: Opt. Commun. **48**, 347–351

FK86   M. Fourrier, A. Kreisler: Appl. Phys. B **41**, 57–60

FM66   G.T. Flesher, W.M. Müller, Proc. IEEE. **54**, 543–546

FMPG64   W.L. Faust, R.A. McFarlane, C.K.N. Patel, C.G.B. Garrett: Phys. Rev. **133**, A1476–A1486
The review article Patel68 gives identification of many lines and clarifies experimental data

FPVSF89   J.A. Facin, D. Pereira, E.C.C. Vasconcellos, A. Scalabrin, C.A. Ferrari, Appl. Phys. B **48**, 245–248 Lines in CHD20H at 83.7, 83.9, 103.0, 246.8 and 385.4 μm appear to have been seen earlier in CD30H,although for the lines reported by PFS86 polarizations do not agree

FR87   M. Fourrier, M. Redon: Opt. Commun. **64**, 534–536

FSP73   H.R. Fetterman, H.R. Schlossberg, C.D. Parker: Appl. Phys. Lett. **23**, 684–686

FSPB67   L. Frenkel, T. Sullivan, M.A. Pollack, T.J. Bridges: Appl. Phys. Lett. **11**, 344–345

FWSGW82   E.M. Frank, C.O. Weiss, K. Siemsen, M. Grinda, G.D. Willenberg: Opt. Lett. **7**, 96–98

GB81   B. Gilbert, R.J. Butcher: IEEE J. QE-**17**, 827–828

GCK86   J.A. Golby, N.R. Cross, D.J.E. Knight: Int. J. Infrared Mmwaves **7**, 1309–1327

GD81   E.B. Gamble, Jr., E.J. Danielewicz: IEEE J. QE-**17**, 2254–2256

GD84   G. Graner, J.-C. Deroche: Reviews of Infrared and Millimeter Waves 2, K.J. Button, M. Inguscio and F. Strumia (eds.) 43–52 (review CD3CL)

GDB84   G. Graner, J.-C. Deroche, C. Betrencourt-Stirnemann: Reviews of Infrared and Millimeter Waves 2, K.J. Button, M. Inguscio and F. Strumia (eds.) 73–80 (review CD3BR)

GDEFJ83   W.G. Gerasimov, S.F. Dyubko, M.N. Efimenko, L.D. Fesenko, W.I. Jarcev: Ukr. Fiz. Zh. **28**, 1323–1327
Five lines in NH3(D) were found to be very close in wavelength to lines of 14NH2D previously reported by Lands80D and were eliminated. They were at 77 μm, 78 μm, 86 μm, 110 μm and 122 μm wavelength

Grane75   G. Graner: Opt. Commun. **14**, 67–69

GRBF84   C. Gastaud, M. Redon, P. Belland, M. Fourrier: Int. J. Infrared Mmwaves **5**, 875–885

GRBKF85   C. Gastaud, M. Redon, P. Belland, A. Kreisler, M. Fourrier: Int. J. Infrared Mmwaves **6**, 63–70

GRF87   C. Gastaud, M. Redon, M. Fourrier: Int. J. Infrared Mmwaves **8**, 1069–1081

GRF88   C. Gastaud, M. Redon, M. Fourrier: Appl. Phys. B **47**, 303–305

GSRF80   C. Gastaud, A. Sentz, M. Redon, M. Fourrier: IEEE J. QE-**16**, 1285–1287

GW78   M. Grinda, C.O. Weiss: Opt. Commun. **26**, 91

Hard69   T.M. Hard: Appl. Phys. Lett. **14**, 130

HC69   J.C. Hassler, P.D. Coleman: Appl. Phys. Lett. **14**, 135–136
Long-pulsed lines (pulse length around 0.001 ms)

Henni77   J.O. Henningsen: IEEE J. QE-**13**, 435–441

Henni78   J.O. Henningsen: IEEE J. QE-**14**, 958–962
Quasi-cw techniques employed although not necessarily for all lines listed

Henni82   J.O. Henningsen: IEEE J. QE-**18**, 313–317
          This paper was used as the reference for two lines, presumed cw, for which
          the given reference was unobtainable
Henni83   J.O. Henningsen: Int. J. Infrared Mmwaves **4**, 707–732
          Long-pulsed lines (pulse length presumably around 0.05 ms)
Henni84   J.O. Henningsen: Unpublished. Data takem from ISH84 in which this paper
          is Ref 71
Henni86   J.O. Henningsen: Int. J. Infrared Mmwaves **7**, 1605–1629
          Long-pulsed lines (pulse length around 0.05 ms)
HFR76   D.T. Hodges, F.B. Foote, R.D. Reel: Appl. Phys. Lett. **29**, 662–664
HHC73   J.C. Hassler, G. Hubner, P.D. Coleman: J. Appl. Phys. **44**, 795–801
          Long-pulsed lines (pulse length around 0.005 ms)
HHCS71   G. Hubner, J.C. Hassler, P.D. Coleman, G. Steenbeckeliers: Appl. Phys.
          Lett. **18**, 511–513
          Long-pulsed lines (pulse length not stated)
HIMS81   J.O. Henningsen, M. Inguscio, A. Moretti, F. Strumia: 6th Int. Conf. Infrared
          Mmwaves, Miami, pM-2-6
HIMS82   J.O. Henningsen, M. Inguscio, A. Moretti, F. Strumia: IEEE J. QE-**18**, 1004–
          1008
HJ67   L.O. Hocker, A. Javan: Phys. Lett. **25**A, 489–490
HJ68A   L.O. Hocker, A. Javan: Phys. Lett. **26**A, 255–256
HJ68B   L.O. Hocker, A. Javan: Appl. Phys. Lett. **12**, 124–125
HJRFS67   L.O. Hocker, A. Javan, D. Ramachandra Rao, L. Frenkel, T. Sullivan: Appl.
          Phys. Lett. **10**, 147–149
HK66   B. Hartmann, B. Kleman, Can. J. Phys. **44**, 1609–1612, Long-pulsed lines
          (pulse length "a few microsec")
HL82   B. Hartmann, L. Lindgren: Int. J. Infrared Mmwaves **3**, 503–515
          Long-pulsed lines (pulse length around 0.1 ms)
HM76   Y. Horiuchi, A. Murai: IEEE J. QE-**12**, 547–549
          Long-pulsed lines (pulse length around 0.01 ms)
HNSM84   Y. Horiuchi, Y. Nishi, N. Sokabe, A. Murai: Jap. J. Appl. Phys. **23**, 62–67
Hodge78   D.T. Hodges: Infrared Phys. **18**, 375–384
HP78   J.O. Henningsen, J.C. Petersen: Infrared Phys. **18**, 475–479
          Quasi-cw techniques employed although not necessarily for all lines listed
HP79   H. Herman, B.E. Prewer: Appl. Phys. **19**, 241–242
HPPJE79   J.O. Henningsen, J.C. Petersen, F.R. Petersen, D.A. Jennings, K.M. Even-
          son: J. Mol. Spec. **77**, 298–309
HRB73   D.T. Hodges, R.D. Reel, D.H. Barker: IEEE J. QE-**9**, 1159–1160
HRJ67   L.O. Hocker, D. Ramachandra Rao, A. Javan: Phys. Lett. **24**A, 690–691
HSJ69   L.O. Hocker, J.G. Small, A. Javan: Phys. Lett. **29**A, 321–322
HW82   H. Herman, M.J. Wigglesworth: Int. J. Infrared Mmwaves **3**, 395–400
HW84   H. Herman, M.J. Wigglesworth: Int. J. Infrared Mmwaves **5**, 29–36
HWP77   J. Heppner, C.O. Weiss, P. Plainchamp: Opt. Commun. **23**, 381–384
IEPSV84   M. Inguscio, K.M. Evenson, F.R. Petersen, F. Strumia, E. Vasconcellos: Int.
          J. Infrared Mmwaves **5**, 1289–1296
IIMMS81   M. Inguscio, N. Ioli, A. Moretti, G. Moruzzi, F. Strumia: Opt. Commun. **37**,
          211–216
IIMSD86   M. Inguscio, N. Ioli, A. Moretti, F. Strumia, F. D'Amato: Appl. Phys. B**40**,
          165–169
IM84   M. Inguscio, K.M. Evenson: Opt. Lett. **9**, 443–444
IMEJ86   M. Inguscio, G. Moruzzi, K.M. Evenson, D.A. Jennings: J. Appl. Phys. **60**,
          R161–R192
IMMS81A   M. Inguscio, S. Marchetti, A. Moretti, F. Strumia: 6th Int. Conf. Infrared
          Mmwaves, W-5-9
IMMS81B   M. Inguscio, A. Moretti, G. Moruzzi, F. Strumia: Int. J. Infrared Mmwaves
          2, 943–986
IMMS82   M. Inguscio, S. Marchetti, A. Moretti, F. Strumia: Int. J. Infrared Mmwaves
          3, 97–116

IMMSD85   N. Ioli, A. Moretti, G. Moruzzi, F. Strumia, F. D'Amato: Int. J. Infrared Mmwaves **6**, 1017–1029

IMPSG89   N. Ioli, A. Moretti, D. Pereira, F. Strumia, G. Garelli, Appl. Phys. B **48**, 299–304

IMS80   M. Inguscio, A. Moretti, F. Strumia: Opt. Commun. **32**, 87–90

IMS89   N. Ioli, A. Moretti, F. Strumia: Appl. Phys. B **48**, 305–309. Note that an incorrect frequency is given for the 127 μm line

IMSD85   N. Ioli, A. Moretti, F. Strumia, F. D'Amato: Int. J. Infrared Mmwaves **7**, 459–485

IMSL85   N. Ioli, A. Moretti, F. Strumia, I. Longo: Opt. Lett. **10**, 330–332

Ingus84   M. Inguscio: Reviews of Infrared and Millimeter Waves 2, K.J. Button, M. Inguscio and F. Strumia (eds.) 193–222 (review CH3CN, CD3CN, CH3NC)

ISH84   M. Inguscio, F. Strumia, J.O. Henningsen: Reviews of Infrared and Millimeter Waves 12, K.J. Button, M. Inguscio and F. Strumia (eds.) 105–150 Their wavelength no. 212 is incorrect (frequency correct)

JD81   H. Jones, P.B. Davies: IEEE J. QE-**17**, 13–14

JDL83   H. Jones, P.B. Davies, W. Lewis-Bevan: Appl. Phys. B **30**, 1–4

JEJ75   D.A. Jennings, K.M. Evenson, J.J. Jimenez: IEEE J. QE-**11**, 637 Measurements on ethyl chloride found to be in error (DK82) and removed from list

JTT82   H. Jones, G. Taubmann, M. Takami: IEEE J. QE-**18**, 1997–1999

JV87   P. Janssen, H. Vanderstraeten: Int. J. Infrared Mmwaves **8**, 415–429 There are errors in Table 1 of JV87 for the wavelengths of the last three methyl bromide lines. The values shown in the diagrams are correct. The 10R38 pump should read 10P38. Reports cw operation of the 660.7 μm and 245.04 μm lines previously seen in pulsed mode by CM76

KFK82   T. Kachi, M. Fukutani, S. Kon: Int. J. Infrared Mmwaves **3**, 401–408

KHYH75   S. Kon, E. Hagiwara, T. Yano, H. Hirose: Jap. J. Appl. Phys. **14**, 731–732 Lines at 41 μm (10R18) and 255 μm (10R36) and ascribed to CD3OD were eliminated; VSPE81 reported that they lased more strongly in CD3OH; their measured wavelengths agree well with the 41.355 μm and 253.720 μm entries

KK82   T. Kachi, S. Kon: Infrared Phys. **22**, 337–341 Quasi-cw techniques employed

KK83   T. Kachi, S. Kon: Int. J. Infrared Mmwaves **4**, 767–777

KKTY84   S. Kon, T. Kachi, Y. Tsunawaki, M. Yamanaka: Reviews of Infrared and Millimeter Waves 2, K.J. Button, M. Inguscio and F. Strumia (eds.) 159–192

KKPPS68   N.V. Karlov, Yu.B. Konev, Yu.N. Petrov, A.M. Prokhorov, O.M. Stel'makh: JETP Lett. **8**, 12–14 It is not clear whether operation was truly cw

KMNT79   V.M. Klement'ev, Yu.A. Matyugin, M.V. Nikitin, B.A. Timchenko: Sov. J. QE-**9**, 1471–1472

Knigh79   D.J.E. Knight: NPL Report no. Qu 45 5th issue March 79

Knigh81   D.J.E. Knight: NPL Report no. Qu 45 (first revision) and supplement

Knigh86   D.J.E. Knight: Metrologia **22**, 252–257

Knigh89   D.J.E. Knight: Handbook of Lasers and Technology, Supplement 1: Gas Lasers, M.J. Weber (ed), (Far-Infrared Gas Lasers)

KR86   D.F. Kroeker, J. Reid: Appl. Opt. **25**, 2929–2933 We used the calculated values of wavelengths rather than measured values in order to be consistent with the earlier paper SRD86

KS80   F.K. Kneubühl, Ch. Sturzenegger: Infrared and Millimeter Waves, Vol. 3, Chapter 5

KW76   G. Kramer, C.O. Weiss: Appl. Phys. **10**, 187–188

KYHH75   S. Kon, T. Yano, E. Hagiwara, H. Hirose: Jap. J. Appl. Phys. **14**, 1861–1862

Lands80a   B.M. Landsberg: Appl. Phys. B **23**, 345–348

Lands80B   B.M. Landsberg: IEEE J. QE-**16**, 684–685

Lands80C   B.M. Landsberg: IEEE J. QE-**16**, 704–706

Lands80D   B.M. Landsberg: Appl. Phys. B **23**, 127–130

LBG85	J.-L. Lachambre, P. Bernard, M. Gagne: IEEE J. QE-21, 282–283
LD79	M.W. Lund, J.A. Davis: IEEE J. QE-15, 537–538
	The 225 μm line was eliminated in favour of the measurement of 220.1 μm by PD85
LJ69	J.S. Levine, A. Javan: Appl. Phys. Lett. 14, 348–350
LM67	D.R. Lide, Jr., A.G. Maki: Appl. Phys. Lett. 11, 62–64
LM84	J.-M. Loutioz, C. Meyer: Reviews of Infrared and Millimeter Waves 2, K.J. Button, M. Inguscio and F. Strumia (eds.) 53–73 (review CF3BR)
LMJ85	R.M. Lees, I. Mukhopadhyay, J.W.C. Johns: Opt. Commun. 55, 127–130
Lourt84	J.-M. Lourtioz: Reviews of Infrared and Millimeter Waves 2, K.J. Button, M. Inguscio and F. Strumia (eds.) 251–260 (review CF2CL2)
LPM81	J.-M. Lourtioz, J. Pontnau, C. Meyer. Int. J. Infrared Mmwaves 2, 525–532
LPMD81	J.-M. Lourtioz, J. Pontnau, M. Morillon-Chapey, J.-C. Deroche: Int. J. Infrared Mmwaves 2, 49–63
LSB81	B.M. Landsberg, M.S. Shafik, R.J. Butcher: IEEE J. QE-17, 828–829
Maki68	A.G. Maki: Appl. Phys. Lett. 12, 122–124
Maki78	A.G. Maki: J. Appl. Phys. 49, 7–11
MC64	L.E.S. Mathias, A. Crocker: Phys. Lett. 13, 35–36
	Comparison with later frequency measurements shows that the accuracy of the wavelength measurements reported in this paper is very high, often better than 0.03%
MCW65	L.E.S. Mathias, A. Crocker, M.S. Wills: Phys. Lett. 14, 33–34
	Seven lines between 21.47 μm and 31.95 μm micrometres wavelength were not included because they were pumped with a laser of less than 0.001 ms pulse length
MCW67	L.E.S. Mathias, A. Crocker, M.S. Wills: IEEE J. QE-3, 170
	Long-pulsed lines (pulse length around 0.002 ms)
MCW68	L.E.S. Mathias, A. Crocker, M.S. Wills: IEEE J. QE-4, 205–208
	Long-pulsed lines (pulse length around 0.003 ms)
MD78	M.V. Moskienko, S.F. Dyubko: Izv. Vuz. Radiofiz. 21, 951–960
	The footnote to their table 5 suggests that most lines were not observed in cw mode
MDT84	W.H. Matteson, F.C. De Lucia, M.S. Tobin: Infrared Phys. 24, 397–401
MF66	W.M. Müller, G.T. Flesher: Appl. Phys. Lett. 8, 217–218
MF67	W.M. Müller, G.T. Flesher: Appl. Phys. Lett. 10, 93–94
MH84A	G. Merkle, J. Heppner: Opt. Lett. 9, 542–543
MH84B	G. Merkle, J. Heppner: Opt. Commun. 51, 265–270
MLJ87	I. Mukhopadhyay, R.M. Lees, J.W.C. Johns: IEEE J. QE-23, 1378–1384
MLL88	I. Mukhopadhyay, R.M. Lees, W. Levis-Bevan: Int. J. Infrared Mmwaves 9, 545–553
MNPC78	E.G. Malk, J.W. Niesen, D.F. Parsons, P.D. Coleman: IEEE J. QE-14, 544–550
	TEA-operation; includes spectroscopic data on PH3
MOS70	A.G. Maki, W.B. Olson, R.L. Sams: J. Mol. Spec. 36, 433–447
MPD83	J. McCombie, J.C. Petersen, G. Duxbury: Q. Electronics & Electr. Opt., P.L. Knight (ed.) (Proc. 5th Nat. Q. Elec. Conf. 1981)
NH80	Y.C. Ni, J. Heppner: Opt. Commun. 32, 459–460
Patel65	C.K.N. Patel: Appl. Phys. Lett. 7, 273–275
Patel68	C.K.N. Patel: Lasers – A Series of Advances, Vol.2, A.K. Levine (ed.), Dekker N.Y. 1–190 (gas lasers)
PBT67	M.A. Pollack, T.J. Bridges, W.A. Tomlinson: Appl. Phys. Lett. 10, 253–256
PCD73	T.K. Plant, P.D. Coleman, T.A. DeTemple: IEEE J. QE-9, 962–963
	Long-pulsed lines (pulse length around 0.1 ms)
	DSF75A claim that many of the lines identified as belonging to ethylene glycol are actually of methanol
PD82A	J.C. Petersen, G. Duxbury: Appl. Phys. B 27, 19–25
PD82B	J.C. Petersen, G. Duxbury: Int. J. Infrared Mmwaves 3, 607–618
	The 677.4 μm 9P14 line was reidentified by GCK86 as 677.96 μm 9P12

PD84   J.C. Petersen, G. Duxbury: Appl. Phys. B **34**, 17–21
PD85   J.C. Petersen, G. Duxbury: Appl. Phys. B **37**, 209–211
     In Table 1 of this paper 9R06 is misprinted as 9R16
PEJS80 F.R. Petersen, K.M. Evenson, D.A. Jennings, A. Scalabrin: IEEE J. QE-**16**, 319–323
PEJWG75 F.R. Petersen, K.M. Evenson, D.A. Jennings, J.S. Wells, K. Goto, J.J. Jimenez: IEEE J. QE-**11**, 838–843
PFMG64A C.K.N. Patel, W.L. Faust, R.A. McFarlane, C.G.B. Garrett: Proc. IEEE **52**, 713
PFMG64B C.K.N. Patel, W.L. Faust, R.A. McFarlane, C.G.B. Garrett: Appl. Phys. Lett. **4**, 18–19
PFS68   M.A. Pollack, L. Frenkel, T. Sullivan: Phys. Lett. **26** A, 381–382
PFS86   D. Pereira, C.A. Ferrari, A. Scalabrin: Int. J. Infrared Mmwaves **7**, 1241–1250
     These authors indicate ten lines for which DSF75A appear to have misidentified the pump transition. Their data, including polarizations, allow one to make four further corrections to DSF75A: the lines become 337.3 μm 10R30, 351.2 μm 10R30, 352.3 μm 9R06, and 472.4 μm 9R18. These were previously assigned to 10R38, 9R and 9R16, respectively. See also the footnote to DSF74C.
     The 858.3 μm 10R20 line was assumed to be identical with the 858.254 μm 10R18 line and was eliminated.
PH84   J.C. Petersen, J.O. Henningsen: Reviews of Infrared and Millimeter Waves 2, K.J. Button, M. Inguscio and F. Strumia (eds.) 151–158 (review 13CH3OH)
PLM79 J. Pontnau, J.-M. Lourtioz, C. Meyer: IEEE, J. QE-**15**, 1088–1090
Polla71 M.A. Pollack: Handbook of Lasers, R.J. Pressley (ed.) Chemical Rubber Co. Ohio 298–349 (molecular gas lasers)
PP65   Yu.N. Petrov, A.M. Prokhorov: JETP. Lett. **1**, 24–25
PRP83 J. Prasad, D.N. Rao, P.N. Prasad: Int. J. Infrared Mmwaves **4**, 15–19
PS87   D. Pereira, A. Scalabrin: Appl. Phys. B **44**, 67–69
PSE80 F.R. Petersen, A. Scalabrin, K.M. Evenson: Int. J. Infrared Mmwaves **1**, 111–115
PVSEP85 D. Pereira, E.C.C. Vasconcellos, A. Scalabrin, K.M. Evenson, F.R. Petersen, D.A. Jennings: Int. J. Infrared Mmwaves **6**, 877–882
Radfo75 H.E. Radford: IEEE J. QE-**11**, 213–214
RBK86 P.A. Rochefort, E. Brannen, Z. Kucerovsky: Appl. Opt. **25**, 3838–3842
RGF79 M. Redon, C. Gastaud, M. Fourrier: IEEE J. QE-**15**, 412–414
RGF84 M. Redon, C. Gastaud, M. Fourrier: Opt. Lett. **9**, 71–72
RPJM77 H.E. Radford, F.R. Peterson, D.A. Jennings, J.A. Mucha: IEEE J. QE-**13**, 92–94
     see also erratum: IEEE J. QE-**13**, 881
RRG84 C. Rolland, J. Reid, B.K. Garside: Appl. Phys. Lett. **44**, 380–382
RWS83 H.P. Röser, R. Wattenbach, G.V. Schultz: Int. J. Infrared Mmwaves **4**, 1–14
SB75   G. Steenbeckeliers, J. Bellet: J. Appl. Phys. **46**, 2620–2626
SBW84 H. Sigg, H.J.A. Bluyssen, P. Wyder: IEEE J. QE-**20**, 616–617
SCLB81 S. Shafik, D. Crocker, B.M. Landsberg, R.J. Butcher: IEEE J. QE-**17**, 115–116
     The 136 μm and 82 μm wavelengths were corrected using calculated values from MNPC78. Several lines may be Raman, see Wille81
SDEF85 A.S. Shevirev, S.F. Dyubko, M.N. Efimenko, L.D. Fesenko: Zh. Prikl. Spektrosk. **42**, 480–481
SDFY86 A.S. Shevyrev, S.F. Dyubko, L.D. Fesenko, V.I. Yartsev: Sov. J. QE-**16**, 568–569
SE79   A. Scalabrin, K.M. Evenson: Opt. Lett. **4**, 277–279
SEJZS87 R.J. Saykally, K.M. Evenson, D.A. Jennings, L.R. Zink, A. Scalabrin: Int. J. Infrared Mmwaves **8**, 653–662
SH84   Z. Solajic, J. Heppner: Appl. Phys. B **33**, 23–27
SHZM85 N. Sokabe, K. Horikawa, N. Zumoto, A. Murai: Int. J. Infrared Mmwaves **6**, 893–907

SIM85   F. Strumia, N. Ioli, A. Moretti: from "Physics of New Laser Sources", N.B. Abraham et al. (eds.) (Plenum Press)

SJFWM87   P.A. Stimson, B.W. James, I.S. Falconer, L.B. Whitbourn, J.C. Macfarlane: Appl. Phys. Lett. **50**, 786–788

SL84   J.P. Sattler, W.J. Lafferty: Reviews of Infrared and Millimeter Waves 2, K.J. Button, M. Inguscio and F. Strumia (eds.) 359–382 (review SO2)
           Lines attributed to this review article are otherwise unpublished and their pulsed or cw nature is not indicated

SMNM83   N. Sokabe, T. Miyatake, Y. Nishi, A. Murai: J. Phys. B: At. Mol. Phys. **16**, 4487–4493

SPEJ80   A. Scalabrin, F.R. Petersen, K.M. Evenson, D.A. Jennings: Int. J. Infrared Mmwaves **1**, 117–126

SRD86   K.J. Siemsen, J. Reid, D.J. Danagher: Appl. Opt. **25**, 86–91
           Although not explicitly stated, it is apparent that this paper presents calculated, rather than measured, wavelength values

SSKYM81   N. Sokabe, T. Sasabe, T. Kimura, Y. Yasuda, A. Murai: Jap. J. Appl. Phys. **20**, 2127–2132

STPVE85   A. Scalabrin, J. Tomaselli, D. Pereira, E.C.C. Vasconcellos, K.M. Evenson, F.R. Petersen, L. Zink, D.A. Jennings: Int. J. Infrared Mmwaves **6**, 973–979

Strum84   F. Strumia: Unpublished. Data taken from ISH84, in which this paper is Ref 64

SWTRD80   J.P. Sattler, T.L. Worchesky, M.S. Tobin, K.J. Ritter, T.W. Daley, W.J. Lafferty: Int. J. Infrared Mmwaves **1**, 127–138

TD80   M.S. Tobin, T.W. Daley: IEEE J. QE-**16**, 592–594
           Quasi-cw techniques employed although not necessarily for all lines listed

TD86   M.S. Tobin, T.W. Daley: Int. J. Infrared Mmwaves **7**, 1649–1652

Telle83   J. Telle: IEEE J. QE-**19**, 1469–1473

TF80   M.S. Tobin, R.D. Felock: Opt. Lett. **5**, 430–432
           Quasi-cw techniques employed although not necessarily for all lines listed. Note the error in Table 1 of this paper: 10R44 should be 10R46 (private communication)

TF81   M.S. Tobin, R.D. Felock: IEEE J. QE-**17**, 825–826

TH86   F. Tang, J.O. Henningsen: IEEE J. QE-**22**, 2084–2087
           Long-pulsed lines (pulse length around 0.020 ms)

TJD86   G. Taubmann, H. Jones, P.B. Davies: Appl. Phys. B **41**, 179–181

TLD83   M.S. Tobin, R.P. Leavitt, T.W. Daley: J. Mol. Spec. **101**, 212–220

TM76   R. Turner, R.A. Murphy: Infrared Phys. **16**, 197–200
           Long-pulsed lines (pulse length around 0.001 ms)

Tobin80   M.S. Tobin: Proc. SPIE **259**, 13–17

Tobin82   M.S. Tobin: Opt. Lett. **7**, 322–324
           and erratum: Opt. Lett. **8**, 509

Tobin84   M.S. Tobin: IEEE J. QE-**20**, 5–8
           and erratum: IEEE J. QE-**20**, 985

TOH88   F. Tang, A. Olafsson, J.O. Henningsen: Appl. Phys. B **47**, 47–54
           Long-pulsed lines (pulse length around 0.020 ms)

TP68   R. Turner, T.O. Poehler: Phys. Lett. **27** A, 479–480
           Long-pulsed lines (pulse length around 0.001 ms)

TSD82   M.S. Tobin, J.P. Sattler, T.W. Daley: IEEE J. QE-**18**, 79–86

TSW79A   M.S. Tobin, J.P. Sattler, G.L. Wood: 4th Int. Conf. Infrared MM Waves, 209–210

TSW79B   M.S. Tobin, J.P. Sattler, G.L. Wood: Opt. Lett. **4**, 384–386

TTMYY74   A. Tanaka, A. Tanimoto, N. Murata, M. Yamanaka, H. Yoshinaga: Jap. J. Appl. Phys. **13**, 1491–1492
           Quasi-cw (N2O laser) and long-pulsed (CO2 laser; pulse length about 0.3 ms) pumping was used

TTMYY77   A. Tanaka, A. Tanimoto, N. Murata, M. Yamanaka, H. Yoshinaga: Opt. Commun. **22**, 17–21
           Quasi-cw techniques employed although not necessarily for all lines listed

272

TW82   F. Temps, H.Gg. Wagner: Appl. Phys. B **29**, 13–14
TYY75   A. Tanaka, M. Yamanaka, H. Yoshinaga: IEEE J. QE-**11**, 853–854
            Quasi-cw techniques employed although not necessarily for all lines listed
VE85   E.C.C. Vasconcellos, K.M. Evenson: Int. J. Infrared Mmwaves **6**, 1157–1167
VHHR86   J.K. Vij, F. Hufnagel, M. Helker, C.J. Reid: IEEE J. QE-**22**, 1123–1130
VJE86   E.C.C. Vasconcellos, D.A. Jennings, K.M. Evenson: Int. J. Infrared Mmwaves
            7, 291–292
VPE81   E.C.C. Vasconcellos, F.R. Petersen, K.M. Evenson: Int. J. Infrared Mmwaves
            2, 705–711
VSPE81   E.C.C. Vasconcellos, A. Scalabrin, F.R. Petersen, K.M. Evenson: Int. J.
            Infrared Mmwaves **2**, 533–539
            Ten wavelength measurements replace those of either HP79 or KHYH75;
            they are given in Table III of VSPE81. In a number of cases polarizations
            differ from those previously reported
VWDP82   A. Vass, R.A. Wood, B.W. Davis, C.R. Pidgeon: Appl. Phys. B **27**, 187–190
VWE87   E.C. Vasconcellos, J. Wyss, K.M. Evenson: Int. J. Infrared Mmwaves **8**,
            647–651
            This paper appeared in preprint form in 1984. The authors were Vasconcel-
            los, Wyss, Evenson and F.R. Petersen
VWPE83   E.C.C. Vasconcellos, J.C. Wyss, F.R. Petersen, K.M. Evenson: Int. J. In-
            frared Mmwaves **4**, 401–406
WDVP80   R.A. Wood, B.W. Davis, A. Vass, C.R. Pidgeon: Opt. Lett. **5**, 153–154
WFGR84   C.O. Weiss, M. Fourrier, C. Gastaud, M. Redon: Reviews of Infrared and
            Millimeter Waves 2, K.J. Button, M. Inguscio and F. Strumia (eds.) 277–336
            [review NH3 (^{14}N and ^{15}N)]
WGS77   C.O. Weiss, M. Grinda, K. Siemsen: IEEE J. QE-**13**, 892
WHH80   G.D. Willenberg, U. Hubner, J. Heppner: Opt. Commun. **33**, 193–196
Wille71   C.S. Willett: Handbook of Lasers, R.J. Pressley (ed.) Chemical Rubber Co.
            Ohio 183–241 (neutral gas lasers)
Wille81   G.D. Willenberg: Opt. Lett. **6**, 372–373
            The 234 μm and 291 μm lines are Raman. The latter was observed in quasi-
            cw mode
WJ75   P. Woskoboinikow, W.C. Jennings: Appl. Phys. Lett. **27**, 658–660
WVPCN80   R.A. Wood, A. Vass, C.R. Pidgeon, M.J. Colles, B. Norris: Opt. Commun.
            33, 89–90
            The reported line at 163.01 μm (10R34) is assumed to be identical with the
            listed 163.034 μm (10R38) line
WZN73   R.J. Wagner, A.J. Zelano, L.H. Ngai: Opt. Commun. **8**, 46–47
            Long-pulsed lines (pulse length around 0.08 ms). Several wavelength mea-
            surements were eliminated; the frequency measurements which replaced them
            carry the footnote 121. Comparison indicates that the wavelength measure-
            ments in WZN73 were exceptionally accurate. If this is true for all lines
            then there are some discrepancies with the pump identifications of certain
            other lines; the lines at 163.01 μm, 418.51 μm, 232.93 μm and 458.43 μm,
            all with R-branch pumping, have been marked as possible errors. However
            some R-branch lines agree with later measurements
XKP84   F. Xi-Sheng, D.J.E. Knight, K.I. Pharaoh: Int. J. Infrared Mmwaves **5**, 421–
            425
YFMF81   N. Yamabayashi, K. Fukai, K. Miyazaki, K. Fujisawa: Appl. Phys. B **26**,
            33–36
YKYSF81   T. Yoshida, M. Kobayashi, T. Yishihara, K. Sakai, S. Fujita: Opt. Commun.
            40, 45–48
YYSF82   T. Yoshida, T. Yoshihara, K. Sakai, S. Fujita: Infrared Phys. **22**, 293–298
            Investigated Stark effect, but all lines were observable at zero electric field
            intensity. This paper was used as the source for several lines for which ref-
            erences were obscure
ZD78   G. Ziegler, U. Duerr: IEEE J. QE-**14**, 708
ZRGB80   T.A. Znotins, J. Reid, B.K. Garside, E.A. Ballik: Opt. Lett. **5**, 528–530
            Pulse length unclear, but lines presented as possibly cw

Subject Index

Springer Series in Optical Sciences

Editorial Board: J.M. Enoch   D.L. MacAdam   A.L. Schawlow   K. Shimoda   T. Tamir